Microbial Biotechnology

Microbial Biotechnology

Uma Shankar Singh
Kiran Kapoor

Oxford Book Company
Jaipur ▪ New Delhi

Published By

Oxford Book Company

Regd. Off. "Bony Residency", Gate No. 2
Opp. Tilak Public School, Vishweswariya Nagar,
Gopalpura Road, Jaipur - 302018, India
Ph: 0141 - 2761280, Telefax: 2761381
email - oxfordbookjpr@gmail.com
website - oxfordbookcompany.com

Delhi Office: 102, 1st Floor, Satyam House,
4327/3, Ansari Road, Darya Ganj,
New Delhi - 110002, India
Ph: 011-45652440

Distributed by
KEYA INTERNATIONAL PRESS LLP
E-mail: keyainternationalpressllp@gmail.com

Published in India, 2025

ISBN : 9789380179247

Typeset by Shivangi Computers, Jaipur

Printed and bound in India by Replika Press Pvt. Ltd.

Preface

Microbes are organisms that are too small to be seen by the unaided eye. They include bacteria, fungi, protozoa, microalgae, and viruses. Microbes live in familiar settings such as soil, water, food, and animal intestines, as well as in more extreme settings such as rocks, glaciers, hot springs, and deep-sea vents. The wide variety of microbial habitats reflects an enormous diversity of biochemical and metabolic traits that have arisen by genetic variation and natural selection in microbial populations. Historically, humans have exploited some of this microbial diversity in the production of fermented foods such as bread, yogurt, and cheese. Some soil microbes release nitrogen that plants need for growth and emit gases that maintain the critical composition of the Earth's atmosphere.

Microbial biotechnology, enabled by genome studies, will lead to breakthroughs such as improved vaccines and better disease-diagnostic tools, improved microbial agents for biological control of plant and animal pests, modifications of plant and animal pathogens for reduced virulence, development of new industrial catalysts and fermentation organisms, and development of new microbial agents for bioremediation of soil and water contaminated by agricultural runoff. Microbial genomics and microbial biotechnology research is critical for advances in food safety, food security, biotechnology, value-added products, human nutrition and functional foods, plant and animal protection, and furthering fundamental research in the agricultural sciences.

The present book on microbial biotechnology provides an essential intellectual link between the breakthroughs of the last two decades in our understanding of the fundamental processes that drive microbial function and the application of this knowledge to the technological challenges faced by society. It is unique in the clarity with which specific industrial problems are delineated and in the cogent description of how current technology provides solutions. The range of subjects covered in this volume is astounding. They extend from microbial metabolites, antibiotics and polymers to recombinant vaccine production and metabolic engineering. In addition enough details and useful references are provided to engage the most sophisticated reader. The book is designed for students pursuing undergraduate or graduate course in biotechnology. It will also be a useful reference tool for all those associated with microbiology, including research workers, microbiologists and regulatory agents.

Uma Shankar Singh
Kiran Kapoor

Contents

1

Introduction to Microbial Biotechnology

Microbes (or microorganisms) are organisms that are too small to be seen by the unaided eye. They include bacteria, fungi, protozoa, microalgae, and viruses. Microbes live in familiar settings such as soil, water, food, and animal intestines, as well as in more extreme settings such as rocks, glaciers, hot springs, and deep-sea vents. The wide variety of microbial habitats reflects an enormous diversity of biochemical and metabolic traits that have arisen by genetic variation and natural selection in microbial populations.

Historically, humans have exploited some of this microbial diversity in the production of fermented foods such as bread, yogurt, and cheese. Some soil microbes release nitrogen that plants need for growth and emit gases that maintain the critical composition of the Earth's atmosphere. Other microbes challenge the food supply by causing yield-reducing diseases in food-producing plants and animals. In our bodies, different microbes help to digest food, ward off invasive organisms, and engage in skirmishes and pitched battles with the human immune system in the give-and-take of the natural disease process. A genome is the totality of genetic material in the DNA of a particular organism. Genomes differ greatly in size and sequence across different organisms. Obtaining the complete genome sequence of a microbe provides crucial information about its biology, but it is only the first step toward understanding a microbe's biological capabilities and modifying them, if needed, for agricultural purposes.

Microbial biotechnology, enabled by genome studies, will lead to breakthroughs such as improved vaccines and better disease-diagnostic tools, improved microbial agents for biological control of plant and animal pests, modifications of plant and animal pathogens for reduced virulence, development of new industrial catalysts and fermentation organisms, and development of new microbial agents for bioremediation of soil and water contaminated by agricultural runoff. Microbial genomics and microbial biotechnology research is critical for advances in food safety, food security, biotechnology, value-added products, human nutrition and functional foods, plant and animal protection, and furthering fundamental research in the agricultural sciences.

HISTORICAL DEVELOPMENT

Biotechnology is technology based on biology, especially when used in agriculture, food science, and medicine. It is often used to refer to genetic engineering technology of the 21st century, however the term encompasses a wider range and history of procedures for modifying biological organisms according to the needs of humanity, going back to the initial modifications of native plants into improved food crops through artificial selection and hybridization. Bioengineering is the science upon which all biotechnological applications are based. With the development of new approaches and modern techniques, traditional biotechnology industries are also acquiring new horizons enabling them to improve the quality of their products and increase the productivity of their systems.

Before 1971, the term, biotechnology, was primarily used in the food processing and agriculture industries. Since the 1970s, it began to be used by the Western scientific establishment to refer to laboratory-based techniques being developed in biological research, such as recombinant DNA or tissue culture-based processes, or horizontal gene transfer in living plants, using vectors such as the bacteria to transfer DNA into a host organism. In fact, the term should be used in a much broader sense to describe the whole range of methods, both ancient and modern, used to manipulate organic materials to reach the demands of food production. So the term could be defined as, "The application of indigenous and/or scientific knowledge to the management of (parts of) microorganisms, or of cells and tissues of higher organisms, so that these supply goods and services of use to the food industry and its consumers.

Biotechnology combines disciplines like genetics, molecular biology, biochemistry, embryology and cell biology, which are in turn linked to practical disciplines like chemical engineering, information technology, and biorobotics. Patho-biotechnology describes the exploitation of pathogens or pathogen derived compounds for beneficial effect.Although not normally thought of as biotechnology, agriculture clearly fits the broad definition of "using a biological system to make products" such that the cultivation of plants may be viewed as the earliest biotechnological enterprise. Agriculture has been theorized to have become the dominant way of producing food since the Neolithic Revolution. The processes and methods of agriculture have been refined by other mechanical and biological sciences since its inception. Through early biotechnology, farmers were able to select the best suited and highest-yield crops to produce enough food to support a growing population. Other uses of biotechnology were required as crops and fields became increasingly large and difficult to maintain specific organisms and organism by-products were used to fertilize, restore nitrogen, and control pests. Throughout the use of agriculture, farmers have inadvertently altered the genetics of their crops through introducing them to new environments and breeding them with other plants—one of the first forms of biotechnology. Cultures such as those in Mesopotamia, Egypt, and India developed the process of brewing beer. It is still done by the same basic method of using malted grains (containing enzymes) to convert starch from grains into sugar and then adding specific yeasts to produce beer. In this process the carbohydrates in the grains were broken down into alcohols such as ethanol. Ancient Indians also used the juices of the plant Ephedra vulgaris and used to call it Soma. Later other cultures produced the process of Lactic acid fermentation which allowed the fermentation and preservation of other forms of food. Fermentation was also used in this time period to produce leavened bread. Although the process of fermentation was not fully understood until Louis Pasteur's work in 1857, it is still the first use of biotechnology to convert a food source into another form.

Combinations of plants and other organisms were used as medications in many early civilizations. Since as early as 200 BC, people began to use disabled or minute amounts of infectious agents to immunize themselves against infections. These and similar processes have been refined in modern medicine and have led to many developments such as antibiotics, vaccines, and other methods of fighting sickness.

In the early twentieth century scientists gained a greater understanding of microbiology and explored ways of manufacturing specific products. In 1917, Chaim Weizmann first used a pure microbiological culture in an industrial process, that of manufacturing corn starch using Clostridium acetobutylicum, to produce acetone, which the United Kingdom desperately needed to manufacture explosives during World War I.The field of modern biotechnology is thought to have largely begun on June 16, 1980, when the United States Supreme Court ruled that a genetically-modified microorganism could be patented in the case of Diamond v. Chakrabarty. Indian-born Ananda Chakrabarty, working for General Electric, had developed a bacterium (derived from the Pseudomonas genus) capable of breaking down crude oil, which he proposed to use in treating oil spills. Revenue in the industry is expected to grow by 12.9% in 2008.

APPLICATIONS OF MICROBIAL BIOTECHNOLOGY

Microbial biotechnology has applications in four major industrial areas, including health care (medical), crop production and agriculture, non food (industrial) uses of crops and other products (e.g. biodegradable plastics, vegetable oil, biofuels), and environmental uses. For example, one application of biotechnology is the directed use of organisms for the manufacture of organic products (examples include beer and milk products). Another example is using naturally present bacteria by the mining industry in bioleaching. Biotechnology is also used to recycle, treat waste, clean up sites contaminated by industrial activities (bioremediation), and also to produce biological weapons.

Whilst there are undoubtedly some who fear all microbes due to the association of some microbes with many human illnesses, many microbes are also responsible for many beneficial processes such as industrial fermentation (e.g. the production of alcohol and dairy products), antibiotic production and as vehicles for cloning in higher organisms such as plants. Scientists have also exploited their knowledge of microbes to produce biotechnologically important enzymes such as Taq polymerase, reporter genes for use in other genetic systems and novel molecular biology techniques such as the yeast two-hybrid system.

Bacteria can be used for the industrial production of amino acids. Corynebacterium glutamicum is one of the most important bacterial species with an annual production of more than two million tons of amino acids, mainly L-glutamate and L-lysine. A variety of biopolymers, such as polysaccharides, polyesters, and polyamides, are produced by microorganisms. Microorganisms are used for the biotechnological production of biopolymers with tailored properties suitable for high-value medical application such as tissue engineering and drug delivery. Microorganisms are used for the biosynthesis of xanthan, alginate, cellulose, cyanophycin, poly(gamma-glutamic acid), levan, hyaluronic acid, organic acids, oligosaccharides and polysaccharide, and polyhydroxyalkanoates.

Microorganisms are beneficial for microbial biodegradation or bioremediation of domestic, agricultural and industrial wastes and subsurface pollution in soils, sediments and marine environments. The ability of each microorganism to degrade toxic waste depends on the nature of each contaminant. Since most sites are typically comprised of multiple pollutant types, the most effective approach to microbial biodegradation is to use a mixture of bacterial species and strains, each specific to the biodegradation of one or more types of contaminants. There are also various claims concerning the contributions to human and animal health by consuming probiotics (bacteria potentially beneficial to the digestive system) and/or prebiotics (substances consumed to promote the growth of probiotic microorganisms).

Pharmacogenomics

Pharmacogenomics is the study of how the genetic inheritance of an individual affects his/her body's response to drugs. It is a coined word derived from the words "pharmacology" and "genomics". It is hence the study of the relationship between pharmaceuticals and genetics. The vision of pharmacogenomics is to be able to design and produce drugs that are adapted to each person's genetic makeup. Pharmacogenomics results in the following benefits:

1. *Development of tailor-made medicines.* Using pharmacogenomics, pharmaceutical companies can create drugs based on the proteins, enzymes and RNA molecules that are associated with specific genes and diseases. These tailor-made drugs promise

not only to maximize therapeutic effects but also to decrease damage to nearby healthy cells.

2. *More accurate methods of determining appropriate drug dosages.* Knowing a patient's genetics will enable doctors to determine how well his/ her body can process and metabolize a medicine. This will maximize the value of the medicine and decrease the likelihood of overdose.

3. *Improvements in the drug discovery and approval process.* The discovery of potential therapies will be made easier using genome targets. Genes have been associated with numerous diseases and disorders. With modern biotechnology, these genes can be used as targets for the development of effective new therapies, which could significantly shorten the drug discovery process.

4. *Better vaccines.* Safer vaccines can be designed and produced by organisms transformed by means of genetic engineering. These vaccines will elicit the immune response without the attendant risks of infection. They will be inexpensive, stable, easy to store, and capable of being engineered to carry several strains of pathogen at once.

Pharmaceutical Products

Most traditional pharmaceutical drugs are relatively simple molecules that have been found primarily through trial and error to treat the symptoms of a disease or illness. Biopharmaceuticals are large biological molecules known as proteins and these usually target the underlying mechanisms and pathways of a malady (but not always, as is the case with using insulin to treat type 1 diabetes mellitus, as that treatment merely addresses the symptoms of the disease, not the underlying cause which is autoimmunity); it is a relatively young industry. They can deal with targets in humans that may not be accessible with traditional medicines. A patient typically is dosed with a small molecule via a tablet while a large molecule is typically injected.

Small molecules are manufactured by chemistry but larger molecules are created by living cells such as those found in the human body: for example, bacteria cells, yeast cells, animal or plant cells.

Modern biotechnology is often associated with the use of genetically altered microorganisms such as E. coli or yeast for the production of substances like synthetic insulin or antibiotics. It can also refer to transgenic animals or transgenic plants, such as Bt corn. Genetically altered mammalian cells, such as Chinese Hamster Ovary (CHO) cells, are also used to manufacture certain pharmaceuticals. Another promising new biotechnology application is the development of plant-made pharmaceuticals.

Biotechnology is also commonly associated with landmark breakthroughs in new medical therapies to treat hepatitis B, hepatitis C, cancers, arthritis, haemophilia, bone fractures, multiple sclerosis, and cardiovascular disorders. The biotechnology industry has also been instrumental in developing molecular diagnostic devices than can be used to define the target patient population for a given biopharmaceutical. Herceptin, for example, was the first drug approved for use with a matching diagnostic test and is used to treat breast cancer in women whose cancer cells express the protein HER2.

Modern biotechnology can be used to manufacture existing medicines relatively easily and cheaply. The first genetically engineered products were medicines designed to treat human diseases. To cite one example, in 1978 Genentech developed synthetic humanized insulin by joining its gene with a plasmid vector inserted into the bacterium Escherichia coli. Insulin, widely used for the treatment of diabetes, was previously extracted from the pancreas of abattoir animals (cattle and/or pigs). The resulting genetically engineered bacterium enabled the production of vast quantities of synthetic human insulin at relatively low cost, although the cost savings was used to increase profits for manufacturers, not passed on to consumers or their healthcare providers. According to a 2003 study undertaken by the International Diabetes Federation (IDF) on the access to and availability of insulin in its member countries, synthetic 'human' insulin is considerably more expensive in most countries where both synthetic 'human' and animal insulin are commercially available: e.g. within European countries the average price of synthetic 'human' insulin was twice as high as the price of pork insulin. Yet in its position statement, the IDF writes that "there is no overwhelming evidence to prefer one species of insulin over another" and "[modern, highly-purified] animal insulins remain a perfectly acceptable alternative.

Modern biotechnology has evolved, making it possible to produce more easily and relatively cheaply human growth hormone, clotting factors for hemophiliacs, fertility drugs, erythropoietin and other drugs. Most drugs today are based on about 500 molecular targets. Genomic knowledge of the genes involved in diseases, disease pathways, and drug-response sites are expected to lead to the discovery of thousands more new targets.

Genetic Testing

Genetic testing involves the direct examination of the DNA molecule itself. A scientist scans a patient's DNA sample for mutated sequences.There are two major types of gene tests. In the first type, a researcher may design short pieces of DNA ("probes") whose sequences are complementary to the mutated sequences. These probes will seek their complement among the base pairs of an individual's genome. If the mutated sequence is present in the patient's genome, the probe will bind to it and flag the mutation. In the second type, a researcher may conduct the gene test by comparing the sequence of DNA bases in a patient's gene to disease in healthy individuals or their progeny.

Some genetic tests are already available, although most of them are used in developed countries. The tests currently available can detect mutations associated with rare genetic disorders like cystic fibrosis, sickle cell anemia, and Huntington's disease. Recently, tests have been developed to detect mutation for a handful of more complex conditions such as breast, ovarian, and colon cancers. However, gene tests may not detect every mutation associated with a particular condition because many are as yet undiscovered, and the ones they do detect may present different risks to different people and populations. Several issues have been raised regarding the use of genetic testing:

1. *Absence of cure.* There is still a lack of effective treatment or preventive measures for many diseases and conditions now being diagnosed or predicted using gene tests. Thus, revealing information about risk of a future disease that has no existing cure presents an ethical dilemma for medical practitioners.

2. *Ownership and control of genetic information.* Who will own and control genetic information, or information about genes, gene products, or inherited characteristics derived from an individual or a group of people like indigenous communities? At the macro level, there is a possibility of a genetic divide, with developing countries that do not have access to medical applications of biotechnology being deprived of benefits accruing from products derived from genes obtained from their own people. Moreover, genetic information can pose a risk for minority population groups as it can lead to group stigmatization.

At the individual level, the absence of privacy and anti-discrimination legal protections in most countries can lead to discrimination in employment or insurance or other misuse of personal genetic information. This raises questions such as whether genetic privacy is different from medical privacy.

1. *Reproductive issues.* These include the use of genetic information in reproductive decision-making and the possibility of genetically altering reproductive cells that may be passed on to future generations. For example, germline therapy forever changes the genetic make-up of an individual's descendants. Thus, any error in technology or judgment may have far-reaching consequences. Ethical issues like designer babies and human cloning have also given rise to controversies between and among scientists and bioethicists, especially in the light of past abuses with eugenics.

2. *Clinical issues.* These center on the capabilities and limitations of doctors and other health-service providers, people identified with genetic conditions, and the general public in dealing with genetic information.

3. *Effects on social institutions.* Genetic tests reveal information about individuals and

their families. Thus, test results can affect the dynamics within social institutions, particularly the family.

4. *Conceptual and philosophical implications regarding human responsibility,* free will vis-à-vis genetic determinism, and the concepts of health and disease.

Gene Therapy

Gene therapy using an Adenovirus vector. A new gene is inserted into an adenovirus vector, which is used to introduce the modified DNA into a human cell. If the treatment is successful, the new gene will make a functional protein.

Gene therapy may be used for treating, or even curing, genetic and acquired diseases like cancer and AIDS by using normal genes to supplement or replace defective genes or to bolster a normal function such as immunity. It can be used to target somatic (i.e., body) or gametes (i.e., egg and sperm) cells. In somatic gene therapy, the genome of the recipient is changed, but this change is not passed along to the next generation. In contrast, in germline gene therapy, the egg and sperm cells of the parents are changed for the purpose of passing on the changes to their offspring.

There are basically two ways of implementing a gene therapy treatment:

1. Ex vivo, which means "outside the body" – Cells from the patient's blood or bone marrow are removed and grown in the laboratory. They are then exposed to a virus carrying the desired gene. The virus enters the cells, and the desired gene becomes part of the DNA of the cells. The cells are allowed to grow in the laboratory before being returned to the patient by injection into a vein.
2. In vivo, which means "inside the body" – No cells are removed from the patient's body. Instead, vectors are used to deliver the desired gene to cells in the patient's body.

Currently, the use of gene therapy is limited. Somatic gene therapy is primarily at the experimental stage. Germline therapy is the subject of much discussion but it is not being actively investigated in larger animals and human beings. As of June 2001, more than 500 clinical gene-therapy trials involving about 3,500 patients have been identified worldwide. Around 78% of these are in the United States, with Europe having 18%. These trials focus on various types of cancer, although other multigenic diseases are being studied as well. Recently, two children born with severe combined immunodeficiency disorder ("SCID") were reported to have been cured after being given genetically engineered cells.

Gene therapy faces many obstacles before it can become a practical approach for treating disease. At least four of these obstacles are as follows:

1. *Gene delivery tools.* Genes are inserted into the body using gene carriers called vectors. The most common vectors now are viruses, which have evolved a way of encapsulating and delivering their genes to human cells in a pathogenic manner. Scientists manipulate the genome of the virus by removing the disease-causing genes and inserting the therapeutic genes. However, while viruses are effective, they can introduce problems like toxicity, immune and inflammatory responses, and gene control and targeting issues. In addition, in order for gene therapy to provide permanent therapeutic effects, the introduced gene needs to be integrated within the host cell's genome. Some viral vectors effect this in a random fashion, which can introduce other problems such as disruption of an endogenous host gene.
2. *High costs.* Since gene therapy is relatively new and at an experimental stage, it is an expensive treatment to undertake. This explains why current studies are focused on illnesses commonly found in developed countries, where more people can afford to pay for treatment. It may take decades before developing countries can take advantage of this technology.
3. *Limited knowledge of the functions of genes.* Scientists currently know the functions of only a few genes. Hence, gene therapy can address only some genes that cause a particular disease. Worse, it is not known exactly whether genes have more than one function, which creates uncertainty as to whether replacing such genes is indeed desirable.
4. *Multigene disorders and effect of environment.* Most genetic disorders involve more than one gene. Moreover, most diseases involve the interaction of several genes and the environment. For example, many people with cancer not only inherit the disease gene for the disorder, but may have also failed to inherit specific tumor suppressor genes. Diet, exercise, smoking and other environmental factors may have also contributed to their disease.

Improving Agricultural Yield

Using the techniques of modern biotechnology, one or two genes may be transferred to a highly developed crop variety to impart a new character that would increase its yield. However, while increases in crop yield are the most obvious applications of modern biotechnology in agriculture, it is also the most difficult one. Current genetic engineering techniques work best for effects that are controlled by a single gene. Many of the genetic characteristics associated with yield (e.g., enhanced growth) are controlled by a large number of genes, each of which has a minimal effect on the overall yield. There is, therefore, much scientific work to be done in this area.

Crops containing genes that will enable them to withstand biotic and abiotic stresses may be developed. For example, drought and excessively salty soil are two important limiting factors in crop productivity. Biotechnologists are studying plants that can cope with these extreme conditions in the hope of finding the genes that enable them to do so and eventually transferring these genes to the more desirable crops. One of the latest developments is the identification of a plant gene, At-DBF2, from thale cress, a tiny weed that is often used for plant research because it is very easy to grow and its genetic code is well mapped out. When this gene was inserted into tomato and tobacco cells, the cells were able to withstand environmental stresses like salt, drought, cold and heat, far more than ordinary cells. If these preliminary results prove successful in larger trials, then At-DBF2 genes can help in engineering crops that can better withstand harsh environments. Researchers have also created transgenic rice plants that are resistant to rice yellow mottle virus (RYMV). In Africa, this virus destroys majority of the rice crops and makes the surviving plants more susceptible to fungal infections.

Modern biotechnology can be used to slow down the process of spoilage so that fruit can ripen longer on the plant and then be transported to the consumer with a still reasonable shelf life. This may improve the taste, texture and appearance of the fruit. More importantly, it could expand the market for farmers in developing countries due to the reduction in spoilage. However, there is sometimes a lack of understanding by researchers in developed countries about the actual needs of prospective beneficiaries in developing countries. For example, engineering soybeans to resist spoilage makes them less suitable for producing tempeh which is a significant source of protein that depends on fermentation. The use of modified soybeans results in a lumpy texture that is less palatable and less convenient when cooking.

Biotechnology in Cheese Production

Enzymes produced by micro-organisms provide an alternative to animal rennet – a cheese coagulant - and an alternative supply for cheese makers. This also eliminates possible public concerns with animal-derived material, although there is currently no plans to develop synthetic milk, thus making this argument less compelling. Enzymes offer an animal-friendly alternative to animal rennet. While providing comparable quality, they are theoretically also less expensive.

About 85 million tons of wheat flour is used every year to bake bread. By adding an enzyme called maltogenic amylase to the flour, bread stays fresher longer. Assuming that 10-15% of bread is thrown away, if it could just stay fresh another 5–7 days then 2 million tons of flour per year would be saved. That corresponds to 40% of the bread consumed in a country such as the USA. This means more bread becomes available with no increase in input. In combination with other enzymes, bread can also be made bigger, more appetizing and better in a range of ways.

Biofertilizers, Biopesticides and other Agrochemicals

Most of the current commercial applications of modern biotechnology in agriculture are on reducing the dependence of farmers on agrochemicals. For example, Bacillus thuringiensis (Bt) is a soil bacterium that produces a protein with insecticidal qualities. Traditionally, a fermentation process has been used to produce an insecticidal spray from these bacteria. In this form, the Bt toxin occurs as an inactive protoxin, which requires digestion by an insect to be effective. There are several Bt toxins and each one is specific to certain target insects. Crop plants have now been engineered to contain and express the genes for Bt toxin, which they produce in its active form. When a susceptible insect ingests the transgenic crop cultivar expressing the Bt protein, it stops feeding and soon thereafter dies as a result of the Bt toxin binding to its gut wall. Bt corn is now commercially available in a number of countries to control corn borer (a lepidopteran insect), which is otherwise controlled by spraying (a more difficult process).

Crops have also been genetically engineered to acquire tolerance to broad-spectrum herbicide. The lack of cost-effective herbicides with broad-spectrum activity and no crop injury was a consistent limitation in crop weed management. Multiple applications of numerous herbicides were routinely used to control a wide range of weed species detrimental to agronomic crops. Weed management tended to rely on preemergence – that is, herbicide applications were sprayed in response to expected weed infestations rather than in response to actual weeds present. Mechanical cultivation and hand weeding were often necessary to control weeds not controlled by herbicide applications. The introduction of herbicide tolerant crops has the potential of reducing the number of herbicide active ingredients used for weed management, reducing the number of herbicide applications made during a season, and increasing yield due to improved weed management and less crop injury. Transgenic crops that express tolerance to glyphosate, glufosinate and bromoxynil have been developed. These herbicides can now be sprayed on transgenic crops without inflicting damage on the crops while killing nearby weeds.

Production of Novel Substances

Biotechnology is being applied for novel uses other than food. For example, oilseed can be modified to produce fatty acids for detergents, substitute fuels and petrochemicals. Potatoes, tomatos, rice, tobacco, lettuce, safflowers, and other plants have been genetically-engineered to produce insulin and certain vaccines. If future clinical trials prove successful, the advantages of edible vaccines would be enormous, especially for developing countries. The transgenic plants may be grown locally and cheaply. Homegrown vaccines would also avoid logistical and economic problems posed by having to transport traditional preparations over long distances and keeping them cold while in transit. And since they are edible, they will not need syringes, which are n only an additional expense in the traditional vaccine preparations but also a source infections if contaminated. In the case of insulin grown in transgenic plants, it is wel established that the gastrointestinal system breaks the protein down therefore this coul not currently be administered as an edible protein. However, it might be produced significantly lower cost than insulin produced in costly, bioreactors. For exampl Calgary, Canada-based SemBioSys Genetics, Inc. reports that its safflower-produce insulin will reduce unit costs by over 25% or more and approximates a reduction the capital costs associated with building a commercial-scale insulin manufacturir facility of over $100 million, compared to traditional biomanufacturing facilities.

Biological Engineering

Biotechnological engineering or biological engineering is a branch of engineering th focuses on biotechnologies and biological science. It includes different disciplines suc as biochemical engineering, biomedical engineering, bio-process engineering, biosyste engineering and so on. Because of the novelty of the field, the definition of a bioengine is still undefined. However, in general it is an integrated approach of fundament biological sciences and traditional engineering principles.

Bioengineers are often employed to scale up bio processes from the laboratory sca to the manufacturing scale. Moreover, as with most engineers, they often deal wi management, economic and legal issues. Since patents and regulation (e.g., U.S. Foc and Drug Administration regulation in the U.S.) are very important issues for biote enterprises, bioengineers are often required to have knowledge related to these issue

The increasing number of biotech enterprises is likely to create a need f bioengineers in the years to come. Many universities throughout the world are no providing programs in bioengineering and biotechnology (as independent programs specialty programs within more established engineering fields).

Bioremediation and Biodegradation

Biotechnology is being used to engineer and adapt organisms especially microorganisr in an effort to find sustainable ways to clean up contaminated environments. Tl elimination of a wide range of pollutants and wastes from the environment is an absolu requirement to promote a sustainable development of our society with lo environmental impact. Biological processes play a major role in the removal contaminants and biotechnology is taking advantage of the astonishing catabo versatility of microorganisms to degrade/convert such compounds. New methodologic breakthroughs in sequencing, genomics, proteomics, bioinformatics and imaging a producing vast amounts of information. In the field of environmental microbiolog enome-based global studies open a new era providing unprecedented in silico views f metabolic and regulatory networks, as well as clues to the evolution of degradation athways and to the molecular adaptation strategies to changing environmental onditions. Functional genomic and metagenomic approaches are increasing our nderstanding of the relative importance of different pathways and regulatory networks carbon flux in particular environments and for particular compounds and they will ertainly accelerate the development of bioremediation technologies and iotransformation processes.

Marine environments are especially vulnerable since oil spills of coastal regions and ıe open sea are poorly containable and mitigation is difficult. In addition to pollution ırough human activities, millions of tons of petroleum enter the marine environment very year from natural seepages. Despite its toxicity, a considerable fraction of etroleum oil entering marine systems is eliminated by the hydrocarbon-degrading tivities of microbial communities, in particular by a remarkable recently discovered oup of specialists, the so-called hydrocarbonoclastic bacteria (HCCB).

Naturally occurring bioremediation and phytoremediation have been used for nturies. For example, desalination of agricultural land by phytoextraction has a long adition. Bioremediation technology using microorganisms was reportedly invented by eorge M. Robinson. He was the assistant county petroleum engineer for Santa Maria, alifornia. During the 1960's, he spent his spare time experimenting with dirty jars and rious mixes of microbes.

Bioremediation technologies can be generally classified as in situ or ex situ. In situ oremediation involves treating the contaminated material at the site while ex situ volves the removal of the contaminated material to be treated elsewhere. Some camples of bioremediation technologies are bioventing, landfarming, bioreactor, mposting, bioaugmentation, rhizofiltration, and biostimulation.Not all contaminants, owever, are easily treated by bioremediation using microorganisms. For example, avy metals such as cadmium and lead are not readily absorbed or captured by ganisms. The assimilation of metals such as mercury into the food chain may worsen atters. Phytoremediation is useful in these circumstances, because natural plants or ansgenic plants are able to bioaccumulate these toxins in their above-ground parts, hich are then harvested for removal. The heavy metals in the harvested biomass may further concentrated by incineration or even recycled for industrial use.

The elimination of a wide range of pollutants and wastes from the environment quires increasing our understanding of the relative importance of different pathways d regulatory networks to carbon flux in particular environments and for particular mpounds and they will certainly accelerate the development of bioremediation chnologies and biotransformation processes.

The use of genetic engineering to create organisms specifically designed for bioremediation has great potential. The bacterium Deinococcus radiodurans (the most radioresistant organism known) has been modified to consume and digest toluene and ionic mercury from highly radioactive nuclear waste.

Mycoremediation is a form of bioremediation, the process of using fungi to return an environment (usually soil) contaminated by pollutants to a less contaminated state. The term mycoremediation was coined by Paul Stamets and refers specifically to the use of fungal mycelia in bioremediation. One of the primary roles of fungi in the ecosystem is decomposition, which is performed by the mycelium. The mycelium secretes extracellular enzymes and acids that break down lignin and cellulose, the two main building blocks of plant fiber. These are organic compounds composed of long chains of carbon and hydrogen, structurally similar to many organic pollutants. The key to mycoremediation is determining the right fungal species to target a specific pollutant. Certain strains have been reported to successfully degrade the nerve gases VX and sarin.

In an experiment conducted in conjunction with Thomas, a major contributor in the bioremediation industry, a plot of soil contaminated with diesel oil was inoculated with mycelia of oyster mushrooms; traditional bioremediation techniques (bacteria) were used on control plots. After four weeks, more than 95% of many of the PAH (polycyclic aromatic hydrocarbons) had been reduced to non-toxic components in the mycelial-inoculated plots. It appears that the natural microbial community participates with the fungi to break down contaminants, eventually into carbon dioxide and water. Wood-degrading fungi are particularly effective in breaking down aromatic pollutants (toxic components of petroleum), as well as chlorinated compounds.

There are a number of cost/efficiency advantages to bioremediation, which can be employed in areas that are inaccessible without excavation. For example, hydrocarbon spills (specifically, petrol spills) or certain chlorinated solvents may contaminate groundwater, and introducing the appropriate electron acceptor or electron donor amendment, as appropriate, may significantly reduce contaminant concentrations after a lag time allowing for acclimation. This is typically much less expensive than excavation followed by disposal elsewhere, incineration or other ex situ treatment strategies, and reduces or eliminates the need for "pump and treat", a common practice at sites where hydrocarbons have contaminated clean groundwater.

REFERENCES

Biotechnology Industry Organization. *Biotechnology in Perspective.* Washington, D.C.: Biotechnology Industry Organization. 1990.

Ginzburg, Lev. R. *Assessing Ecological Risks of Biotechnology.* Boston: Butterworth-Heinemann. 1991.

Margaret G. Mellon. *Biotechnology and the Environment : A Primer on the Environmental Implications of Genetic Engineering.*

2

Genetic Engineering of Microorganisms

Genetic engineering, recombinant DNA technology, genetic modification/manipulation (GM) and gene splicing are terms that apply to the direct manipulation of an organism's genes. Genetic engineering is different from traditional breeding, where the organism's genes are manipulated indirectly. Genetic engineering uses the techniques of molecular cloning and transformation to alter the structure and characteristics of genes directly. Genetic engineering techniques have found some successes in numerous applications. Some examples are in improving crop technology, the manufacture of synthetic human insulin through the use of modified bacteria, the manufacture of erythropoietin in hamster ovary cells, and the production of new types of experimental mice such as the oncomouse (cancer mouse) for research.

There are a number of ways through which genetic engineering is accomplished. Essentially, the process has five main steps:

1. Isolation of the genes of interest
2. Insertion of the genes into a transfer vector
3. Transfer of the vector to the organism to be modified
4. Transformation of the cells of the organism
5. Selection of the genetically modified organism (GMO) from those that have not been successfully modified

Isolation is achieved by identifying the gene of interest that the scientist wishes to insert into the organism, usually using existing knowledge of the various functions of genes. DNA information can be obtained from cDNA or libraries, and amplified using techniques. If necessary, i.e. for insertion of eukaryotic genomic DNA into prokaryotes, further modification may be carried out such as removal of introns or ligating prokaryotic promoters. Insertion of a gene into a vector such as a plasmid can be done once the gene of interest is isolated.

Once the vector is obtained, it can be used to transform the target organism. Depending on the vector used, it can be complex or simple. For example, using raw DNA with gene guns is a fairly straightforward process but with low success rates, where the DNA is coated with molecules such as gold and fired directly into a cell. Other more complex methods, such as bacterial transformation or using viruses as vectors have higher success rates. After transformation, the GMO can be selected from those that have failed to take up the vector in various ways. One method is screening with DNA probes that can stick to the gene of interest that was supposed to have been transplanted. Another is to package genes conferring resistance to certain chemicals such as antibiotics or herbicides into the vector. This chemical is then applied ensuring that only those cells that have taken up the vector will survive.

The first genetically engineered medicine was synthetic human insulin, approved by the United States Food and Drug Administration in 1982. Another early application of genetic engineering was to create human growth hormone as replacement for a compound that was previously extracted from human cadavers. In 1987 the FDA approved the first genetically engineered vaccine for humans, for hepatitis B. Since these early uses of the technology in medicine, the use of GM has gradually expanded to supply a number of other drugs and vaccines. One of the best-known applications of genetic engineering is the creation of genetically modified organisms (GMOs) such as foods and vegetables that resist pest and bacteria infection and have longer freshness than otherwise. Although there has been a revolution in the biological sciences in the past twenty years, there is still a great deal that remains to be discovered. Expedient and inexpensive access to comprehensive genetic data has become a reality with billions of sequenced nucleotides already online and annotated.

Loss of function experiments, such as in a gene knockout experiment, in which an organism is engineered to lack the activity of one or more genes. This allows the experimenter to analyze the defects caused by this mutation, and can be considerably useful in unearthing the function of a gene. It is used especially frequently in developmental biology. Another method, useful in organisms such as (fruitfly), is to induce mutations in a large population and then screen the progeny for the desired mutation. Gain of function experiments, the logical counterpart of knockouts. These are sometimes performed in conjunction with knockout experiments to more finely establish the function of the desired gene. The process is much the same as that in knockout engineering, except that the construct is designed to increase the function of the gene, usually by providing extra copies of the gene or inducing synthesis of the protein more frequently. Tracking experiments, which seek to gain information about the localization and interaction of the desired protein. One way to do this is to replace the wild-type gene with a 'fusion' gene, which is a juxtaposition of the wild-type gene with a reporting element such as Green Fluorescent Protein (GFP) that will allow easy visualization of the products of the genetic modification. While this is a useful technique, the manipulation can destroy the function of the gene, creating secondary effects and possibly calling into question the results of the experiment. More sophisticated techniques are now in development that can track protein products without mitigating their function, such as the addition of small sequences that will serve as binding motifs to monoclonal antibodies.

Expression studies aim to discover where and when specific proteins are produced. In these experiments, the DNA sequence before the DNA that codes for a protein, known as a gene's promoter, is reintroduced into an organism with the protein coding region replaced by a reporter gene such as GFP or an enzyme that catalyzes the production of a dye. Thus the time and place where a particular protein is produced can be observed. Expression studies can be taken a step further by altering the promoter to find which pieces are crucial for the proper expression of the gene and are actually bound by transcription factor proteins.

GENETICALLY MODIFIED ORGANISMS

A genetically modified organism (GMO) or genetically engineered organism (GEO) is an whose material has been altered using genetic engineering techniques. These techniques, generally known as recombinant DNA technology, use DNA molecules from different sources, which are combined into one molecule to create a new set of genes. This DNA is then transferred into an organism, giving it modified or novel genes. Transgenic organisms, a subset of GMOs, are organisms which have inserted DNA that originated in a different species. Some GMOs contain no DNA from other species and are therefore not transgenic but cisgenic.

The general principle of producing a GMO is to add new genetic material into an organism's genome. This is called genetic engineering and was made possible through the discovery of DNA and the creation of the first recombinant bacteria in 1973, i.e., E .coli expressing a salmonella gene. This led to concerns in the scientific community about potential risks from genetic engineering, which were thoroughly discussed at the . One of the main recommendations from this meeting was that government oversight of recombinant DNA research should be established until the technology was deemed safe. Herbert Boyer then founded the first company to use recombinant DNA technology, Genentech, and in 1978 the company announced creation of an *E. coli* strain producing the human protein insulin.

In 1986, field tests of bacteria genetically engineered to protect plants from frost damage (ice-minus bacteria) at a small biotechnology company called Advanced Genetic Sciences of California, were repeatedly delayed by opponents of biotechnology. In the same year, a proposed field test of a microbe genetically engineered for a pest resistance protein by Monsanto was dropped. GMOs have widespread applications. They are used in biological and medical research, production of pharmaceutical drugs, experimental medicine (e.g. gene therapy), and agriculture (e.g. golden rice). The term "genetically modified organism" does not always imply, but can include, targeted insertions of genes from one species into another. For example, a gene from a jellyfish, encoding a fluorescent protein called GFP, can be physically linked and thus co-expressed with mammalian genes to identify the location of the protein encoded by the GFP-tagged gene in the mammalian cell. Such methods are useful tools for in many areas of research, including those who study the mechanisms of human and other diseases or fundamental biological processes in eukaryotic or prokaryotic cells.To date the broadest application of GMO technology is patent-protected food crops which are resistant to commercial herbicides or are able to produce pesticidal proteins from within the plant, or stacked trait seeds, which do both. The largest share of the GMO crops planted globally are owned by Monsanto according to the company. In 2007, Monsanto's trait technologies were planted on 246 million acres (1,000,000 km^2) throughout the world, a growth of 13 percent from 2006.

In the corn market, Monsanto's triple-stack corn – which combines Roundup Ready 2 weed control technology with YieldGard Corn Borer and YieldGard Rootworm insect control – is the market leader in the United States. U.S. corn farmers planted more than 17 million acres (69,000 km^2) of triple-stack corn in 2007, and it is estimated the product could be planted on 45 million to 50 million acres (200,000 km^2) by 2010. In the cotton market, Bollgard II with Roundup Ready Flex was planted on nearly 3 million acres (12,000 km^2) of U.S. cotton in 2007.Rapid growth in the total area planted is measurable by Monsanto's growing share. On January 3, 2008, Monsanto Company (MON.N) said its quarterly profit nearly tripled, helped by strength in its corn seed and herbicide businesses, and raised its 2008 forecast.to the International Service for the Acquisition of Agri-Biotech Applications (ISAAA), of the approximately 8.5 million farmers who grew biotech crops in 2005, some 90% were resource-poor farmers in developing countries.

TRANSGENIC MICROBES

A transgenic microorganism is a microbe, usually a bacterium, into which genetic information has been introduced from the outside and which possesses the ability to pass that information on to subsequent generations in astable manner. This is not an entirely novel idea in microorganisms, since bacteria have been practicing and perfecting this art over billions of years of evolution. We, on the other hand, have only recently learned to duplicate this phenomenon and turn it to our own purposes. Genetic engineering is the field that has developed as a consequence of research into this process. Its commercial application forms the basis of the biotechnology industry today.

The process by which scientists introduce new genetic material into a microorganism is called molecular or gene cloning. It involves the isolation of DNA from a source other than the microorganism itself. Source organisms span the world of living things, from microbes to plants to animals, including humans. Scientists obtain source DNA in several different ways: by disrupting cells of the target microbe (or plant or animal) and fragmenting it into small pieces, by synthesizing it from an RNA template using an enzyme called reverse transcriptase, or by knowing the specific gene sequence and synthesizing it directly in the laboratory. Once obtained, the pieces of DNA are inserted into a small genetic component that has the ability to make copies of itself (replicate) independently from the microbial genome. This self-replicating unit is called a cloning vector. Although these genetic elements exist naturally in the form of plasmids and bacterial viruses, many of the ones used today have been altered to improve their properties for transferring genes. Restriction enzymes, which nick the donor DNA and the cloning vector at specific sites, and DNA ligase, which attaches the donor DNA to the cloning vector, allow the source genes of interest to be inserted into the cloning vector without disrupting its ability to replicate.

The next step in the process is the introduction of the cloning vector with its segment of new DNA into a living cell. Bacteria have the ability to transport DNA into their cells in a process called transformation, and this ability is commonly exploited to achieve this goal. Getting the DNA into the cell, however, is only the beginning. No transformation is 100 percent efficient, and so the bacteria that receive the gene(s) of interest must be separated from those that did not. One of the best studied and most commonly used cloning vectors, pBR322, is especially useful for this purpose, as it contains several genes for antibiotic resistance. Hence, any cell transformed with DNA containing pBR322 will be antibiotic resistant, and thus can be isolated from similar cells that have not be so transformed by merely growing them in the presence of the appropriate drugs. All that remains is to identify bacteria that are producing the product of the desired gene(s), and cloning is a success.

The introduction of human genes into bacteria has several complicating wrinkles that make cloning them even more challenging. For example, a bacterial gene codes for a protein from start to finish in one long string of nucleotides, whereas human cells have stretches of noncoding nucleotides called introns within their genes. Bacteria do not have the same ability as human cells to remove these introns when producing proteins from the gene, and if the introns are not removed, the intended protein cannot be produced. This, along with other complications, has been overcome using many of the tools of genetic engineering.

Transgenic microbes have many commercial and practical applications, including the production of mammalian products. A company called Genentech was among the earliest and most successful commercial enterprises to use genetically engineered bacteria to produce human proteins. Their first product was human insulin produced by genetically engineered Escherichia coli. A variety of other human hormones, blood proteins, and immune modulators are now produced in a similar fashion, in addition to vaccines for such infectious agents as hepatitis B virus and measles.

Another promising application of genetically engineered microbes is in environmental cleanup, or biomediation. Scientists have discovered many naturally occurring genes that code for enzymes that degrade toxic wastes and wastewater pollutants in bacteria. Examples include genes for degrading chlorinated pesticides, chlorobenzenes, naphthalene, toluene, anilines, and various hydrocarbons. Researchers are using molecular cloning to introduce these genes from several different microbes into a single microbe, creating "super microbes" with the ability to degrade multiple contaminants.

Ananda Chakrabarty created one of the first microbes of this nature in the early 1970s. He introduced genes from several different bacteria into a strain of Burkholderia cepacia, giving it the ability to degrade toxic compounds found in petroleum. This microbe offered a potential alternative to skimming and absorbing spilled oil. Chakrabarty's genetically modified bacterium has never been used, however, due to public concerns about the release of genetically engineered microbes into the environment. The microbe did, on the other hand, play an important role in establishing the biotechnology industry. The U.S. Patent Office granted Chakrabarty the first patent ever for the construction and use of a genetically engineered bacterium. This established a precedent allowing biotechnology companies to protect their "inventions" in the same way chemical and pharmaceutical companies have done in the past. Bacteria were the first organisms to be modified in the laboratory, due to their simple genetics. These organisms are now used for several purposes, and are particularly important in producing large amounts of pure human for use in medicine.

Genetically modified bacteria are used to produce the protein insulin to treat diabetes. Similar bacteria have been used to produce clotting factors to treat haemophilia,and human growth hormone to treat various forms of diesases. These recombinant proteins are safer than the products they replaced, since the older products were purified from cadavers and could transmit diseases. Indeed the human-derived proteins caused many cases of AIDS and hepatitis C in haemophilliacs and Creutzfeldt-Jakob disease from human growth hormone. In addition to bacteria being used for producing proteins, genetically modified viruses allow gene therapy, which is a relatively new idea in medicine. A virus reproduces by injecting its own genetic material into an existing cell. That cell then follows the instructions in this genetic material and produces more viruses. In medicine, this process is engineered to deliver a gene that could cure disease into human cells. Although gene therapy is still relatively new, it has had some successes. The bacteria which cause tooth decay are called Streptococcus mutans. These bacteria consume leftover sugars in the mouth, producing that corrodes and ultimately causes cavities. Scientists have recently modified Streptococcus mutans to produce no lactic acid. These transgenic bacteria, if properly colonized in a person's mouth, could reduce the formation of cavities.Transgenic microbes have also been used in recent research to kill or hinder tumors, and to fight Crohn's disease. Genetically modified bacteria are also used in some soils to facilitate crop growth, and can also produce chemicals which are toxic to crop pests.

APPLICATIONS OF GEMS

Genetically Engineered Microorganisms (GEMS)—A large number of human genes encoding pharmaceutically valuable proteins have been cloned and expressed in microorganisms. Initially, E. coli was used as the host for obvious reasons of ease in cloning. But yeast is fast becoming the host of choice for production of recombinant proteins. Several of the recombinant proteins are used for treatment of diabetes mellitus (protein: insulin), dwarfism [protein: human growth hormone (hGI I)], cancer (proteins: interferons, interleukins, granulocyte macrophage colony stimulating factor), thrombosis (streptokinase), and AIDS (e.g., interferons, granulocyte macrophage colony stimulating factor). Many other useful recombinant proteins are in advanced stages of development. Human insulin is a dimer comprising one chain of 21 amino acids (A chain) and the other of 30 amino acids (B chain). Both chains A and B are derived from a single polypeptide chain, and are held together by two disulphide bridges.

The genes (DNA sequences) for chains A and B of insulin were synthesized separately as early as 1978. These genes were integrated separately in a pBR322 type vector translational fusion with lacZa. The A and B chains were separated by a methionine residue from the ß -galactosidase sequence encoded by lacZa. Therefore, the insulin sequences were separated from the ß-galactosidase sequence by treating the fusion proteins with cyanogen bromide. The purified chains A and B were then attached to each other by disulphide bonds induced in vitro; this, however, turned out to be an inefficient reaction.Subsequently, a gene representing B, e and A chains was synthesized and expressed in E. coli; in this case, the intervening chain is removed proteolytic ally following spontaneous folding of the proinsulin molecule.

Live Recombinant Vaccines

An approach holding considerable promise employs a live vector for the delivery of immunogen encoding gene into the vaccinated individuals. The vectors are, vaccinia viruses, adenoviruses, *E. coli,* Salmonella typhimurium, etc.The concerned gene is introduced into the genome of selected viral/ bacterial vector, which is suitably attenuated, and the live microorganism is used for vaccination. Of the various vectors studied, vaccinia virus appears to be the most promising. Vaccinia virus (VV) is a close relative of the variola virus causing small pox and was used as the vaccine to generate protection against small pox. Generally, antigen encoding genes are inserted within its thymidine kinase (TK) locus, which makes the virus TK- and attenuates its pathogenicity. Further attenuation of VV can be achieved by integration, in its genome, of lymphokine genes like interferon gamma or interleukin-2 (IL-2). A large number of genes encoding antigenic proteins have been integrated into the VV genome, which was then used for vaccination.

Recently, a highly effective vaccine against rinderpest virus has been developed by inserting the viral genes Hand F in the VV genes TK and HA. Cattle immunized with the recombinant VV vaccine were completely protected even when they were challenged by a more than 1000 times the normally lethal inoculum. A VV expressing rabies virus glycoprotein (G protein) is widely used vaccine.

Recombinant VV

(i) is not transmitted from vaccinated to contact animals,

(ii) and induces both humoral and cellular immune responses. But

(i) individuals previously immunised or exposed to infection by VV may respond poorly to recombinant VV vaccines. In addition,

(ii) children and adults with congenital or acquired immunodeficiency may run the risk of severe infections; this could be resolved by incorporating into the VV genome the gene I L-2. DNA vaccines offer the following advantages:

(i) purification and preparation of DNA for vaccines is easier, cheaper and more rapid,

(ii) they are safer and more specific because of high purity, and

(iii) they elicit a more potent immune response than purified protein vaccines.

Disease Diagnosis

An accurate diagnosis of the disease and its casual organism is critical to its effective management and cure. Conventionally, disease diagnosis is based on the following.

1. Microscopy. The specimen (tissue, body fluid, excreta, pus, exudates etc.) are subjected to microscopic examination for detection of the causal organisms, e.g., stool examination for ova and cyst.

2. Culture of the specimen on specific and selective media to allow specific pathogens to grow, which are then tested for their susceptibility to various therapeutic agents, e.g., antibiotics.

3. Immunological assays for specific antigens present on the surface of pathogens.
4. Detection and measurement of the pathogen specific antibodies produced by the patient in response to the invasion by pathogen.

These tests are often tedious, take long time, may yield ambiguous results, and some of them can not be applied in certain cases. Novel diagnostic approaches have been developed by biotechnology, which are precise and very rapid, viz.,

(i) probes and

(ii) monoclonal antibodies.

Probes

Probes—Probes are small (15-30 bases long) nucleotide (DNA/RNA) sequences used to detect the presence of complementary sequences in nucleic acid samples. Both DNA and RNA are used as probes. The probes can be prepared in many ways, and are either radioactively or non radioactively labelled.

Probes are being used in clinical diagnosis for the detection of microorganisms in various samples, e.g., tissues, excreta, body fluids, etc. Use of probes for disease diagnosis offers several advantages over the conventional diagnostic tools, which are briefly listed below.

1. They are highly specific, relatively rapid and much simpler.
2. They are extremely powerful especially when combined with PCR; even a single molecule in the test sample can be detected.
3. Since culture of microbes is not required, the risk of accidental infection to laboratory personnel is eliminated.
4. It is applicable to even such organisms, which can not be cultured.
5. Probes detect even latent viral infections, which do not lead to an increase in antibody titre in the blood.
6. A single species specific probe can identify all the serotypes of a pathogen.
7. Pure probe preparations are relatively easily obtained.

However, since probes are usually radioactively labelled, they present a health hazard in handling and disposal. Therefore, the emphasis is shifting to non radioactively labelled probes. Probes are available for the detection of a variety of pathogenic micro organisms. Probes can be used for hybridization (dot blot, Southern, in situ) or ligase chain reaction (LCR).

Hybridization

DNA may be isolated from the test samples and subjected to Southern blot or dot blot hybridization with the probe. For dot blot analyses, test samples like blood are generally lysed directly on the nitrocellulose filter. A probe can hybridize with a test DNA sample only when the latter contains the complementary sequence. The probes used in diagnostic assays are highly specific to the concerned pathogenic microorganisms. Therefore, a positive hybridization signal of a test DNA sample with a given probe reveals the presence of concerned microorganism. Probes are used for hybridization assays using microscopic preparations of tissues. Generally, the tissues are fixed in formalin, embedded in paraffin, sectioned and stained with conventional stains like eosin and haematoxylin for routine examination. Subsequently, probes are used for in situ hybridization to detect the presence of concerned pathogens. This approach has proved quite useful for the detection of viral pathogens.

Obtaining Foetal Cells

Earlier foetal cells were obtained by amniocentesis, i.e., withdrawal of amniotic fluid (which has free cells of foetal origin) with the help of a hypodermic syringe. But amniocentesis is applicable only 18 weeks or later after the pregnancy, which is rather late for abortion. Therefore, foetal cells are now obtained from biopsies of trophoblastic villi, which are an external part of human embryo and later form a part of the placenta The biopsy is performed during 6th or 8th week of pregnancy (using an endoscope passed through the cervix of uterus); it usually provides 100 μg of pure foetal DNA.

Disease Detection

The foetal cells are used for detection of the genetic disorders in one of the following ways.

1. Determination of karyotype of cells provides information on various syndromes produced by gross chromosomal aberrations.
2. Most of the genetic diseases produce defective proteins/enzymes or no enzymes many of these proteins have been identified and some of them (at least 35 genetic diseases) can be assayed. The foetal cells are used to assay the concerned enzyme activities to detect such genetic diseases.
3. In case of some genetic diseases, the concerned gene mutation may alter the recognition site for a restriction enzyme. The RFLP so produced can be detected by Southern hybridization; a sequence of the concerned gene is used as probe. For example, in case of sickle cell anaemia, the mutation from GAG to GTG eliminates a recognition site for the restriction enzyme MstII (CCTNAGG) in the ß- globin gene

of haemoglobin. DNAs from a normal (ßA) and the test individuals (including foetal samples) are digested with MstII, subjected to gel electrophoresis and probed with it sequence of ß -globin gene. If the test individuals have normal ß-globin gene, the bands detected in their Southern blots will be comparable to those of normal DNA. A sickle cell mutant of the ß -globin (ßs) gene will change this pattern in a detectable manner. Heterozygotes will show the bands present in both normal and sickle cell DNAs. This approach is applicable to only those disorders in which the gene mutation changes the restriction pattern. As a result, this approach is not of general application.

4. A more general approach utilizes oligonucleotide probes representing the sequence altered by the gene mutation causing the genetic disease. Typically, a set of two separate probes are used for each disease: one probe is complementary to the normal sequence, while the other is complementary to the mutant sequence.

The probes are radiolabelled and used to probe Southern blots; under appropriate conditions, the probes can distinguish the normal and mutant DNA samples.A set of two 19-mer (19 base long) oligonucleotide probe has been successfully used to detect sickle cell anemia. One of the two probes (ßA probe) contains the sequence complementary to that changed by the sickle cell mutation, while the other (pt probe) is complementary to the same segment of the normal allele. Similarly, other mutant genes (e.g. a-antitrypsin gene implicated in pulmonary emphysema) differing from the normal allele for a single base could be detected using this approach. But this assay can be used only in such cases where the base sequence of the gene segment containing the mutation (for both normal and mutant alleles) is known to allow the synthesis of the two oligonucleotide probes.

Disease Treatment

Treatment of diseases utilizes a wide variety of preparations of both biological and a biological origins. The preparations of biological origin may either be crude (e.g., ayurvedic medicines, some allopathic drugs, etc.) or purified to various degrees. A large number of such compounds originate from microorganisms, cultured cells and recombinant organisms.

Products from Nonrecombinant Organisms

Micro-organisms

A large number of pharmaceuticals originate from microorganisms; they range from whole microorganisms, e.g., spores of Lactobacillus sporogenes, through biomass, used as food/feed supplements, e.g., single cell proteins, to a variety for highly valuable compounds like antibiotics, vitamins, enzymes, organic acids, etc.

Plant Cell Cultures

Some biochemicals of pharmaceutical value are produced by cultured plant cells, e.g., shikonin, berberine, ginseng biomass

Animal Cell Cultures

Cultured animal cells are the source of several compounds used in treatment of diseases, e.g., angiogenic factor, interleukin-2, ß -interferon, etc.

Drug Delivery and Targetting

Drugs are normally delivered either orally or parenterally (by injection). They become distributed in the whole body tissues and fluids, and only a small portion reaches the diseased tissue/organ. This necessitates a much larger dose of expensive drugs, and may often produce. severe undesirable effects in other organs/tissues. Further, oral route of drug administration is much more desirable than that by injection for obvious reasons. But this route is unsuitable for the new class of protein/peptide drugs due to poor uptake; this is because of proteolytic degradation in the gastrointestinal tract and poor permeability of the intestinal mucosa to these high molecular weight therapeutic agents. The following approaches are being developed for a more efficient and/or targetted delivery of drugs.

1. Peptide/protein drugs may be delivered by other routes e.g., nasal, buccal, rectal, ocular, pulmonary and vaginal routes; of these the nasal route appears to be the most promising.

   The epithelial membranes present barriers to drug uptake, which can be overcome by using compounds that enhance permeability of these membranes. Some commonly used permeability enhancers are, sodium glycocholate, sodium deoxycholate, dimetlyl-p-cyclodextrin etc.

2. A variety of drugs can encapsulated in liposomes but liposomes become concentrated in liver and spleen. Tissue selectivity of liposomes can be greatly in eased by attaching to their surface specific ligands, e.g., monoclonal antibodies (Mabs). Liposomes hold great promise as a DNA delivery system in gene therapy, but they have to be injected into the subjects.

3. Polymers have been used as drug delivery systems; the drug is generally released by cleavage of the drug from the polymer, swelling of the polymer (for drugs trapped within the polymeric chains), through osmotic pressure generated pores, or simple diffusion.Polyesters are the widely studied biodegradable products; their hydrolysis yields nontoxic alcohols and organic acids.

Polyesters of lactic acid and glycolic acid are the most widely used polyesters for slow release of drugs having large molecules; e.g., proteins, polysaccharides and oligonucleotides. The drugs being delivered by the polymer systems include insulin growth factors, steroids, anticancer drugs etc.

4. The most effective system for site directed delivery of drugs and other moieties (called drug targetting) is based on monoclonal antibodies. Immunotoxins serve as a good example of this approach.

The application of this technology is limited by the low availability of Mabs specific to a given tumor cell type and also by the changing 'surface decorations' (antigens displayed on the cell surface) of tumour cells.

Bioremediation

During the past 20 years, recombinant DNA techniques have been studied intensively to improve the degradation of hazardous wastes under laboratory conditions. Only one field test has been successfully implemented. Recombinant bacteria can be obtained by genetic engineering techniques or by natural genetic exchange between bacteria. Applications for genetically engineered microorganisms (GEM) in bioremediation have received a great deal of attention, but have largely been confined to the laboratory environment. This has been due to regulatory risk assessment concerns, and to a large extent the uncertainty of their practical impact and delivery under field conditions. There are at least four principal approaches to GEM development for bioremediation application.These include:

(1) modification of enzyme specificity and affinity,

(2) pathway construction and regulation,

(3) bioprocess development, monitoring, and control, and

(4) bioaffinity bioreporter sensor applications for chemical sensing, toxicity reduction, and end point analysis.

These genetically engineered microorganisms have higher degradative capacity and have been demonstrated successfully for the degradation of various pollutants under defined conditions. However, ecological and environmental concerns and regulatory constraints are major obstacles for testing GEM in the field. These problems must be overcome before GEM can provide an effective clean-up process at lower cost.

The use of genetically engineered microorganisms has been applied to bioremediation process monitoring, strain monitoring, stress response, end point analysis, and toxicity assessment. The range of tested contaminants includes chlorinated compounds, aromatic hydrocarbons, heavy metals, and nonpolar toxicants, etc.

Development of effective and cost-efficient bioremediation processes is the goal for environmental biotechnology. The combination of microbiological and ecological knowledge, biochemical mechanisms, and field engineering designs are essential elements for successful in situ bioremediation using GEM.

Degradative (Catabolic) Genes

Pseudomonas spp.

The TOL plasmid pathway of Pseudomonas putida has been manipulated extensively to expand its catabolic capability on different branched compounds. RAMOS et al. modified the TOL metabolic pathway to use 4-ethylbenzoate as a substrate. These restructured strains were mutants isolated by either altering a pathway regulator XylS or by modifying substrate specificity of enzymes (catechol-2,3-dioxygenase). The complete upper TOL operon of plasmid pWW0 with its regulator gene, xylR, was reconstructed as a single gene cassette and yielded a hybrid mini-Tn5 [upp TOL] transposon.

Chlorobenzoates

Pseudomonas sp. B13 - Hybrid Pathways

The plasmid pWR1 encoding 3-chlorobenzoate and 4-chlorophenol degradation genes in strain B 13 have been modified extensively to generate hybrid pathways that extend vertical degradation of haloaromatic compounds. REINEKE and KNACKMUSS first reported the introduction of TOL plasmid pWW0, of P. putida mt-2 into strain B13 via conjugation. The transconjugants demonstrated the capability to utilize mono- and dichlorobenzoates as growth substrates. The inactivation of the TOL plasmid encoded m-pathway (by loss of catechol-2,3-dioxygenase activity) was an obligatory requirement for the hybrid strain, Pseudomonas sp. WR211, to utilize 4-chlorobenzoate as the sole carbon source.

Schwien and Schmidt reported that a benzoate and phenol degrader, Alcaligenes sp. A7, acquired chlorocatechol degrading capacity from strain B13 through conjugation. The recombinant strain, Alcaligenes sp. A7-2, exhibited the ability to utilize all three isomeric chlorophenols as the sole source of carbon and energy.The xylXYZ (toluate-1,2-dioxygenase) and xylL (carboxylate dehydrogenase) genes from TOL plasmid (pWW0-161) of P. putida and nahG (salicylate hydroxylase) from NAH7 plasmid were cloned into strain B13 to extend its substrate range.A hybrid plasmid, including genes (xylXYZL), positive regulator (xylS), and their native promoter (Pm), was introduced into strain B13 by conjugation. The resulting Pseudomonas sp. B13(TOL) was capable of utilizing 3-chloro-, 4-chloro-, and 3,5-dichlorobenzoate.The transconjugants containing the nahG gene were able to degrade salicylate, 3-, 4-, and 5-chlorosalicylate.ROJO et al. reported a modified ocleavage pathway, in strain B13, for simultaneous degradation of chloro- and methylaromatic compounds.This hybrid pathway included:

(1) toluate 1,2-dioxygenase and carboxylate dehydrogenase from the TOL pathway to degrade 4-chlorobenzoate and transform methylbenzoates into methyl-2-enelactones (strain FR1),

(2) 4-methyl-2-enelactone isomerase from Ralstonia eutropha (formerly Alcaligenes eutrophus) JMP134 to transform 4-methyl-2-enelactones to 3-methyl-2- enelactones to complete the o-pathway for 4-methylbenzoate [strain FR1(pRFC20P)],

(3) phenol hydroxylase from B13 to mineralize chloro- and methylphenols (cresols) to corresponding catechols [strain FR1(pFRC20P)-1].

A hybrid plasmid, pFRC4P (Tn5::xylXYZLS), containing Pm promoter was introduced into strain B13 through triparental mating using E. coli HB101(pFRC4P) as a donor, and E. coli HB101(pRK2013) as a helper.The hybrid Tn5 transposon that carried TOL genes was transposed into the chromosome of B13 and yielded strain FR1. Gene encoded 4-methyl-2-enelactone isomerase from R. eutropha JMP134 was cloned and transferred by conjugation into FR1.The resulting strain containing a hybrid plasmid, pFRC20P, was designated as B13FR1(pFRC20P) and was capable of growing on 4-methylbenzoate.

Spontaneous mutants of FR1(pFRC20P), such as FR1-(pFRC20P)-1 and -2, were able to use 4-methylphenol as a sole source of carbon and energy. The engineered bacterium Pseudomonas sp. B13FR1(pFRC20P)-1 was capable of growing on and mineralizing mixtures of 3-chloro-, 4-chloro-, and 4-methylbenzoate and 4-chloroand 4-methylphenol via the modified o-pathway.

Pseudomonas aeruginosa AC869(pAC31)

Strain P. putida, containing plasmid pAC25 that encodes genes for the degradation of 3-chlorobenzoate, has been shown to be structurally homologous to pWR1 (CHATTERJEE and CHAKRABARTY, 1983). Genetic rearrangements occurred between plasmids pAC25 and TOL under chemostat maintenance and yielded pAC31. The new plasmid contained xylDGEF genes located on the chromosome. The resulting strain AC869(pAC31) showed the capability to degrade 3,5-dichlorobenzoate and 3- and 4-chlorobenzoate.

Pseudomonas sp. US1 ex.

A monochlorobenzoate and 2,4-D degrader, strain US1 ex. containing pJP4 was obtained through conjugation. The *E. coli* JMP397, harboring plasmid pJP4 (no expression), was used as a donor and the 4-chlorobenzoate degrader, US1, was the recipient. The new strain released stoichiometric amounts of chloride when grown on respective chloroaromatics as carbon source. There are reports on the horizontal transfer of 3-chlorobenzoate catabolic plasmid (pBRC60) in a freshwater ecosystem and in an activated sludge unit.

Polychlorinated Biphenyls (PCB) and Chlorobiphenyls

Ralstonia eutropha (formerly Alcaligenes eutrophus)

Chromosomally located PCB catabolic genes of R. eutropha A5, Achromobacter sp. LBS1C1, and Alcaligenes dentrificans sp. JB1 were transferred into a heavy metal resistant strain R. eutropha CH34 through natural conjugation. All donor strains degraded biphenyl and monochlorobiphenyls to corresponding benzoate and chlorobenzoates. Benzoate was further metabolized via the o-cleavage pathway in strains A5 and LBS1C1, and via the m-cleavage pathway in strain JB1. Strain A5 harbored a catabolic transposon, Tn4317, which carried biphenyl and 4-chlorobiphenyl degradation genes. The dehalogenase activity was mediated by plasmid pSS50 in strain A5.

Transfer of PCB degradation genes from A5 to the heavy metal resistant strain CH34 was carried out by conjugation.The constructed strain,AE707, exhibited phenotype of BphcCbpc, and degraded 4-chlorobiphenyl to 4-chlorobenzoate in the presence of heavy metals. In resting cell assays (grown on biphenyl), strain AE707 cometabolized di- and trichlorinated congeners of Aroclor 1242T in the presence of heavy metals.The PCB catabolic chromosomal genes of strain JB1 were transferred into CH34 through RP4::Mu3A mediated R-prime plasmid formation. A transconjugant, strain AE1216, utilized 2-, 3- and 4-chlorobiphenyl and exhibited properties of metal resistance.

Pseudomonas sp. - Hybrid Strains

Pseudomonas putida BN10 grew on biphenyl and accumulated metabolites of 2-, 4-chlorobenzoate and 3-chlorocatechol from corresponding mono-chlorobiphenyls. 3-Chlorobiphenyl degraders were obtained from conjugation between Pseudomonas sp. B13 and strain BN10.The resulting Pseudomonas strain BN210 (gained chlorocatechol degradation genes) and B131 (acquired biphenyl degradation genes) were able to grow on 3-chlorobiphenyl. Both strains exhibited the capability to degrade monochlorobiphenyls and 2 of the dichlorobiphenyls found in Aroclor 1221T.

Another hybrid Pseudomonas sp. strain UCR2 was isolated from multi-chemostat mating between a chlorobenzoate degrader, *P. aeruginosa* JB2, and a 2-chlorobiphenyl utilizer, Arthrobacter sp. strain B1Barc. Strain UCR2 exhibited ability to mineralize 2-chloro- and 2,5-dichlorobiphenyl. The UCR2 showed higher phenotypic similarity and higher genomic DNA homology to strain JB2. No hybridization was observed when the parental strains were probed against each other.

Recombinant Pseudomonas sp. strain CB15 was obtained by multi-chemostat mating by mixing Pseudomonas sp. strain HF1 (3-chlorobenzoate utilizer) and Acinetobacter sp. strain P6 (biphenyl utilizer) on ceramic beads (ADAMS et al., 1992). 3-chlorobiphenyl, 3-chlorobenzoate, and biphenyl could be utilized as growth substrates by strain CB15. Results of DNA hybridization suggested strain CB15 was closely related to parent strain HF1. An in vivo recombinant plasmid, pDD530, containing the bphABCD genes from Burkholderia sp. strain LB400 was isolated from P. putida KT2442. The bph genes on pDD530 were further cloned into a pUT transposon vector to yield plasmid pDDPCB. Plasmid pDDPCB was mobilized into Pseudomonas sp. strain B13, and its genetically engineered derivative B13FR1 via conjugation. A transconjugant, Pseudomonas sp. strain B13FR1::bph, with a chromosomally integrated bph gene removed approximately 90% of added 4- chlorobiphenyl after 5 d in lake sediment microcosms. The plasmid, pDDPCB, was later transferred into a rhizosphere pseudomonad, Pseudomonas fluorescens

F113, by conjugation to generate a genetically modified strain F113pcb.The bph operon was chromosomally located and was stable in non-sterile soil microcosms for 25 d after inoculated onto sugar beet seeds. Strain F113pcb gained the ability to utilize biphenyl as a sole carbon source.

Pseudomonas cepacia JHR22

Havel and Reineke reported a hybrid bacterium, Pseudomonas cepacia JHR2, was constructed by filter mating with a biphenyl-grown donor and a chlorobenzoate-grown recipient (P. cepacia JH230). Strain 230 is a hybrid strain that originated from the transfer of chlorocatechol degradative genes from strain B13 into Pseudomonas sp. WR401. Strain JHR2 was able to grow on 3- and 4-chlorobiphenyl. A 2-chlorobiphenyl degrader, JHR22, was obtained by growing JHR2 with 4-chlorobenzoate (1 mM) in the presence of 2-chlorobiphenyl (3 mM). The new hybrid strain, P. cepacia JHR22, showed capability to utilize 2-chloro-, 3-chloro-, 4-chloro-, 2,4-dichloro-, and 3,5-dichlorobiphenyl as sole source of carbon and energy.The strain JHR22 also exhibited ability to degrade all monochlorobiphenyls in Aroclor 1221T when tested with soils.

Pseudomonas acidovorans M3GY

Recombinant bacterium, strain M3GY, was produced within a multi-chemostat culture by mixing *P. acidovorans* CC1 (a chloroacetate and biphenyl degrader) and Pseudomonas sp. strain CB15 (a hybrid strain) on ceramic beads that were coated with 3,3b-dichlorobiphenyl. Strain M3GY expressed catabolic ability to utilize 3,4b-dichlorobiphenyl as a sole carbon and energy source.The recipient strain was determined by phenotypic similarity and genetic homology between strains M3GY and CC1. In a recent study, 2,3b-dichloro- and 2,4-bdichlorobiphenyl were mineralized by a twomember consortium, Burkholderia (formerly Pseudomonas) sp. strain LB400 and *P. putida* mt-2a. The strain mt-2a was obtained by intergeneric mating with the chlorocatechol genes transferred from LB400 to mt-2 (TOL). Strain mt-2a exhibited the ability to grow on 3-chloro- and 4-chlorobenzoate.

Pseudomonas putida (pDA261)

Different biphenyl degradative genes (bphABCD) from Comamonas testosteroni B-356 were subcloned into *P. putida* and *E. coli* separately, and their degradative capabilities were examined. The hphC and bphD genes were expressed well in both cells, however, bphA and bphB in *E. coli* were poorly expressed even though located downstream of the tac promoter. A review of the strain B-356 was published earlier.

Comamonas testosteroni VP44(pE43)/VP44(pPC3)

Introduction of dehalogenase genes into the biphenyl degrading strain C. testosteroni VP44 resulted in complete mineralization of o- and p-substituted monochlorobiphenyls. Plasmid pE43, containing the ohbAB gene, encodes iron-sulfur protein (ISPOHB) of the o-halobenzoate-1,2-dioxygenase in P. aerugrinosa 142. This plasmid, pE43, was transformed into strain VP44. The resulting recombinant strain VP44(pE43) mineralized 2-chlorobiphenyl and 2-chlorobenzoate. chlorobenzoate. The other recombinant strain VP44-(pPC3) contained the fcbABC genes (from Arthrobacter globiformis KZT1) that catalyze hydrolytic p-dechlorination of 4-chlorobenzoate. This strain demonstrated the capability to grow on 4-chlorobiphenyl and 4-chlorobenzoate.

Escherichia coli JM109 (pSHF1003)/(pSHF1007) - Hybrid Biphenyl Dioxygenase

The bphA1 genes of biphenyl dioxygenase in Pseudomonas pseudoalcaligenes KF707 and Burkholderia cepacia LB400 were recombined randomly by DNA shuffling. The shuffled bphA1 genes were cloned into p ;HF18 (MulI), upstream of bphA2A3A4BC, to replace the disrupted bphA1 genes in th ; pJHF18. The resulting plasmids containing bphA1 (shuffled) and bphA2A3A4BC (KF707) were transformed into *E. coli* JM109 by electroporation. Some chimeric biphenyl dioxygenases in *E. coli* cells exhibited enhanced degradation of different biphenyl compounds. Biphenyl dioxygenase (pSHF 1003) showed 3 amino acid substitutions in KF707 bphA1 (H255Q, V258I, and D303E), derived from strain LB400, and acquired degradation capabilities not only for PCB, but also for benzene and toluene. Plasmid pSHF1007 showed 4 amino acid substitutions (H255Q, V258I, G268A, and T376N) in KF707 bphA1 gene and increased substrate affinities for some PCB congeners. The above results suggested the substitutions of His with Gln at

255 and of Val with Ile at 258 leading to the differences in substrate specificity and mode of oxygenation between the two enzymes. 4 amino acids in LB400 were converted to the corresponding KF707 sequence. The modified biphenyl dioxygenase exhibited broader substrate specificity with PCB.

Pseudomonas putida IPL5 (Field Application Vectors)

Field application vectors (FAV) are a combination of a selective substrate that can be used easily by host (notindigenous) microorganisms, and a cloning vector to provide a temporary niche for the host bacterium in harsh environments. FAV can stabilize and enhance the expression of foreign genes in contaminated sites. The chromosomally encoded PCB catabolic genes (bphABC) from Pseudomonas sp. strain ENV307 were cloned into broad host range plasmid pRK293. The resulting plasmid was transferred to the host Sphingomonas paucimobilis 1IGP that utilizes non-ionic surfactant Igepal CO-720T (IGP) as selective substrate. The recombinant strain 1IGP4(pCL3) exhibited ability to degrade individual PCB congeners in Aroclor 1242T without biphenyl as an inducer. The transposon encoded PCB degradative genes (bphABC) were more stable than plasmid encoded after insertion into the surfactant utilizing strain, Pseudomonas putida IPL5.

Trichloroethylene (TCE)

Escherichia coli HB101/pMY402 and FM5/pKY287

A XhoI fragment (4.7 kb), containing toluene monooxygenase (TMO) genes from Pseudomonas mendocina KR-1, was subcloned into a broad host range vector, pMMB66EH, which contains an *E. coli* tac promoter to yield plasmid pMY402. The same fragment was also inserted into another *E. coli* expression vector, pCFM1146, containing the temperature inducible *E. coli* phage lPL promoter, to yield plasmid pKY287. Recombinant *E. coli* strains, *E. coli* HB101/pMY402 and FM5/pKY287, were able to oxidize both toluene and TCE. Recently, TMO genes in pMMB503EH, a broad host range vector, were introduced into *P. putida* KT2440 by electroporation.

Pseudomonas pseudoalcaligenes KF707-D2 and Pseudomonas putida KF715-D5

Hrose et al. reported a hybrid dioxygenase gene cluster between the tod and the bph operons in *E. coli* JM109. Plasmid pJHF101, containing todC1::bphA2orf3A3A4, was constructed by deleting a 1.3 kb PpuMI fragment from bphB and bphC genes from pJHF10. The vector pUC118 was used for the construction of pJHF10. This recombinant strain, *E. coli* (pJHF101), degraded TCE at an initial rate of 1.8 µg mLP1 hP1 which was much faster than *E. coli* cells carrying the toluene dioxygenase genes (todC1C2BA) or the biphenyl dioxygenase genes.

The hybrid gene cluster, todC1: :phA2orf3A3A4, was further inserted into the chromosomal bph operons by SUYAMA et al. The plasmid pJHF101 was subcloned into suicide vector, pSUPB30, to yield plasmid pASF101. The pASF101 was inserted into *E. coli* S17-1 (chromosomally integrated RP4-2-Tc::Mu-Km::Tn7) cells via transformation, and the resulting strain was used as a donor in the mating with the recipient biphenyl utilizing *P. pseudoalcaligenes* KF707 and *P. putida* KF715. The resulting double crossover strains, KF707-D2 and KF715-D5, maintained todC1 in LB broth under no selective pressure. Both single and double crossover strains carrying todC1 on chromosomes of KF707 and KF715 degraded TCE efficiently, and grew on toluene and benzene.

Pseudomonas sp. strain JR1A::ipb - Hybrid Strains

A multi-component isopropylbenzene (IPB) dioxygenase from strain Pseudomonas sp. JR1, growing on IPB exhibited the capability to co-oxidize TCE. Organization of ipb genes encodes IPB dioxygenase (ipbA1A2A3A4), 2,3-dihydro- 2,3-dihyroxy-IPB dehydrogenase (ipbB), and 3-IPB-2,3-dioxygenase (ipbC). Recently, a recombinant, Pseudomonas sp. JR1A, exhibited constitutive TCE oxidation activity was reported. A transposon vector pC8 (pUT/miniTn5Km::ipbABC) was constructed, then transformed into E. coli S17.1(pir).The pC8 vector in E. coli was conjugatively transferred into different recipients. The ipb genes were detected in following transconjugants: strain JR1A (spontaneous IPB negative mutants of strain JR1), *P. putida* 548, and Pseudomonas sp. strain CBS-3. Two of these hybrid strains, JR1A::ipb and CBS-3::ipb, were stable for more than 120 generations in antibiotic free medium and degraded TCE and IPB constitutively. Further studies showed that strain JRlA::ipb can oxidize TCE without inducer.

Escherichia coli JM109(pDTG601)

Toluene dioxygenase from *P. putida* F1 has been reported as an enzyme responsible for degradation of TCE. The structural genes (todC1C2BA) of toluene dioxygenase were cloned from plasmid pKK223-3 then transformed into strain JM109. The resulting strain *E. coli* JM109 (pDTG601) degraded TCE at a slower initial rate when compared to *P. putida* F39/D, a mutant strain of strain F1 that does not contain cis-toluene dihydrodiol dehydrogenase.

Pseudomonas putida G786(pHG-2)

A recombinant *P. putida* G786(pHG-2), containing two multicomponent enzyme systems, cytochrome P450cam, and toluene dioxygenase, capable of degrading pentachloroethane and TCE was reported by WACKETT et al.. The toluene dioxygenase genes (todC1C2BA) and tac promoter (Ptac) were cloned into plasmid pKT230 to form plasmid pHG-1. The lacIQ gene cassette from plasmid pMMB24 was cloned into plasmid pHG-1 to yield pHG-2, then transformed into *E. coli* DH5. Plasmid pRK2073 was used as a helper to transconjugate plasmid pHG-2 into *P. putida* G786 containing cytochrome P450cam genes on the CAM plasmid. The new hybrid strain, G786(pHG-2), transformed pentachloroethane to TCE by the cytochrome P450cam, then TCE was further mineralized by the toluene dioxygenase.

Pseudomonas putida F1/pSMMO20

Slow growth rate, copper repression of the smmo locus, and strong competition between TCE and methane for soluble methane monooxygenase (sMMO) are restrictions to the use of strain OB3b to degrade TCE. To overcome these impediments, the complete smmo cluster of Methylosinus trichosporium OB3b was cloned into a wide host range vector, pMMB277, to form pSMMO20 (JAHNG and WOOD, 1994). pSMMO20 is a 14.7 kb plasmid and contains the IPTG-inducible tac promoter upstream of the smmo cluster.

Plasmid pSMMO20 was transformed into *P. putida* F1, *P. putida* KT2440, *P. mendocina* KR1, *B. cepacia* G4, and B. cepacia G4 PR1 through electroporation. *P. putida* F1/pSMMO20 was the only bacterium that was able to degrade TCE. However, inconsistent sMMO activity was a major drawback in the recombinant strain. Furthermore, the sMMO protein bands were detected on SDS-PAGE gels only when TPTG was present to induce the tac promoter. The constructed strain, *P. putida* F1/pSMMO20, showed a lower TCE degradation rate and a much higher growth rate than strain OB3b. This recombinant strain also demonstrated the capability to mineralize chloroform.

Burkholderia cepacia G4 5223-PR1

An aerobic bacterium, *Burkholderia cepacia* G4, cometabolically degrades TCE to CO2 and non-volatile products by toluene o-monooxygenase (SHIELDS et al., 1989). Tn5 mutants of strain G4 were constructed via triparental mating among G4, *E. coli* C600(pRZ102), and E. coli HB101(pRK2013). The resulting Tn5 mutants could not express toluene o-monooxygenase activity, and were unable to degrade TCE, toluene, and phenol. A mutant spontaneously reverted to express toluene omonooxygenase constitutively and was designated as G4 5223-PR1.This strain metabolized TCE and m-trifluoromethyl phenol without induction.

Ralstonia eutropha AEK301/pYK3021

Strain AEK301, a Tn5 mutant that lost phenol hydroxylase activity, was derived from strain JMP134. Plasmid pYK3021 encoding phenol hydroxylase was subcloned into pMMB67EH (vector from pJP4) and exhibited TCE degradation capability without phenol induction. Triparental mating was used to transfer plasmid pYK3021 from *E. coli* to *R. eutropha* AEK301 with the helper plasmid pRK2013. The recombinant strain AEK301/ pYK3021 expressed phenol hydroxylase activity constitutively and degraded TCE efficiently. The removal rate of TCE by the strain could be influenced by growth substrates.

2,4-Dichlorophenoxyacetic Acid (2,4-D)

Pseudomonas putida PPO300(pRO101) and PPO301(pRO103)

Plasmid pJP4 encodes the degradation genes of 2,4-D and 3-chlorobenzoate in R. eutropha JMP134. GEM containing catabolic genes for the degradation of 2,4-D to 2-chloromaleylacetate were reported by Harker et al. Plasmid pRO101 (pJP4::Tn1721) was constructed by insertion of transposon Tn1721 into pJP4, then further transferred by conjugation to different Pseudomonas strains. One of the resulting strains, *P. putida* PPO300(pRO101), with chromosomally encoded phenol hydroxylase, also degrades phenoxyacetate in the presence of an inducer (2,4-D or 3-chlorobenzoate) of the 2,4-D pathway.

A mutant plasmid, pRO103, derived from pRO101 by the spontaneous deletion of negative regulatory gene (tfdR) was isolated from strain PPO300-(pRO101).The deletion of tfdR resulted in the constitutive expression of the tfdA gene (encodes 2,4-D monooxygenase), and enabled mutant strain PPO300(pRO103) to grow on phenoxyacetate as the sole carbon source. Strain PPO301, derived from PPO300 (ATCC 17514) and resistant to nalidixic acid, was used to harbor plasmid pRO103 to yield a constitutive 2,4-D degrader, PPO301(pRO103), that was used in agricultural soil studies. A dual substrate (2,4-D and succinate) chemostat study with P. cepacia DBO1-(pRO101) indicated that succinate can act as repressor of the 2,4-D catabolic pathway. But, this repression can be relieved with appropriate adjustments, such as lower 2,4-dichlorophenol accumulation or reduced succinate concentration (~2,4-D) in the media.

Pseudomonas cepacia RHJ1 Burkholderia cepacia (formerly P. cepacia)

RH1 is a recombinant strain created by performing conjugation between *R. eutropha* JMP134 (2,4-D degrader) and *B. cepacia* AC1100 (2,4,5-trichlor-ophenoxyacetate degrader). The self-transmissible 2,4-D degradative plasmid, pJP4, in *A. eutrophus* JMP134 was transferred into a 2,4,5-trichlorophenoxyacetic acid (2,4,5-T) degrader, *B. cepacia* AC1100. This new strain, designated RHJ1, was capable of degrading mixtures of 2,4-D and 2,4,5-T simultaneously.

USE OF GENETICALLY MODIFIED MICROBES IN AGRICULTURE

The widespread deployment of GM microbes has been going on for at least six years with little or no public awareness and input into the reviews on environmental impact. The widely deployed GM microbes include the following: Sinorhizobium meliloti a bacterium which is added to soil or used as a seed innoculant to enhance nodulation and nitrogen fixation in legumes, it has seen wide application since its release for commercial production in 1997. The other commercial GM microbes are designated to be biopesticides. These include GM Agrobacterium radiobacter k1026 that is used to treat fruit and vegetable plants to prevent crown gall tumors from appearing. The bacterium Pseudomonas flourescens has been modified with a number of different Cry delta-endotoxin genes from different subspecies of *Bacillus thruingiensis* (Bt) the resulting biopestide is produced by killing modified *P. flouresens,* the modified *P. flourescns* provides a persistent biopesticide that degrades much slower in sunlight than Bt used to control insect pests of fruits and vegetables. Information indicating that the commercial microbial preparations are genetically modified is not widely recognized by those selling or using the preparations and the preparations can easily be employed by organic farmers.

The legume symbiont, Sinorhizobium meliloti, is tremendously important in fixing nitrogen from the air into plants and soil. Legumes signals the bacterium by exuding flavonoids from its roots, activating expression of nodulation genes in the bacterium resulting in the production of Nod factors that regulate formation of nitrogen fixing root nodules. The sequence of S.meliloti has been fully determined, the bacterial genome is unusual in that it contains three chromosomes (or a chromosome and two very large plasmids), all three of the bacterial replicons contribute to symbiosis. The genetically modified commercial strain (RMBPC-2) has added genes that regulate nitrogenase enzyme (for nitrogen fixation) along with genes that increase the organic acid delivered from the plant to the nodule bacterium and finally the antibiotic resistance to streptomycin and spectomycin are added to the commercial bacterium.

The commercial release was permitted in spite of concerns about the impact of the GM microbe on the environment. Evidence supporting the initial concerns has accumulated but has not deterred the spread of the GM microbe. For example, a recent review showed that GM *S. meliloti* strains persisted in the soil for six years, even in the absence of legume hosts. Horizontal gene transfer to other soil bacteria and microevolution of plasmids was observed.

Other studies showed that a soil micro arthropod ingested *S. meliloti* and rescience in the arthropod gut facilitated gene transfer to a range of bacteria. There is little doubt that the antibiotic markers of GM. *S. meliloti,* streptomycin and spectomycin will be transfer to soil bacteria from which they can be transferred to a range of animal pathogens. For example, the resistance genes for streptomysin and spectomysin were found to be transferred from their insertion as transgenes in plant chloroplast into the infecting bacterium Actinobacter sp. so the mobility of the transgnes are well established.

It is clear that the antibiotics, spectinomycin and streptomycin, are used extensively in human and animal medicine. For example spectinomycin is used to treat human gonnnorhea and bovine pneumonia Streptomycin is used to treat human tubeculosis and Meniere's disease and it is used as a pesticide on fruits and vegetables. In conclusion, the commercial release of GM Sinorhizobium meliloti resulted in the soil establishment of the GM microbe in millions of acres of cropland and spread antibiotic resistance genes for antibiotics used extensively in medicine and agriculture.

Agrobacterium radiobacter k1026 is a biopesticide derived from A. radiobacter k84 a natural bacterium used to control the crown gall disease of fruits and ornamental trees and shrubs. It is used to treat germinating seeds or roots and stems of the plants. Crown gall disease is caused by the bacterium Agrobacterium tumefaciens that causes tumors to form on the plant stems (the bacterium is the one used in genetic engineering). GM Agrobacterium radiobacter is used a great deal to control crown gall disease in fruits and ornamental trees. The GM A. radiobacter releases a chemical warfare agent against disease causing A. tumefaciens, the chemical is called bacteriocin (agrocin)a novel nucleic acid derivative that prevents the crowwn ball tumors from forming in crop plants.The GM A. radiobacter has an engineered deletion in the genes controlling plasmid transfer so that the male bacterium cannot transfer its plasmid but can act as a female to receive a plasmid transfer.The strain is used a great deal to treat fruits and ornamental shrubs and trees, the bacterium appears to be very stable and persist for years in the soil around treated plants.

The other GM biopestides commercially released include a number of Pseudomonas flourescens strains that have been modified with Cry delta endotoxin genes from Bacillus thuringiensis then the transformed P. flourescens strains are killed before being marketed. The killed GM bacteria are more persistent used as foliar sprays than are the B. thuringiensis sprays. The main fallacy in the approval of these biopesticides is the fact that bacteria enjoy sex after death. Soil bacteria are easily transformed with cell lysates (squashed dead cells)and function in soil microcosms. P. flourescens and A. tumefacians both are transformed in soil. The soil Pseudomonas and Actinobacter also easlly take up genes from transgenic plants so that combination of transgenic crops and GM biopesticides can create genetic combinations cabable of devastating the soil microflora and microfauna.

REFERENCES

Cookson, J. T. *Bioremediation Engineering: Design and Application*. McGraw-Hill. New York. 1994.

Environmental Defense Fund. *Genetically Engineered Foods: Who's Minding the Store?* New York. NY: Environmental Defense Fund. 1995.

Olson, Steve. *Biotechnology: An Industry Comes of Age*. Washington. D.C.: National Academy Press. 1986.

3

Microbial Genomics and Proteomics

The era of genomics (the study of genes and their function) began a scant dozen years ago with a suggestion by James Watson that the complete DNA sequence of the human genome be determined. Since that time, the human genome project has attracted a great deal of attention in the scientific world and the general media; the scope of the sequencing effort, and the extraordinary value that it will provide, has served to mask the enormous progress in sequencing other genomes.

Genomics was established by Tattersol Smith when he first sequenced the complete genomes of a virus and a mitochondrion. His group established techniques of sequencing, genome mapping, data storage, and bioinformatic analyzes in the 1970-1980s. A major branch of genomics is still concerned with sequencing the genomes of various organisms, but the knowledge of full genomes has created the possibility for the field of functional genomics, mainly concerned with patterns of gene expression during various conditions. The most important tools here are microarrays and bioinformatics. Study of the full set of proteins in a cell type or tissue, and the changes during various conditions, is called proteomics. A related concept is materiomics, which is defined as the study of the material properties of biological materials (e.g. hierarchical protein structures and materials, mineralized biological tissues, etc.) and their effect on the macroscopic function and failure in their biological context, linking processes, structure and properties at multiple scales through a materials science approach. The actual term 'genomics' is thought to have been coined by Dr. Tom Roderick, a geneticist at the Jackson Laboratory over beer at a meeting held in Maryland on the mapping of the human genome in 1986.

In 1972, Walter Fiers and his team at the Laboratory of Molecular Biology of the University of Ghent (Ghent, Belgium) were the first to determine the sequence of a gene: the gene for Bacteriophage MS2 coat protein. In 1976, the team determined the complete nucleotide-sequence of bacteriophage MS2-RNA. The first DNA-based genome to be sequenced in its entirety was that of bacteriophage F-X174; sequenced by Frederick

Sanger in 1977. The first free-living organism to be sequenced was that of Haemophilus influenzae in 1995, and since then genomes are being sequenced at a rapid pace. A rough draft of the human genome was completed by the Human Genome Project in early 2001, creating much fanfare.

As of September 2007, the complete sequence was known of about 1879 viruses, 577 bacterial species and roughly 23 eukaryote organisms, of which about half are fungi. Most of the bacteria whose genomes have been completely sequenced are problematic disease-causing agents, such as Haemophilus influenzae. Of the other sequenced species, most were chosen because they were well-studied model organisms or promised to become good models. Yeast (Saccharomyces cerevisiae) has long been an important model organism for the eukaryotic cell, while the fruit fly Drosophila melanogaster has been a very important tool (notably in early pre-molecular genetics). The worm Caenorhabditis elegans is an often used simple model for multicellular organisms. The zebrafish Brachydanio rerio is used for many developmental studies on the molecular level and the flower Arabidopsis thaliana is a model organism for flowering plants. The Japanese pufferfish (Takifugu rubripes) and the spotted green pufferfish (Tetraodon nigroviridis) are interesting because of their small and compact genomes, containing very little non-coding DNA compared to most species. The mammals dog (Canis familiaris), brown rat (Rattus norvegicus), mouse (Mus musculus), and chimpanzee (Pan troglodytes) are all important model animals in medical research.

BACTERIOPHAGE GENOMICS

Bacteriophages have played and continue to play a key role in bacterial genetics and molecular biology. Historically, they were used to define gene structure and gene regulation. Also the first genome to be sequenced was a bacteriophage. However, bacteriophage research did not lead the genomics revolution, which is clearly dominated by bacterial genomics.

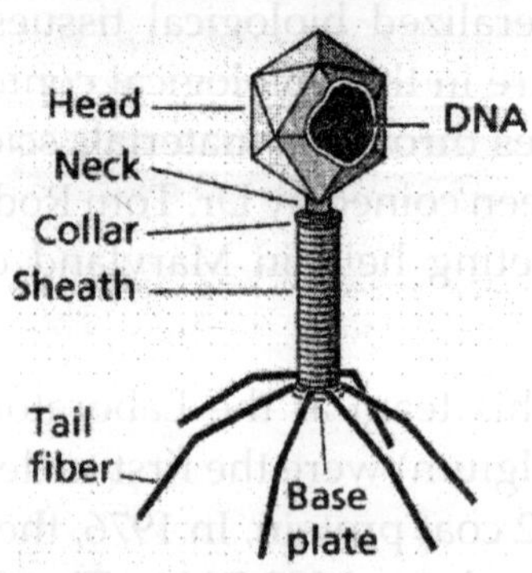

Figure 1. Structure of Bacteriophage

Only very recently has the study of bacteriophage genomes become prominent, thereby enabling researchers to understand the mechanisms underlying phage evolution. Bacteriophage genome sequences can be obtained through direct sequencing of isolated bacteriophages, but can also be derived as part of microbial genomes. Analysis of bacterial genomes has shown that a substantial amount of microbial DNA consists of prophage sequences and prophage-like elements. A detailed database mining of these sequences offers insights into the role of prophages in shaping the bacterial genome.

CYANOBACTERIA GENOMICS

At present there are 24 cyanobacteria for which a total genome sequence is available. 15 of these cyanobacteria come from the marine environment. These are six Prochlorococcus strains, seven marine Synechococcus strains, Trichodesmium erythraeum IMS101 and Crocosphaera watsonii WH8501. Several studies have demonstrated how these sequences could be used very successfully to infer important ecological and physiological characteristics of marine cyanobacteria.

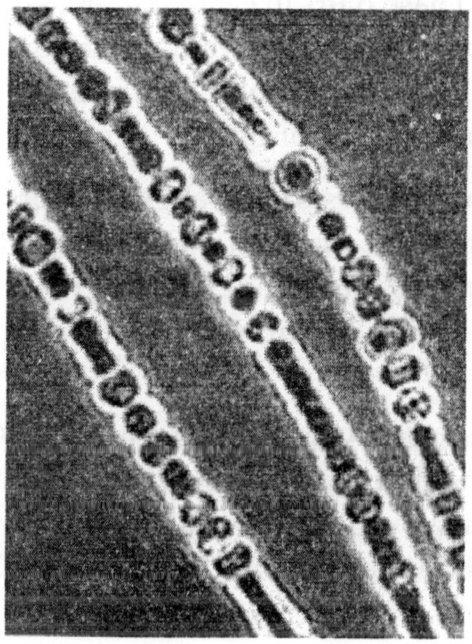

Figure 2. Cyanobacteria

However, there are many more genome projects currently in progress, amongst those there are further Prochlorococcus and marine Synechococcus isolates, Acaryochloris and Prochloron, the N2-fixing filamentous cyanobacteria Nodularia spumigena, Lyngbya aestuarii and Lyngbya majuscula, as well as bacteriophages infecting marine cyanobaceria. Thus, the growing body of genome information can also be tapped in a more general way to address global problems by applying a comparative approach. Some new and exciting examples of progress in this field are the identification of genes for regulatory RNAs, insights into the evolutionary origin of photosynthesis, or estimation of the contribution of horizontal gene transfer to the genomes that have been analyzed.

FULL GENOME SEQUENCING (FGS),

Full genome sequencing (FGS), also known as whole genome sequencing, complete genome sequencing, or entire genome sequencing, is a laboratory process that determines the complete DNA sequence of an organism's genome at a single time. This entails sequencing all of an organism's chromosomal DNA as well as DNA contained in the mitochondria or chloroplast, depending respectively on whether the organism is an animal or plant. Almost any biological sample—even a very small amount of DNA or ancient DNA—can provide the genetic material necessary for full genome sequencing. Such samples may include saliva, epithelial cells, bone marrow, hair (as long as the hair contains a hair follicle), seeds, plant leaves, or anything else that has DNA-containing cells. Because the sequence data that is produced can be quite large (for example, there are approximately three billion base pairs in each human genome), genomic data is stored electronically and requires a large amount of computing power and storage capacity. Full genome sequencing would have been nearly impossible before the advent of the microprocessor, computers, and the Information Age.

Full genome sequencing only refers to the laboratory process of deducing a person's entire genetic code and, on its own, may not contain any clinical assessment or useful clinical information. However, this may change over time as a large number of scientific studies continue to be published detailing clear associations between specific genetic variants and disease.

The first full genome sequenced was J. Craig Venter's. Although it was not known during the race to sequence the first entire human genome during the 1990s, it was later revealed that Celera Genomics' sequence was that of Venter, the company's chief executive officer. Since that time, many other full genomes have been sequenced, such as James Watson's, although as of March 2009 true commercialization of full genome sequencing has yet to be achieved.

New Techniques

One possible way to accomplish the cost-effective high-throughput sequencing necessary to accomplish full genome sequencing is by using Nanopore technology, which is a patented technology held by Harvard University and Oxford Nanopore and licensed to biotechnology companies. To facilitate their full genome sequencing initiatives, Illumina licensed nanopore sequencing technology from Oxford Nanopore and Sequenom licensed the technology from Harvard University. Another possible way to accomplish cost-effective high-throughput sequencing is by utilizing fluorophore technology. Pacific Biosciences is currently using this approach in their SMRT (single molecule real time) DNA sequencing technology. Complete Genomics is developing DNA Nanoball (DNB) technology that are arranged on self-assembling arrays. Pyrosequencing is a method of DNA sequencing based on the sequencing by synthesis principle. The technique was developed by Pål Nyrén and his student Mostafa Ronaghi at the Royal Institute of Technology in Stockholm in 1996, and is currently being used by 454 Life Sciences in their effort to deliver an affordable, fast and highly accurate full genome sequencing platform.

Older Techniques

Full genome sequencing of the entire human genome was first accomplished in 2000 partly through the use of shotgun sequencing technology. While full genome shotgun sequencing for small (4000–7000 base pair) genomes was already in use in 1979, broader application benefited from pairwise end sequencing, known colloquially as double-barrel shotgun sequencing. As sequencing projects began to take on longer and more complicated genomes, multiple groups began to realize that useful information could be obtained by sequencing both ends of a fragment of DNA. Although sequencing both ends of the same fragment and keeping track of the paired data was more cumbersome than sequencing a single end of two distinct fragments, the knowledge that the two sequences were oriented in opposite directions and were about the length of a fragment apart from each other was valuable in reconstructing the sequence of the original target fragment.

The first published description of the use of paired ends was in 1990 as part of the sequencing of the human HPRT locus, although the use of paired ends was limited to closing gaps after the application of a traditional shotgun sequencing approach. The first theoretical description of a pure pairwise end sequencing strategy, assuming fragments of constant length, was in 1991. In 1995 Roach et al.introduced the innovation of using fragments of varying sizes, and demonstrated that a pure pairwise end-sequencing strategy would be possible on large targets. The strategy was subsequently adopted by The Institute for Genomic Research (TIGR) to sequence the entire genome of the bacterium Haemophilus influenzae in 1995, and then by Celera Genomics to sequence the entire fruit fly genome in 2000, and subsequently the entire human genome. Applied Biosystems, now called Life Technologies, manufactured the shotgun sequencers utilized by both Celera Genomics and The Human Genome Project.

While shotgun sequencing was one of the first approaches utilized to successfully sequence the full genome of a human, it is too expensive and requires too long of a turn-around-time to be utilized for commercial purposes. Because of this, shotgun sequencing technology, even though it is still relatively 'new', is being displaced by technologies like pyrosequencing, SMRT sequencing, and nanopore technology.

Race to Commercialization

In October 2006, the X Prize Foundation, working in collaboration with the J. Craig Venter Science Foundation, established the Archon X Prize for Genomics, intending to award US$10 million to "the first Team that can build a device and use it to sequence 100 human genomes within 10 days or less, with an accuracy of no more than one error in every 100,000 bases sequenced, with sequences accurately covering at least 98% of the genome, and at a recurring cost of no more than $10,000 per genome." However, a higher accuracy rate will be needed before this technology has clinical utility. An error rate of 1 in 100,000 bases, out of a total of three billion base pairs (six billion bases total) in the human genome, would mean about 30,000 errors per genome, which is a highly significant number of false positives and false negatives, especially since it will not be known where the errors occurred (for example, did the error occur while sequencing the BRCA1 gene, which may then be associated with a false breast cancer risk analysis). Instead, the error rate will have to be better than 1 in 10 million bases, which will provide 300 inaccurate base pairs per genome or less and will then make the data applicable for global, widespread clinical use, such as in Predictive Medicine. As of February 2009, the Archon X Prize for Genomics remains unclaimed.

In 2007, Applied Biosystems started selling a new type of sequencer called SOLiD System, with the first sale to Helicos Biosciences in 2008. Helicos stated that, utilizing the new sequencers, they will attempt to provide a full genome sequencing service with a target price of $72,000 per sample. However, this price point is still too high for true, global commercialization as is not competitive to DNA arrays, whose price point is now under $500 per sample.

In 2008 and 2009, both public and private companies have emerged that are now in a competitive race to be the first mover to provide a full genome sequencing platform that is commercially viable for both research and clinical use, including Illumina, Sequenom, 454 Life Sciences, Pacific Biosciences, Complete Genomics, Intelligent Bio-Systems, Genome Corp., and Helicos BioScience. These companies are heavily financed and backed by venture capitalists, hedge funds, investment banks and, in the case of Illumina, Sequenom and 454, heavy re-investment of revenue into research and development, mergers and acquisitions, and licensing initiatives.

In the race to commercialize full genome sequencing, companies have made claims about being able to offer a service at a specific time for a specific price that have turned out to not be true. Intelligent Bio-Systems stated in November 2007 that by the end of 2008 they would release a platform capable of a providing a $5,000 full genome sequence, but, as of March 2009, no such platform has yet to be released.Pacific Biosciences stated that they will start selling their full genome sequencers in early 2010. While they didn't disclose the cost to sequence a single genome, they did state they may not release their second-generation machine capable of a $1,000 genome until 2013. Complete Genomics, however, stated that they'll be able to provide a $5,000 full genome sequencing service by the summer of 2009. The accuracy, precision, and reproducibility of both Pacific Biosciences and Complete Genomics technology, however, is still unknown.

A personal genomics company located in Massachusetts, Knome.com, currently provides genome sequencing services but the cost is about $99,500 per genome (down from $350,000 per genome initially), the turn-around time is unknown, the accuracy is unknown, and the number of people is reportedly limited to 20 so this does not qualify as the true commercialization of full genome sequencing.As of January 2009, there are no indications that any of these companies have been hindered by the global recession. And thus, the race appears to be proceeding forward at full speed.

At the end of February 2009, Complete Genomics released a full sequence of a human genome that was sequenced using their service. The data indicates that Complete Genomics' full genome sequencing service accuracy is just under 99.99%, meaning that there is an error in one out of every ten thousand base pairs. This means that their full sequence of the human genome will contain approximately 80,000-100,000 false positive errors in each genome. However, this accuracy rate was based on Complete Genomics' sequence that was completed utilizing a 90x depth of coverage (each base in the genome was sequenced 90 times) while their commercialized sequence is reported to be only 40x, so the accuracy may be substantially lower unless they can find some way to improve it before their first service release planned for the summer 2009. This accuracy rate may be acceptable for research purposes but is still way too high for clinical use. In March 2009, it was announced that Complete Genomics has signed a deal with the Broad Institute to sequence cancer patient's genomes and will be sequencing five full genomes to start.

Disruptive Technology

Full genome sequencing provides information on a genome that is orders of magnitude larger than that provided by the current leader in sequencing technology, DNA arrays. For humans, DNA arrays currently provides genotypic information on up to one million genetic variants, while full genome sequencing will provide information on all three billion base pairs in the human genome, or 3,000 times more data. Because of this, full genome sequencing will be disruptive to the DNA array markets once the accuracy (> 99.998%), precision (> 99.998%), and consistent reproducibility of extremely high-quality results (> 99.998%) is demonstrated and its cost becomes competitive with DNA arrays (less than $500 per sample). It is unknown whether Agilent, another established DNA array manufacturer, is also working on full genome sequencing technologies or services. It is thought that Affymetrix, the pioneer of array technology in the 1990s, has fallen behind due to significant corporate and stock turbulence and is currently not working on any known full genome sequencing approach. It is unknown what will happen to the DNA array market once full genome sequencing becomes commercially viable, especially when companies and laboratories providing this disruptive technology starts to realize economies of scale. It is postulated, however, that this new technology may significantly diminish the total market size for arrays and any other sequencing technology once it becomes commonplace for individuals and newborns to have their full genomes sequenced.

Societal Impact

Inexpensive, time-efficient full genome sequencing will be a major accomplishment not only for the field of Genomics, but for the entire human civilization because, for the first time, individuals will be able to have their entire genome sequenced. Utilizing this information, health care professionals, such as physicians and genetic counselors, will be greatly empowered by this information and be able to use it to predict what diseases a person may get in the future and attempt to either minimize the impact of that disease or avoid it altogether through the implementation of personalized, preventive medicine. Full genome sequencing will allow health care professionals to analyze the entire human genome of an individual and therefore detect all disease-related genetic variants, regardless of the genetic variant's prevalence or frequency. This will enable the rapidly emerging medical fields of Predictive Medicine and Personalized Medicine and will mark a significant leap forward for the clinical genetic revolution.

COMPUTATIONAL GENOMICS

Computational genomics is the study of deciphering biology from genome sequences using computational analysis., including both DNA and RNA. Computational genomics focuses on understanding the human genome, and more generally the principles of how DNA controls the biology of any species at the molecular level. With the current abundance of massive biological datasets, computational studies have become one of the most important means to biological discovery.

Computational genomics began in spirit, if not in name, during the 1960s with the research of Margaret Dayhoff and others at the National Biomedical Research Foundation, who first assembled a database of protein sequences. Their research developed a that determined the evolutionary changes that were required for a particular protein to change into another protein based on the underlying amino acid sequences

This led them to create a scoring matrix that assessed the likelihood of one protein being related to another.

Beginning in the 1980s, databases of genome sequences began to be recorded, but this presented new challenges in the form of searching and comparing the databases of gene information. Unlike text-searching algorithms that are used on websites such as google or Wikipedia, searching for sections of genetic similarity requires one to find strings that are not simply identical, but similar. This led to the development of the Needleman-Wunsch algorithm, which is a dynamic programming algorithm for comparing sets of amino acid sequences with each other by using scoring matrices derived from the earlier research by Dayhoff. Later, the algorithm was developed for performing fast, optimized searches of gene sequence databases. BLAST and its derivatives are probably the most widely-used algorithms for this purpose.

The first meeting of the Annual Conference on Computational Genomics was in 1998, providing a forum for this speciality and effectively distinguishing this area of science from the more general fields of Genomics or Computational Biology. The first use of this term in scientific literature, according to MEDLINE abstracts, was just one year earlier in Nucleic Acids Research.

The development of computer-assisted mathematics (using products such as Mathematica or Matlab) has helped engineers, mathematicians and computer scientists to start operating in this domain, and a public collection of case studies and demonstrations is growing, ranging from whole genome comparisons to gene expression analysis. This has increased the introduction of different ideas, including concepts from systems and control, information theory, strings analysis and data mining. It is anticipated that computational approaches will become and remain a standard topic for research and teaching, while students fluent in both topics start being formed in the multiple courses created in the past few years.

APPLICATIONS OF MICROBIAL GENOMICS

Bacteria and their viruses represent only part of the vast interconnected web of life that make up the global ecosystem, in numbers they make up the majority. The variety of environments in which they live, the strategies they use to survive and grow, and the substrates they transform in that service lead to a wealth of forms and functions, the extent of which we are only beginning to understand. The growth and death of subpopulations of microbes in response to environmental change and their invasion into new niches can lead to large changes in the balance of a local ecosystem and can lead to interference with human operation - with effects ranging from the corrosion of oil lines to increasing the prevalence of and the introduction of new pathogenic strains.

Beneficial aspects of microbial populations include their contribution to important geochemical cycles, their ability to buffer environmental change through bioremediation, and the possibility that they can provide a wealth of new functions for energy conversion, catalysis and natural product synthesis. High-throughput sequencing and advances in DNA cloning and amplification technology, coupled with genomic tools, are enabling holistic views into the composition and dynamics of predominantly unculturable microbial communities.

This emerging field, termed 'metagenomics', offers new discoveries into the capabilities of microbes that allow them to collaborate and compete to survive in a wide range of environments. Genomic investigations into the diversity of environmental bacteria are leading to insights into ecological dynamics, the evolution of new forms of biological systems, and the discovery of new functions that might be exploited for biotechnological and biomedical purposes.

Efforts to understand the biological composition of environments and the nature of engendered ecologies and their place in regional and global geochemistry are considerably aided by the identification of the constituent organisms. Modern approaches, made possible by genomic technologies, provide a much broader ability to access this diversity than traditional microscopic and culturing techniques. Current estimates indicate that less than 1% of microbial species are amenable to growth in isolation under standard laboratory conditions. Instead, sequencing and other techniques for identifying DNA from environmental samples can yield a far more complete picture of the organisms involved in a community and, ultimately, the placement of those organisms into their ecological roles.

Identification and classification of both well-known and novel organisms is greatly aided by phylogenetic marker genes, which result from the commonality of certain tasks such as transcription and translation. Although conserved protein-coding genes can be used to identify bacterial lineages, the most commonly used phylogenetic markers are genes for RNA subunits of the ribosome, most frequently the small subunit (16S rRNA gene); these have been used in numerous studies to determine the presence and relative abundance of taxonomic groups within environmental samples. These studies have either sequenced rRNA genes directly, used PCR amplification to scan rRNA genes in large-insert clones built from environmental samples or employed sequencing followed by computational identification of the 16S rRNA gene.

One key benefit has been the identification of novel species, clades and divisions, which guides future research into a more balanced understanding of the tree of life. Additionally, these studies have revealed that the diversity of different communities can vary dramatically from just a few species to thousands, often to a much greater extent than had been expected, and possess members that had not been previously identified.

For example, one study using the 16S rRNA gene discovered a new clade that is one of the most abundant members of bacterioplankton communities in the ocean, but which had escaped detection by other approaches.

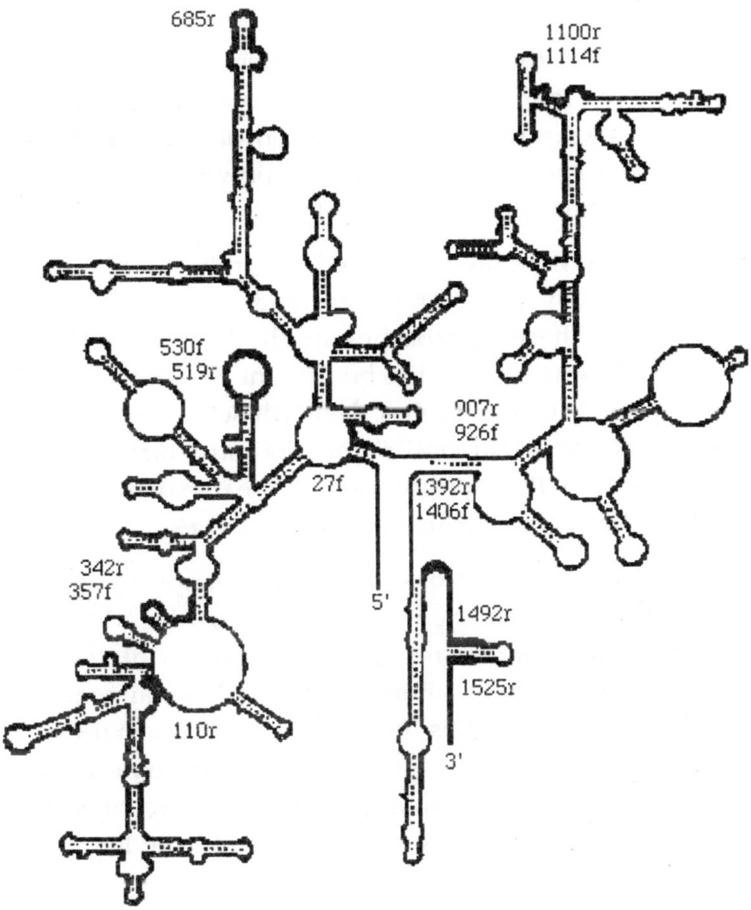

Figure 3. 16S rRNA Secondary Structure

Subsequent culturing and genomic sequencing of one of the members of this clade, Pelagibacter ubique, revealed a streamlined genome with fewer genes than any free-living bacteria to date, adding to its interest and highlighting the value of initial survey studies in discovering important organisms for further investigation. Temporal and spatial studies of microbial population structure beyond initial surveys are also of great importance, and will facilitate comparative analyses of community composition that will yield insight into the relationship of the ecology with the conditions that favor one population structure over another. In light of this, we expect less labor-intensive approaches than sequencing that capture the presence of organisms in environmental samples, such as those based on the hybridization of probes to the 16S rRNA gene, to prove useful as researchers attempt to more rapidly characterize populations, potentially even in the field. One such approach takes advantage of fluorescence in situ hybridization to ribosomal genes within the sample, whereas others make use of microarray technology.

In the former approach, identification of only a small number of types is possible in a given experiment, owing to the uniform application of the probes and the limited number of fluorescent dyes that can be utilized at once. In the latter approach, the separation of the probes allows for the identification of far more types. Additionally, probes of varying taxonomic specificity can be used, allowing for rapid classification, and perhaps quantification, of the organisms present within a sample.

However, one drawback of techniques based on probes is that such investigations are limited to the identification of known groups and will fail to capture the presence of truly novel organisms. These approaches might be best applied once a better understanding of the organisms expected to be present is achieved by an initial survey to allow for the synthesis of custom probes. Nevertheless, it is the expectation of the authors that as such spatial and temporal population structure studies are one of the essential directions for the field, the rapidity and relative inexpensiveness of probe-based approaches will lead to their frequent use in future studies.

In contrast to phylogenetic marker-based studies that survey the microbes present in an environment, DNA sequencing of environmental samples addresses the functional capabilities of the constituent organisms through analysis of the community gene complement. Recent studies utilizing 'shotgun' environmental sequencing reflect the challenges associated with these studies and the conclusions that can be drawn from them. The primary challenge to 'piece together' fragmentary sequences to determine the genetic content of each species in a community is greatly affected by the complexity of the sample, the comprehensiveness of the sequencing, and the length of the fragments themselves.

For example, the shotgun approach taken in studies of the Sargasso Sea, soil and whale carcasses yielded hundreds to thousands of unique species, but reads that were possible to group together were primarily from those species that already had sequenced genomes. Additionally, although these undertakings were huge in scope, ultimately the sequencing was not comprehensive over all regions of the genomes of each species, and could not be comprehensive owing to the large amount of strain variation. However, the nature of microbial genomes (generich with small genes) permitted the quantitative assessment of gene repertoires in each of the sampled environments. These 'environmental gene tags' could be used to distinguish environments using differences in the inferred metabolic activities and functional roles of each microbial community.

In contrast to the challenges presented by more complex communities, the relatively low species complexity of an acid mine drainage (AMD) biofilm lent itself more readily to the assembly of two nearly complete genomes and the partial assembly of three additional genomes. This achievement allowed for the assignment of the roles played by the members of the community, such as the appearance that only one of the members possesses the capacity for essential nitrogen fixing, and highlights the potential of metagenomic studies to reveal the keystone organisms within a given ecology and the interactions between community members.

Given the cost and complications of genome assembly associated with environmental shotgun sequencing, it is often practical and informative to sequence large-insert fosmid or BAC (bacterial artificial chromosome) clones to meet research goals. The ability to retrieve archived DNA in the form of a large-insert library greatly aids phylogenetic identification of the clones, assembly of metagenome DNA sequence, linking of genes with organisms, and the linking of genes and organisms to environmental function. For example, Hallam et al. sequenced a marine sediment-derived fosmid library enriched for archaeal DNA to demonstrate that an uncultured archaea contains the gene complement required to oxidize methane. Similarly, DeLong et al. sequenced fosmid clones derived from marine microbial communities isolated from multiple depths at a single site in the Pacific Ocean to link microbial diversity to oceanic parameters like nutrients, salinity, temperature and the availability of light.

Despite the insight into microbial functional capacity provided by environmental DNA sequences, these data in isolation are typically not sufficient to determine gene function. This is highlighted by the large number of both conserved and non-conserved genes with unknown function in individual bacterial genomes and metagenomes. The discrepancy between our ever-increasing sequencing capacity and our inability to systematically determine gene function is exasperated by the realization that a comprehensive understanding of microbial life requires the elucidation of complex interactions and dynamics between genes, organisms and their environment. It is clear that omics level technologies derived from primary sequence information are necessary to make the transition from gene and genome catalogues to functional significance.

Microarray-based gene expression profiling provides a quantitative assessment of transcript abundance and can be used to predict gene function based on the hypothesis that functionally related genes are more likely to be transcriptionally coregulated. In natural microbial communities, microarray technology can be applied both as a tool to monitor critical gene activities across a diverse spectrum of genomes or to access the transcriptome of single microbial strains in a complex community.

Regardless of the nature of the study, substantial challenges (e.g. efficient RNA extraction, detection of signal above background noise for complex samples, and crosshybridization) need to be overcome before environmental gene expression studies approach the reproducibility of similar laboratory-based analyses. Finally, compared with large-scale sequencing, gene expression studies are more amenable to time-course studies. The significance of this is that the dynamics of a complex microbial community in a changing environment can be tracked with a single, portable experimental tool.

In addition to gene expression, environmental proteomics are enabled by the availability of near-complete microbial metagenomes. Through alignment of mass-spectrometrygenerated peptide signatures to the assembled AMD biofilm metagenome, high-confidence detection was achieved for ~2000 proteins including ~50% of the predicted proteins from the high-abundance Leptospirillum group II strain. Although DNA sequence illustrates the metabolic and functional potential of an organism, the detection of expressed proteins in a community provides critical insight into the important cellular activities at temporal and spatial environmental resolution.

In the AMD proteomics study, many genes with a role in oxidative stress and protein folding were highly expressed, potentially reflecting the challenge in maintaining cellular integrity in a harsh environment. The classic approach to assess gene function is to identify which genes are required for fitness in a given condition through gene disruption. One attractive mutagenesis technology that can be employed directly in the environment is the 'tagging' of individual mutants in an approach analogous to bacterial signature tagged mutagenesis and the parallel phenotypic analysis of the yeast deletion collection. In these strategies, each tag is a unique DNA sequence that serves to mark a single mutant strain. The presence of common PCR priming sites surrounding the unique tags enables the amplification of all tags in a complex pool of mutants in a single reaction. The relative abundance of each mutant can then be assessed by hybridization of the tags to a microarray containing the tag complements. In this manner, all pooled mutants that did not survive an experimental selection can be identi-fied in parallel.

For environmental studies, the main advantage of tagged mutagenesis is that the tag signals can potentially be PCR-amplified from the environmental 'noise' and quantified using a microarray without the need for culturing the pooled mutants after addition to the environment. Such experiments would identify genes required for survival in a natural environment. Groh et al. applied the signature tagged mutagenesis approach in the metal-reducing bacteria Shewanella oneidensis and Desulfovibrio desulfuricans. Pools of 60 tagged mutants were analysed for survival in an artificial anaerobic sediment environment using a custom microarray. Simulation of the natural environment will identify genes required for fitness under more natural conditions when studies cannot be performed in the field. As more environmental microbes are cultured, often enabled by the blueprint of the genome sequence, tagged mutagenesis will become increasingly applicable.

The biotechnology applications derived from microbial diversity range from the isolation of genes encoding novel functionality for industrial or biomedical applications to the cleanup of environmental pollutants using engineered microbes. Both companies and academic groups have constructed libraries of environmental DNA from diverse sources such as soil and seawater to identify genes with particular characteristics, such as those conferring antibiotic resistance or encoding specific enzymatic activities, or more generally to gain a better understanding of the variety and range of a protein family of interest. The challenges associated with this approach include potential problems with expressing heterologous DNA in a surrogate host (typically *Escherichia coli*), insufficient homology to identify clones using PCR, and the laborious task of screening through thousands of clones for rare 'hits'.

Methods to rapidly screen or select clones of interest from the thousands in a standard metagenome library are required to bring this technology to the average research laboratory lacking high-throughput infrastructure. One promising development is the substrate-induced gene expression screening (SIGEX) technique. Taking advantage of the observation that most catabolic genes are induced by their substrates, a groundwater metagenome library was cloned in an operon-trap vector driving expression of the gene encoding green fluorescent protein gfp). Upon induction by a hydrocarbon substrate, gfpexpressing clones (presumably containing catabolic genes involved in the degradation of the hydrocarbon) were identified and separated from non-induced clones using fluorescence-activated cell sorting. The use of individual microbes for complex environmental tasks such as bioremediation of contaminated and polluted sites represents a great challenge for environmental biotechnology on several levels. Foremost, there is substantial discord between the laboratory conditions where the organism is manipulated and the in situ environment that is targeted by the microbe. Consequently, it is not surprising that genetically modified bacteria rarely function in a natural environment. How do we cope with the laboratory-environment discrepancy and how do we design laboratory experiments that adequately represent natural conditions? One solution is to take a global, systems biology approach by examining the numerous stress responses, regulatory systems, and genes critical for the desired biological activity such as bioremediation.

The key to this approach will be the integration of gene expression, proteomics, physiological, mutant phenotype, and metabolic data into working cellular models that can accurately predict the response of the organism to a given environment. Meeting these goals of microbial systems biology will additionally require the development of computational resources and infrastructure that link services such as data storage and integration into coherent, testable models. Finally, the functionality of an environmentally introduced, engineered microbe(s) will be aided by the culture-independent technologies described previously to determine the impact of the endogenous microbial population, track the activity and progress of the engineered microbe over time, and to understand the ecological impact of the human intrusion.

Genomic-based Analyses

Genomic-based analyses in environmental microbiology are in their infancy. Meeting the challenges associated with applying experimental techniques in the environment, analysing complex data, and meshing biogeochemical cycles with the relevant microorganism will go a long way towards realizing the biotechnological potential of natural microbial diversity. Currently, because of the complexity and cost of technologies necessary for environmental genomics, these projects are often accomplished through the formation and collaborative effort of large research teams such that cost and expertise are distributed. However, we anticipate that technological innovations will lead to a severe cost reduction in DNA sequencing and other related technologies and make large-scale metagenomics more accessible to the individual researcher. Similar to the development of standard laboratory genomics, the availability of the necessary tools to a larger number of researchers will spur future discoveries in environmental genomics.

As environmental microbiology data accumulates we can begin to move from 'What is there' and 'What is it doing' towards higher order questions regarding the generation and maintenance of genetic diversity and the impact of environmental change on microbial evolution. These questions will be aided by analyses into the prevalence and function of viruses, transposable DNA elements, plasmids, and horizontally transferred genes within and across communities. It is becoming increasing clear, both from comparative studies of whole microbes and from studies of sequences obtained in environmental samples, that the horizontal transfer of genes plays a large role in the spread of functional abilities within communities and in enabling the adaptation of organisms to changing niches. It appears that, in addition to the measures taken by bacteria and archaea to confer fitness upon their brethren, phage might also provide a means for the transfer of useful genetic elements between microbes, and perhaps even contribute to the evolution of novel functions. Therefore, metagenomic studies of viruses are an important, and perhaps essential, complement to genomic studies of microbial communities.

MICROBIAL PROTEOMICS

Proteomics is the large-scale study of proteins, particularly their structures and functions. Proteins are vital parts of living organisms, as they are the main components of the physiological metabolic pathways of cells. The term "proteomics" was first coined in 1997 to make an analogy with genomics, the study of the genes. The word "proteome" is a blend of "protein" and "genome", and was coined by Prof Marc Wilkins in 1994 while working on the concept as a PhD student. The is the entire complement of proteins, including the modifications made to a particular set of proteins, produced by an organism or system. This will vary with time and distinct requirements, or stresses, that a cell or organism undergoes.

After genomics, proteomics is often considered the next step in the study of biological systems. It is much more complicated than genomics mostly because while an organism's is more or less constant the differs from cell to cell and from time to time. This is because distinct genes are expressed in distinct cell types. This means that even the basic set of proteins which are produced in a cell needs to be determined.

In the past this was done by mRNA analysis, but this was found not to correlate with protein content. It is now known that mRNA is not always translated into protein, and the amount of protein produced for a given amount of mRNA depends on the gene it is transcribed from and on the current physiological state of the cell. Proteomics confirms the presence of the protein and provides a direct measure of the quantity present.

More importantly though, any particular protein may go through a wide variety of alterations which will have critical effects to its function. For example during many and structural proteins can undergo . The addition of a phosphate to particular amino acids—most commonly serine and threoninemediated by serine/threonine kinases, or more rarely tyrosine mediated by tyrosine kinases—causes a protein to become a target for binding or interacting with a distinct set of other proteins that recognize the phosphorylated domain.

Because protein phosphorylation is one of the most-studied protein modifications many "proteomic" efforts are geared to determining the set of phosphorylated proteins in a particular cell or tissue-type under particular circumstances. This alerts the scientist to the signaling pathways that may be active in that instance.

Listing all the protein modifications that might be studied in a "Proteomics" project would require a discussion of most of biochemistry; therefore, a short list will serve here to illustrate the complexity of the problem. In addition to and , proteins can be subjected to methylation, etc. Some proteins undergo ALL of these modifications, which nicely illustrates the potential complexity one has to deal with when studying protein structure and function.Even if one is studying a particular cell type, that cell may make different sets of proteins at different times, or under different conditions. Furthermore, as mentioned, any one protein can undergo a wide range of post-translational modifications.Therefore a "proteomics" study can become quite complex very quickly, even if the object of the study is very restricted. In more ambitious settings, such as when a for a tumor is sought - when the proteomics scientist is obliged to study sera samples from multiple cancer patients - the amount of complexity that must be dealt with is as great as in any modern biological project.

Rationale for Proteomics

The key requirement in understanding protein function is to learn to correlate the vast array of potential protein modifications to particular phenotypic settings, and then determine if a particular post-translational modification is required for a function to occur.Scientists are very interested in proteomics because it gives a much better understanding of an organism than genomics. First, the level of transcription of a gene gives only a rough estimate of its level of expression into a protein. An mRNA produced in abundance may be degraded rapidly or translated inefficiently, resulting in a small amount of protein. Second, as mentioned above many proteins experience that profoundly affect their activities; for example some proteins are not active until they become phosphorylated. Methods such as phosphoproteomics and glycoproteomics are used to study post-translational modifications. Third, many transcripts give rise to more than one protein, through alternative splicing or alternative post-translational modifications. Fourth, many proteins form complexes with other proteins or RNA molecules, and only function in the presence of these other molecules. Finally, protein degradation rate plays an important role in protein content.

One way in which a particular protein can be studied is to develop an antibody which is specific to that modification. For example, there are antibodies which only recognize certain proteins when they are tyrosine-; also, there are antibodies specific to other modifications. These can be used to determine the set of proteins that have undergone the modification of interest.

For sugar modifications, such as glycosylation of proteins, certain lectins have been discovered which bind sugars. These too can be used. A more common way to determine post-translational modification of interest is to subject a complex mixture of proteins to electrophoresis in "two-dimensions", which simply means that the proteins are electrophoresed first in one direction, and then in another... this allows small differences in a protein to be visualized by separating a modified protein from its unmodified form. This methodology is known as "two-dimensional gel electrophoresis". Recently, another approach has been developed called PROTOMAP which combines SDS-PAGE with shotgun proteomics to enable detection of changes in gel-migration such as those caused by proteolysis or post translational modification.

Classically, antibodies to particular proteins or to their modified forms have been used in biochemistry and cell biology studies. These are among the most common tools used by practicing biologists today. For more quantitative determinations of protein amounts, techniques such as ELISAs can be used. For proteomic study, more recent techniques such as Matrix-assisted laser desorption/ionization have been employed for rapid determination of proteins in particular mixtures.

Most proteins function in collaboration with other proteins, and one goal of proteomics is to identify which proteins interact. This is especially useful in determining potential partners in cascades. Several methods are available to probe protein-protein interactions. The traditional method is yeast two-hybrid analysis. New methods include protein microarrays, immunoaffinity chromatography followed by, and experimental methods such as phage display and computational methods.

STRATEGIES OF MICROBIAL PROTEOMICS

Proteomics may be defined as the analysis of the entire protein complement expressed in a cell or any biological sample at a given time under specific conditions. The field can be split into two areas, expression proteomics and functional proteomics, the former aims to measure differential expression of proteins within a cell under varying conditions while the latter seeks to characterise the components of cellular compartments, multiprotein complexes and signalling pathways. Unlike DNA microarray analysis, proteomics currently does not have the equivalent of the polymerase chain reaction to enhance the signal, making proteins of low copy number difficult to detect. Developments in the ability to study gene expression at the genome level have been complemented by the development of high throughput multi-dimensional methods for proteome analysis. Mass spectrometry has greatly enhanced research in the field of microbial proteomics. In the areas of global microorganism identification through intact cell mass spectrometry; identification of membrane, cellular, periplasmic and extracellular proteins; full proteome expression in organisms (2D-PAGE coupled to MS and 2D LC coupled to MS); differential protein expression levels under stress and non stress conditions and identification of posttranslational modifications of proteins within organisms.

Top Down Strategy

The top down strategy was first introduced by McLafferty and colleagues utilising the immense analytical power of FT-ICR MS. The goal of this methodology is to identify intact proteins utilising mass spectrometry, without the need for prior proteolytic digestion of the sample. Significantly, the protein also need not be purified to homogeneity. Initially proteins are introduced into the mass spectrometer in the gas phase and are then fragmented. The fragmentation profile generated is then analysed and compared with a specifically designed database in order to identify the proteins present.

The methodology is not as widely used as peptide fragmentation and usually requires a high resolution mass spectrometer such as FT-ICR, Maldi/TOF-TOF or Q-TOF. This methodology has, however, been used successfully for microbial proteomics in the analysis of *Bacillus* spores in order to ascertain the species that the spore was derived from. In addition it has also been used for the identification of pathogenicity biomarkers from a comparison of 12 strains of *Enterobacter sakazakii*. It should be noted that the classification of top-down proteomics has recently been widened to include the multidimensional separation (gel based or LC based) of undigested protein samples followed by tryptic digestion of isolated proteins and subsequent analysis of peptides by MS.

Bottom up Strategy

This approach refers to any methodology that identifies proteins from the analysis of peptides derived from the proteolytic digestions of those proteins. The resultant peptide mixture is fractionated by chromatography before being subjected to tandem mass spectrometry. The fragmentation pattern from each peptide produces a peptide sequence tag and the resultant data is analysed by bioinformatics tools and searched through amino acid or protein databases in order to identify the protein. The simplest form of this approach is known as 'shotgun proteomics'. This refers to the direct analysis of a complex protein mixture without fractionation.

The complex mixture is enzymatically digested to produce peptides, this peptide mixture is then fractionated on a reverse phase C18 column before analysis on the mass spectrometer. This methodology gives a rapid large scale global analysis of the protein mixture, however, it gives limited penetration into the proteome. The effectiveness and proteome coverage of shotgun analysis has been greatly enhanced by coupling it with multidimensional separation techniques.

Until recently, the study of global protein expression was performed nearly exclusively using two-dimensional gel electrophoresis (2D PAGE), a technique developed in the 1970s with significant advances in the intervening decades. For a detailed description of the current status of this technology the reader is directed to the excellent review by Gorg et al. The strength of 2D PAGE is that it can separate up to 10,000 proteins in one gel. Every component is fractionated on the first dimension, by isoelectric focusing and then further resolved according to molecular weight in the second dimension.

At this point in the proteomic workflow a snapshot of the organism/cell may be visualized. An emerging trend is to deposit images of these 2D gels with databases such as Swiss-2D PAGE or Gelbank as reference material. The usefulness of such repositories is yet to be demonstrated. A limitation of 2D PAGE is typically many more spots are resolved on the gel than are actually identified by the researchers involved. This is as a result of a second analytical step that must be employed in order to identify the proteins present. Proteins are excised from the gel, subjected to proteolytic digestion, and identified or sequenced; this step is usually carried out manually and is very time-consuming although the advent of computerised gel visualisation and robotic spot excision equipment has gone some way to alleviate these 'bottle necks'.

The 2-D PAGE methodology has traditionally had a number of practical limitations that the researcher should be aware of with the main issue being the wide dynamic range of proteins present within a biological sample thus proteins present in low copy numbers, and therefore low concentration, are often not visualised on 2-D PAGE gels. A number of additional limitations can be encountered such as: most isoelectric focusing gels can only focus proteins between the pI ranges 3–10, so proteins with extreme pI will not be seen on the gels; however protocols have been developed to allow separation and then visualisation of highly alkaline proteins with a pI up to 12; Most 2-D PAGE gels cannot resolve proteins smaller than 10 kDa and above 200 kDa.

Due to the nature of the buffers used in isoelectric focusing the range of solubilising detergents that can be used in this methodology are restricted, thus making it difficult to solubilise certain membrane proteins, however the inclusion of amidosulfobetaines can enhance solubilisation of certain membrane proteins.

Despite its limitations 2D-PAGE is still used as a standard tool in the analysis of microbial proteomes. The idea being to first identify the protein complement of the microbe under normal conditions, then subject the organism to a stress stimulus so that the differential expression of proteins can be visualised by either an increase or decrease in spot intensity or by the appearance/disappearance of spots on the gel.

Protein Identification Technologies

An alternative to the traditional 2-D PAGE technology for microbial proteome analysis is the high throughput approach of multidimensional liquid chromatography coupled to tandem mass spectrometry. In its early stage of development this process was used very successfully for the proteome analysis of the *Saccharomyces cerevisiae* ribosome allowing the identification of more than 100 proteins in a single 24-hour run. The process was further developed and led to the multidimensional protein identification technology (MUDPIT).

A MUDPIT experiment entails the following: A reduced, alkylated and tryptically digested mixture of proteins are separated by first running the peptide mixture on a strong cation exchange (SCX) chromatography column. This solution is then separated into several discrete fractions by a series of wash steps with an increase in salt molarity at each step. The peptides eluted at each salt wash step are then run onto a reverse phase C18 column where they are further separated and resolved. The resolved mixtures are then passed directly into the mass spectrometer where tandem mass spectrometry profiles are generated for each peptide; this data is automatically trawled against protein databases for identification. Finally, any novel peptides not in the database can be subjected to *de novo* sequencing.

This process, whilst seemingly complicated, is highly automated with high throughput achieved in a short time. Washburn and co-workers using this process were able to identify1484 proteins from the *Saccharomyces cerevisiae* proteome in a single twenty-seven hour run. MUDPIT can be seen as complimentary to 2D PAGE as it overcomes many of the problems and limitations of this technique, identifying proteins with extreme pI, integral membrane proteins and low abundance proteins.

Intact Cell Mass Spectrometry

Intact cell mass spectrometry (ICMS) can be employed in microbiology for the rapid analysis, identification and subtyping of specific microorganisms. The use of MALDI-TOF-MS allows the examination of specific peptides or proteins that desorb from intact viruses, bacteria and microbial spores, thus generating peptide mass fingerprints that are unique to the individual microorganisms. Walker *et al* assessed ICMS for the identification and subtyping of methicillin-resistant *Staphylococcus aureus* (MRSA) investigating the effects of different culture media and the intra- and inter-laboratory reproducibility of their results in previously characterised isolates of staphylococcal species. Shah *et al* used MALDI-TOF-MS analysis on intact cells of human pathogens to give specific spectral profiles which could be used to delineate bacterial species. Cells were then lysed and subjected to Surface-enhanced laser desorption/ionisation time of flight mass spectrometry (SELDI-TOF-MS): this is a modification of MALDI-TOF-MS in which the stainless steel target plate is replaced by a protein chip array. The chip has a number of sample wells each containing a different chemistry, thus specific classes of molecules may be captured from cell lysates and selectively analyzed. Using this process several toxigenic and nontoxigenic strains of *Bacteroides fragilis* were analyzed revealing potential biomarkers specific to the toxigenic strains in the mass range 3.5–18.5 kDa.

Expressional Proteomics

Whilst techniques described thus far provide the microbiologist with an invaluable snapshot of the processes occurring within a biological system, assessing the quantitative change in protein expression patterns remains the focus for those interested in the fundamental analysis of microbial systems. There are presently several methodologies that attempt to provide quantifiable expressional analysis. These include the label free emPAI technique; the label based ICAT, iTRAQ and metabolic labelling as well as the gel-based differential in gel electrophoresis (DIGE).

The exponentially modified protein abundance index (emPAI) is a label free methodology for estimating absolute protein abundance in a sample. This methodology is a simple calculation that utilises the output information obtained when tandem mass spectrometry data is processed through database search engines.The aim of any labelling strategy is to derivatize all proteins/peptides in a sample to allow their analysis. Gygi and co-workers were the first to utilise isotope coded affinity tagging (ICAT) for differential protein expressional analysis of *Saccharomyces cerevisiae* when utilising either galactose or ethanol as a carbon source. The original ICAT reagent consisited of an affinity tag (biotin), to allow labelled peptides to be removed from a mixture by attachment to an avidin column; an isotopically labelled linker region which was either 'light' containing eight hydrogen atoms d0 or 'heavy' containing eight deuterium atoms d8; and a thiolate-reactive group that allowed labelling of cysteinyl groups.

Protein mixtures from the two states were labelled separately one with the light reagent and one with the heavy. The two samples were then mixed, tryptically digested and the labelled peptides were separated from the unlabelled by running the sample on an avidin column which binds to the biotin tag. The biotin is then removed and the sample separated on a C18 column before analysis on a mass spectrometer. The relative abundance of the light and heavy versions of the peptides can then be compared and information on the protein expressional changes can be identified. The present form of the ICAT reagent differs slightly form the original. It contains the biotin affinity tag which is attached to an acid cleavable linker, making it easier to remove; the light and heavy isotopically labelled region contains either nine C12 or nine C13 atoms (Figure 4), these overcome slight differential elution problems that were observed when using hydrogen and deuterium; the thiol-specific labelling group remains the same.

The latest reagent for use in protein labelling, which was utilised by Ross and co-workers to analyse the global protein expression in a wild type *Saccharomyces cerevisiae* and two isogenic mutant strains, is the amine reactive isobaric tag for relative and absolute quantitation (iTRAQ). The iTRAQ reagent has several advantages over ICAT; four or, in the most recent version, eight states rather than two can be measured; and free amine groups rather than reduced cysteines, which are only present in 95% of proteins, are labelled. The 4-plex reagent contains an amine specific reactive group, a balance group and a reporter group that can have a mass of 114, 115, 116 or 117.

Samples of proteins from up to four states are first trypsinised resulting in a peptide mixture with each cleaved peptide having a free amine group. Each sample then is labelled with one of the specific reagents by attachment of the label via the amine specific reactive group. All four samples are then mixed, separated by liquid chromatography and introduced into the mass spectrometer. During tandem mass spectrometry of the labelled peptides the reporter group is released, and measurement of the peak areas of these resultant ions gives an assessment of the abundance of that particular peptide under each condition (Figure4).

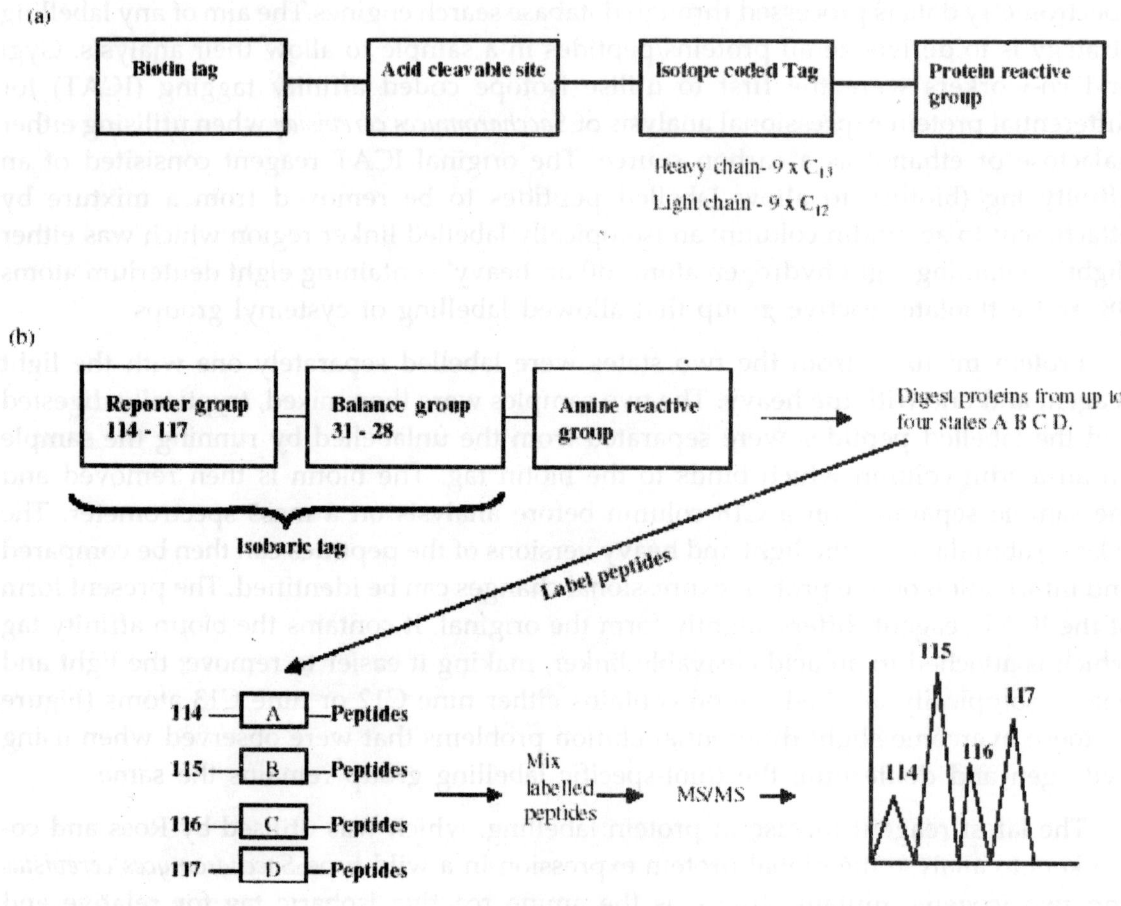

Figure 4. (a) Structure of the isotope coded affinity tag (ICAT). (b) Simplified version of the structure and protocol using the isobaric tag for relative and absolute quantitation (iTRAQ).

Metabolic labelling offers one of the most comprehensive methods of investigating microbial proteomes. Unlike other labelling technologies samples to be analysed can be combined before protein extraction thus removing the main source of sample variation, which is the protein extraction process itself. The simplest methodology involves comparison of 'normal' and 'stress' states by growth of the microorganism on media enriched with N15 for one state and on media containing the naturally abundant isotope N14 for the other state. The ratio of N14/N15 containing proteins from the two conditions are measured and changes in protein expression levels can be identified. This methodology has been successfully utilised by Washburn and co-workers when working on *Saccharomyces cerevisiae.* Additional isotopic variants such as deuterium and C13 enriched media can also be utilised for metabolic labelling.Conventional comparative 2D-PAGE requires the production of two gels, one for each condition being compared. Several inherent problems with this approach include the fact that there can be slight variation in the composition of the gels thus giving rise to slight variations in the proteomic profile observed, also sample loading errors can give rise to similar misleading results. Subsequently several replicate gels must be produced in order to obtain as accurate a picture of the proteome as possible.

The most efficient way to overcome these problems would be to analyse the different protein samples on one gel. This can now be achieved utilising a technique known as differential in gel electrophresis (DIGE). In this methodology each protein sample is labelled with one of three structurally similar but spectrally distinct fluorphores that are N-hydroxy-succinymidyl esters of the cyanin dyes Cy2, Cy3 and Cy5. The samples to be investigated are mixed together and run on a 2D-PAGE gel. The resultant gel is then imaged using filters specific to each fluorphor. The ratio of the different signal intensities can be used to determine changes in observed protein expression patterns.

Post--translational Modifications

Genomic data alone gives no information on post-translational modification events that occur within many proteins as they are converted to their mature forms. There are currently over 200 reported post-translational modifications the vast majority of these are found in eukaryotic systems. Many of these modifications are regulatory in nature as exemplified by phosphorylation events. Phosphorylation of proteins is a key post-transaltional modification that governs the activity of a number of biochemical pathways and enzymatic activities, although these phosphorylation cascades are widespread in eukaryotic systems, prokaryotic proteins are also phosphorylated, markedly in the phosphorylation cascade of bacterial two-component signal transduction systems.

Therefore detection and identification of this post-translational modification within microbial proteomic investigations is highly desirable as it allows further understanding and elucidation of the processes occurring within a given system. Various methodologies have been developed for investigation of protein phosphorylation. Phosphoproteins can be identified on 2 D PAGE by the incorporation of radiolabelled orthophosphate into proteins with identification by autoradiography. This method has limitations; it can only be carried out *in vivo* and background staining of DNA and RNA occurs. An alternative to this method is the staining of immunoblots from 2D PAGE with phosphoamino specific antibodies. These antibodies work well for phosphotyrosine but are less effective in the identification of phosphoserine and phosphothreonine.

An advantage of 2D PAGE over gel-free approaches is that post-translational modifications will cause a mass change and subsequently a shift in the pI of proteins that have been modified, thus the modified and parent proteins usually appear on the gel as horizontal or vertical sets of spots. This has been used to identify parallel profiles of phosphoylated and total proteins within a 2D PAGE gel; using sequential staining the gel is first stained with Pro Q™ Diamond phosphoprotein stain and then imaged. The gel is then stained with Sypro Ruby which reacts with all proteins present, the gel is then re-imaged and the two images compared.

The phosphorylated proteins, with their increased negative charge, migrate to a more acidic region of the gel compared to the parent protein. This pattern can then be visualised and proteins identified. A multiplexing approach can then be taken by running several gels and comparing the profiles. Another useful method for identification of phosphorylation sites by 2D PAGE is to run samples before and after chemical or enzymatic removal of phosphate groups. In this case spots relating to phosphorylated proteins will disappear from the gel. Various gel-free mass spectrometry based protocols have also been developed for the identification of phosphoylation sites on peptides and proteins. However, the analysis of protein phosphorylation is complicated by the fact that these proteins are present in low concentrations and are poorly ionisable. In order to better study these post-translationally modified proteins from a proteomic sample one must try to either enrich the intact phosphoproteins or their derived phosphopeptides form the sample to be investigated.

Phosphoprotein enrichment is usually achieved by immunoprecipitation using antibodies directed against phosphotyrosine. However, a more commonly used approach, which gives a more global overview in studying the phosphoproteome, is the enrichment of phosphopeptides. Enrichment can be achieved with IMAC as demonstrated by Ficarro et al studying the phosphoproteome of *Saccharomyces cerevisiae* and can also be achieved utilising graphite and titanium oxide. This approach both reduces the complexity of the sample to be analysed and allows you to gain specific information on the sites of phosphorylation.Once the phosphoproteome has been enriched mass spectrometry techniques can be used to identify the sites of phosphorylation. These include neutral loss scanning for the loss of the 98 Da phosphoric acid moiety, which is lost from phosphoserine and phosphothreonine during CID in ion trap mass spectrometers. This method cannot be used to detect phosphotyrosine as there is no loss of a phosphoric acid moiety during CID. However, another technique, parent ion scanning in positive ion mode searching for the immonium ion of phosphotyrosine at m/z 216.043 can be used for its identification. This technique utilised in negative ion mode for an ion of m/z 79 (corresponding to PO_3) can also be used for the identification of phosphoserine and phosphothreonine.

Replicate Injections

Due to the complexity of peptide mixtures within a proteomic sample the separation capabilities of LC-MS systems are often exceeded. This, coupled to the limitations of the data dependent acquisition for the selection of peptides for MS/MS, requires that samples be run more than once in order to gain as wide a proteome coverage as possible. During a proteomic investigation of *E. coli* by Taoka and colleagues ten repeated injections of the same sample was carried out and showed that the number of new proteins identified in each run increased until the third and fourth run, where further injections did not greatly increase the number of proteins identified. This suggests that researchers must perform at least triplicate analysis of the same fraction and recent studies have shown that such an approach leads to an increase in the number of proteins identified by up to 40%.

Applications of Microbial Proteomics

One of the most promising developments to come from the study of human genes and proteins has been the identification of potential new drugs for the treatment of disease. This relies on genome and proteome information to identify proteins associated with a disease, which computer software can then use as targets for new drugs. For example, if a certain protein is implicated in a disease, its 3D structure provides the information to design drugs to interfere with the action of the protein. A molecule that fits the active site of an enzyme, but cannot be released by the enzyme, will inactivate the enzyme. This is the basis of new drug-discovery tools, which aim to find new drugs to inactivate proteins involved in disease. As genetic differences among individuals are found, researchers expect to use these techniques to develop personalized drugs that are more effective for the individual.

A computer technique which attempts to fit millions of small molecules to the three-dimensional structure of a protein is called "virtual ligand screening". The computer rates the quality of the fit to various sites in the protein, with the goal of either enhancing or disabling the function of the protein, depending on its function in the cell. A good example of this is the identification of new drugs to target and inactivate the HIV-1 protease. The HIV-1 protease is an enzyme that cleaves a very large HIV protein into smaller, functional proteins. The virus cannot survive without this enzyme; therefore, it is one of the most effective protein targets for killing HIV.

Understanding the proteome, the structure and function of each protein and the complexities of protein-protein interactions will be critical for developing the most effective diagnostic techniques and disease treatments in the future. An interesting use of proteomics is using specific protein biomarkers to diagnose disease. A number of techniques allow to test for proteins produced during a particular disease, which helps to diagnose the disease quickly. Techniques include western blot, immunohistochemical staining, enzyme linked immunosorbent assay (ELISA) or mass spectrometry. The following are some of the diseases that have characteristic biomarkers that physicians can use for diagnosis.

In Alzheimer's disease, elevations in beta secretase create amyloid/beta-protein, which causes plaque to build up in the patient's brain, which is thought to play a role in dementia. Targeting this enzyme decreases the amyloid/beta-protein and so slows the progression of the disease. A procedure to test for the increase in amyloid/beta-protein is immunohistochemical staining, in which antibodies bind to specific antigens or biological tissue of amyloid/beta-protein.

Heart disease is commonly assessed using several key protein based biomarkers. Standard protein biomarkers for CVD include interleukin-6, interleukin-8, serum amyloid A protein, fibrinogen, and troponins. cTnI cardiac troponin I increases in concentration within 3 to 12 hours of initial cardiac injury and can be found elevated days after an acute myocardial infarction. A number of commercial antibody based assays as well as other methods are used in hospitals as primary tests for acute MI.

REFERENCES

Ouellette, Robert P., and Paul N. Cheremisinoff. *Essentials of Biotechnology*. Lancaster. Pennsylvania: Technomic Publishing Company. 1985.

Panem, Sandra. (ed.) *Biotechnology: Implications for Public Policy*. Washington. D.C.: The Brookings Institution. 1985.

Peters, Pamela. *Biotechnology: A Guide to Genetic Engineering*. Dubuque. IA: Wm. C. Brown Publishers, 1993.

Rittmann, B.E., and P.L. McCarty. *Environmental Biotechnology: Principles and Applications*. Boston, MA: McGraw-Hill. pp. 720-721. 2001.

Turner A. P. F. (ed.). *Biosensors: Fundamentals and Applications*. Oxford University Press. Oxford. 1987.

Walker P.M.B. (ed.) *Chambers Science and Technology Dictionary*. Chambers, Edinburgh. 1992.

4

Integrated Microbial Genomes System

Problems related to biological data management systems have been examined extensively over the past decade. These problems are usually discussed in terms of novel methods and technologies needed for developing biological data management systems. For example, a recent report discusses the need for extending database technology to support biological data types, provenance, evolution, and integration. In practice, hundreds of commercial and public biological database have been developed using existing data management technology. Most problems with these database regard effective use of, rather than deficiencies with, existing technologies.

One of the key goals for biological data management systems is to provide support for data analysis, which often involves exploring data across multiple heterogeneous data sources. Data warehousing and data federation technologies have been employed for handling syntactic heterogeneity, that is, differences in data structure and formats, across diverse biological data sources, as discussed. Effective data analysis, however, also needs to support seamless flow (composition) of analysis operations, while addressing semantic heterogeneity, that is, differences in the meaning of related data items (objects). Providing such support presents a significant challenge for biological data management systems, especially for those developed in academic settings.

Biological data management systems in academic settings were originally confined to relatively small individual scientific groups or laboratories: these systems were often limited to specialized data sets and analysis operations and were developed without considering data analysis workflows, heterogeneity, evolution, and scalability issues. Addressing such problems requires a systematic process for analyzing the data structure and operations for the application domain. This process entails substantial documentation which is especially difficult to maintain for biological data whose semantics are complex and tend to evolve. These data are generated via processes that involve multiple transformations between different levels of data granularity and are based on evolving technology platforms and computational methods. In spite of this complexity, a systematic application domain analysis process and comprehensive documentation are essential for providing effective data analysis support and thus address the frustration scientists often encounter in dealing with public biological data management systems. In this chapter we discuss how this challenge has been addressed in the development of the Integrated Microbial Genomes (IMG) system.

The development process for IMG is based on established practices and starts with application domain analysis, followed by abstract data model definition, system design and implementation. Application domain analysis is based on requirements gathered from biologists and entails detailed use case scenarios that serve as a vehicle for bridging the rather steep communication gap between these scientists and data management system developers. Application domain analysis is used for defining an abstract microbial genome data model in terms of data types and operations. This data model then serves as the foundation for the design and development of the data management system.

IMG is the result of the collaboration between the scientists of the Microbial Genome Analysis Program (MGAP) at the Joint Genome Institute (JGI) and members of the Biological Data Management and Technology Centre (BDMTC) at Lawrence Berkeley National Laboratory. The IMG case study is instructive since it deals with genomic sequence data generated using established technologies and methods. Systems that deal with data generated using newer technology platforms and methods, such as gene expression and proteomic data, are likely to encounter similar or more complex challenges. Furthermore, MGAP scientists and BDMTC engineers had prior experience in developing both academic and commercial large scale biological data management systems. Their combined experience was not enough to avoid the communication problems mentioned above, but was essential in following the process required to address these problems.

A public version of IMG that supports microbial genome data analysis was released in March 2005. An enhanced version of IMG, with additional support for genome data creation (editing) is used at JGI for improving the quality of annotations for newly sequences microbial genomes. In the following sections we present a brief overview of the microbial genome data application, and then discuss gathering and analyzing application requirements for IMG.

MICROBIAL GENOME APPLICATION

According to the Genomes On-line Database, about two hundred microbial genomes have been sequences to date with 530 other projects ongoing and more in the process of being launched. Microbial genome analysis is a growing area that is expected to lead to advances in healthcare, environmental cleanup, agriculture, industrial processes, and alternative energy production.

Microbial Genome Data Types

Microbial genome data captures information about raw DNA sequence data, along with genes characterized in terms of functions and pathways.

A gene represents an ordered sequence of nucleotides located on a particular chromosome that encodes a specific product (i.e., a protein or RNA molecule). Characterizing a gene consists of determining its biological context, including its location on a chromosome within a (species specific) genome, and its associated functional roles in cellular pathways. A key characteristic for genome is its taxonomic (phylogenetic) lineage, including its domain, phylum, class, order, family, genus, species and strain.

Pathways can be viewed as ordered lists of reactions, whereby each reaction involves compounds which are reactants (substrates, products), catalysed by enzymes. Pathways can be combined in pathway networks, whereby pathways can be associated via reactions that share common components. Pathways are associated with genes via gene products that function as enzymes that serve as catalysts for individual reactions of metabolic pathways. Accordingly, pathways provide a biologically meaningful framework for examining functional relationships between genes, rather than individual gene functions.

Microbial Genome Annotation

Microbial genome annotation generally refers to a process of assigning biological meaning to the raw sequence data by identifying gene regions or functional features and determining their biological functions. Gene annotation is a combination of automated methods that generate a "preliminary" annotation in terms of predicted genes (also called Open Reading Frames or ORFs, which represent the sequence of DNA or RNA located between the startcodon and stop-codon sequence) and associated functions and pathways based on sequence similarity or profile searches.

The result of a preliminary (baseline) annotation is often sparse, with numerous genes not having associated functions or pathways. Consequently, several techniques are employed for further annotating genes as well as validate baseline annotations. The most effective annotation techniques involve comparative multi-genome analysis based on observed biological evolutionary phenomena: pairs of genes with related (coupled) functions:

(1) are often both present or both absent within genomes;

(2) tend to be collocated (on chromosomes) in multiple genomes;

(3) might be fused into a single gene in some genomes; or

(4) are components of an operon (a set of genes transcribed as a unit under the control of an operator gene).

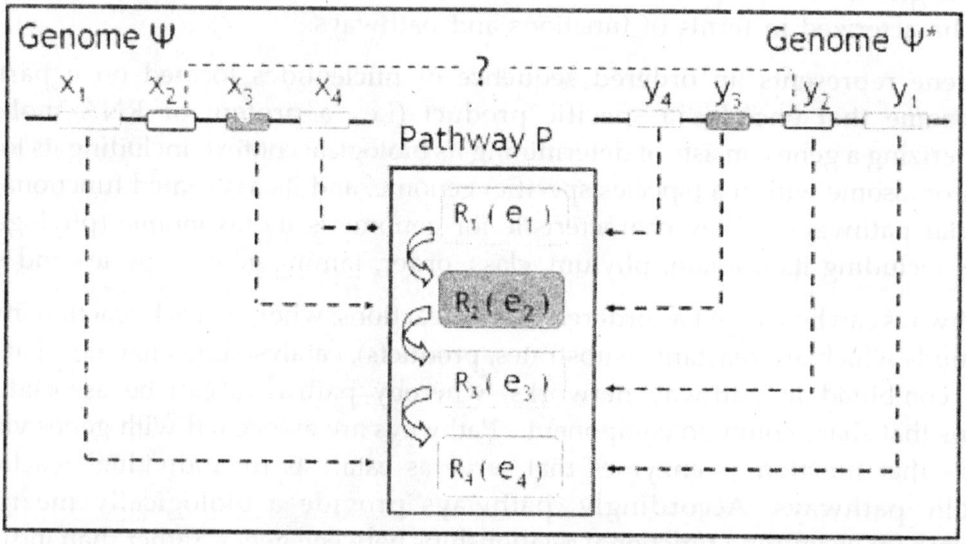

Figure 1. Sketch of Genomes Associated with a Pathway.

Consider the example shown in Figure 1, where pathway P involves reactions R_1,R_2,R_3, and R_4: genes x_1, x_3 and x_4 of genome Ψ are associated with pathway P via enzymes e_1, e_3 and e_4, respectively; genes y_1, y_2, y_3 and y_4 of genome Ψ* are associated with pathway P via enzymes e_1, e_2, e_3 and e_4 respectively; if gene x_2 is similar (i.e., determined to be related via significant sequence similarity) to gene y_2, then, following rules above, x_2 may be associated with P via enzyme e_2.

Microbial Genome Data Sources

Microbial genomes are sequences by organizations worldwide, follow an annotation process similar to that mentioned above, and end up in one of several microbial genome data sources, such as EBI Genome Reviews, CMR, and RefSeq. Furthermore, additional genome annotation details such as protein families and pathways reside in multiple specialized data sources such as UniProt (protein sequences and functions), InterPro (protein families and domains), COG (clusters of orthologous genes), and KEGG (pathway maps). Consequently, analyzing microbial genome data entails integration of data from diverse, usually heterogeneous, data sources.

It is important to distinguish between "shallow" and "deep" integration of biological data. The former amounts to "data sorting and collating" and does not address semantic problems between individual data items, while the latter involves identifying and matching data items (objects) in different data sources that may represent the same underlying biological objects, such as genes. Resolving semantic heterogeneity between diverse biological data sources is a complex problem. For example, a protein sequence is represented in data sources such as GenBank and SwissProt using different accession numbers to identify it and different terms to characterize it. Consequently, mapping objects across data sources may require expert scientific review of individual objects.

Effective comparative analysis of microbial genome data requires a coherent view of biological data and therefore involves "deep" data integration. Different microbial genome data sources provide a variety of alternative or fragmented views of an inherently incomplete and imprecise data domain. These sources share common goals but contain different collections of genomes or data with different degrees of resolution regarding the same genomes. These differences are the result of diverse annotation methods, creation techniques, and functional characterization employed across microbial genome data sources. An additional problem in dealing with these sources is the difficulty of determining the coherence and completeness of their data.

Data coherence regards the quality of annotations: although inherently imprecise, these annotations can be qualified in terms of "biological coherence" rules. For example, predicted genes with overlapping sequences often indicate errors in gene prediction and need to be manually reviewed and corrected. Problems related to data coherence are caused by the high cost in terms of time and expertise needed to validate and correct annotations manually.

Data completeness regards the extent and coverage of functional characterization and depends on the diversity of the genomes included in a data source and the depth of integration of genome annotations collected from diverse sources. Problems related to data completeness are caused by the complexity of "deep" integration, which often requires complex expert scientific reviews to resolve semantic heterogeneity problems.

JGI Microbial Genome Data

The Joint Genome Institute (JGI) is one of the key sources of microbial genome sequence data, covering about 22% of the reported number of microbial genome projects worldwide. Individual microbial genomes are sequences and assembled at JGI's production facility, producing data files with so called "draft" genome sequences. Draft genomes are subsequently completed ("finished") by JGI's partners at Los Alamos National Lab and Stanford. Both draft and finished genomes pass through the automatic

Genome Analysis Pipeline at Oak Ridge National Lab which identifies genes using gene prediction methods, and associates them with preliminary functional annotations, such as InterPro protein families and domains, COG categories, and KEGG pathway maps. Finished genomes and their annotations are eventually published on individual genome portals.

Before publication, scientific groups interested in a specific genome further review and curate the microbial genome data in collaboration with JGI's Microbial Genome Analysis Program. The genome annotation and creation processes are greatly enhanced when individual microbial genomes can be analyzed in the comparative context of other genomes. Providing such a context is the main purpose of the Integrated Microbial Genomes (IMG) system. IMG aims at providing high levels of data diversity in terms of the number of genomes integrated in the system from public sources, data coherence in terms of the quality of the gene annotations, and data completeness in terms of breadth of the functional annotations. IMG also aims at providing a high level of comprehensibility in terms of documenting its data structural and operational semantics.

MICROBIAL GENOME SYSTEM REQUIREMENTS

We discuss below the process of analyzing system and application requirements for developing a biological data management system in the context of the Integrated Microbial Genomes (IMG) data management system.

Developing a biological data management system starts with the analysis of application domain requirements. This analysis is one of the most difficult problems for biological data management systems, and involves domain scientists who outline what they need in abstract, potentially ambiguous or vague, domain-specific terms. The key challenge is to translate the "what" of abstract application domain views into the "how" of data management system components. This process is prone to misinterpretation, may require reconciling conflicting views, and often involves numerous iterations. Furthermore, this process is time consuming and requires a reliable mechanism for clarifying questions between individuals who have different views of the application.

Data Content Requirements

Gathering and analysing requirements for IMG first involved its data content. A prototype database that included a representative set of microbial genome sequences and associated annotations from a variety of sources was developed for this purpose.

The key question addressed in analyzing data content requirements for IMG was finding a primary source of public microbial genomes with annotations that are not only extensive and accurate, but also amenable for integration with additional annotations available in other data sources. For example, the source initially considered for public microbial genome data, NCBI's RefSeq, had only sparse annotations (e.g., in terms of gene names, symbols, etc.), and poor cross references with additional sources of annotations, such as UniProt and InterPro. EBI's Genome Reviews had better annotations and cross references than RefSeq, and therefore was selected as IMG's main source for public microbial genome data. It is worth noting that the quality of and issues with cross references between multiple biological data sources is not well documented and often requires extensive experimentation in collecting and integrating data from these sources. This problem is compounded by changes in the structure of biological data sources which range from occasional minor extensions to restructuring that may affect the semantics of the data. Furthermore, although correlated through mutual cross references, biological data sources tend to evolve on different schedules, which is another source of potential semantic inconsistencies.

Application Requirements

A second, equally important, aspect of analysing requirements for IMG regarded microbial genome data analysis. A prototype analytical tool was devised for examining, validating, refining, and documenting these requirements. This prototype was developed in the framework provided by the Apollo tool, and includes in addition to Apollo's native viewers additional visualization capabilities, for example for displaying genes on multiple genomes in a comparative context and for aligning DNA sequences. A key component of this prototype is a generic query constructor that allows experimenting with a variety of analysis workflows involving composition of individual operations.

For example, consider a typical microbial genome analysis that involves identifying and grouping genes that may belong to a particular protein family. Such an analysis entails:

(1) finding the genes associated with a specific protein family, such as "fusA";

(2) identifying and eliminating so called "duplicate" genes associated with individual genomes - such genes may be paralogs, that is genes that result from gene duplication events and variation within the same species;

(3) finding genes that have strong similarity with genes found in the previous steps - such genes may be orthologs, that is genes in different species that have the same evolutionary origin;

(4) removing ortholog genes whose similarities are determined to be "false positives", by examining their aligned protein sequences.

MICROBIAL GENOME DATA SPACE

Requirements analysis provides the basis for specifying an abstract data model for microbial genome data. For IMG, data warehouse constructs were employed for specifying its data model in order to allow reasoning about genome data in an established framework that also provides helpful analogies to well understood traditional data applications. Consequently, the microbial genome data space is modelled in terms of primary (also known as fact) objects characterized in the context of other (also known as dimension) objects. Each dimension is further characterized by one or several category attributes which are sometimes organized in a classification hierarchy. Operations in such a framework can be then defined in a multidimensional data space.

Microbial Genome Data Model

Microbial genome data can be viewed as an abstract multidimensional data space, whereby genes form the primary class of objects and are characterized in the context of other classes of objects, in particular individual genomes, functions and pathways.

The definition for each class of objects must include specifications for the semantics of component objects and for the operations that can be applied on them. Defining the semantics of biological data objects is a daunting task and requires a thorough understanding of the process involved in their generation. Unlike traditional (e.g., financial) data, biological data are imprecise, generated via processes that involve transformations between different levels of data granularity and are based on evolving technology platforms and computational methods. Consequently, the semantics of biological objects often cannot be fully characterized without information about their generation (i.e., provenance), such as experimental conditions, methods, data transformation parameters.

For example, class Gene models objects that may represent hypothetical genes that are predicted using gene prediction methods, such as Glimmer or Critica or experimentally validated genes. Genomes are annotated using a variety of gene prediction methods that yield results with different precision and reflecting different (e.g., either under or over prediction) biases. Functional characterizations for genes are also generated using diverse methods with different degrees of confidence (e.g. sequence homology based methods, experimental evidence based methods etc.) and often employ different terms for identifying functions. Ontology terms have been traditionally used to specify the functions of gene products. For example, the Gene Ontology provides terms to describe the attributes of gene products in three domains of molecular biology: molecular function, biological process, and cellular component. The Gene Ontology is not the only controlled vocabulary used for this purpose, nor is it used consistently for annotating different genomes.

Every class of objects is associated with basic operations, such as comparing data items of the same type (e.g., two DNA sequences associated with genes), as well as complex operations, such as searches for certain patterns across sets of data items. For example, a typical microbial genome analysis operation involves comparing distribution patterns (known as gene occurrence profiles or simply gene profiles) for genes associated with a specific genome, across other genomes. This operation is described in more detail in the next section.

Relationships between biological classes of objects are usually specified using operations associated with each individual class. For example, associating a gene with a pathway may involve matching (via sequence similarity) the protein sequence of the gene with the protein sequence considered as representative for the enzyme that serves as catalyst for reactions involved in pathways. Alternatively, a gene may be associated with a pathway based on its functional coupling with another gene that is already associated with the pathway. Specific methods employed for associating genes with pathways affect the precision of the functional characterization for genes. Operations that are based on these characterizations, such as grouping genes based on their association with pathways, will be also affected by the choice of these methods.

Microbial Genome Data Analysis

Microbial genome data analysis is set mainly in the comparative context of multiple microbial genomes. Comparative analysis is essential in the identification of similar or unique genes among different, potentially phylogenetically related, genomes, which provides the foundation for characterizing microbial genomes.

Analysis operations allow navigating the microbial genome data space along one or several dimensions and are often set in the context of specific genomes, pathways, and genes. Setting this context corresponds to reducing the dimensionality of the data for on-line analytical processing (OLAP) operations for traditional (e.g., financial) applications. Genome (organism) selections help focus the analysis on a subset of interest, especially in terms of phylogenetic relationships. For example, a set of interest may include the organisms for all the strains within a specified species. Similarly, pathway selections focus the analysis on a subset of interest, such as pathways involved in lipid metabolism. Gene selections reduce the scope of analysis to genes with certain properties, such as genes sharing a common gene symbol, function, or pathway.

Aggregation operations (usually called summaries or statistics) involving groups of objects, such as organisms, pathways, and genes, are commonly used over microbial genome data and are similar to analogous OLAP operations for traditional applications. For example, genes can be grouped over one level of the phylogenetic classification hierarchy associated with organisms. Such a grouping is then associated with a count of the group members and is employed to assess the extent of annotations over a selected set of organisms or genes. For example, the number of genes with functional characterization for a given set of organisms provides an assessment of the functional characterization across these organisms. Ranking (sorting) is another OLAP like operation that is employed over genome data summaries.

Additional operations in the microbial genome data space are domain-specific and do not have counterparts in the traditional data domain. Consider part of the genome data space that involves three dimensions: Genes, Genomes, and (sequence similarity) Methods. For a specific genome Ψ and specific method (e.g, sequence similarity comparison of genes, associated with a specific precision), the data space consists of either "p" or "a" for every pair (x, $Ψ_i$), whereby x is a Ψ gene (the Genes dimension), $Ψ_i$ is a genome (on the Genomes dimension), and "p" / "a" indicates presence or absence of a $Ψ_i$ gene that is similar to x, where the similarity is determined using the selected method. Figure 2 shows an example of the genome data space projected over $Ψ_i$ and a specific method, with presence/ absence for a set of selected Ψ genes (x_1 to x_4) shown across a set of selected genomes ($Ψ_2$ to $Ψ_8$).

A typical operation in this simplified data space involves examining (computationally predicted) genes of a specific genome, Ψ, in the context of other related genomes, $Ψ_1$,..., Ψk : this operation allows determining what genes genome Ψ may have in "common" with $Ψ_1$,..., Ψk. Such an operation is sometimes called gene profile (also called phylogenetic profile and occurrence profile in), and involves first selecting a specific method (i.e., projection on the Method dimension), and then selecting the genomes of interest ($Ψ_1$,..., Ψk), for example based on their phylogenetic relationship (on the Genomes dimension). Then Ψ genes with specific "profiles" can be examined. For the example shown in Figure 2, genome Ψ has gene x_4 in "common" with genomes $Ψ_1$ to $Ψ_8$; and genes x_1 and x_2 of Ψ have the same profile across genomes $Ψ_1$ to $Ψ_8$. Note that although this operation seems to be similar to (and is often confused with) a set operation (e.g., intersection) over genes, it is not an operation over sets of genes associated with different genomes: from a data point of view each gene is unique although it may be similar (but not identical) to other genes in terms of their associated protein sequences

$Ψ$	$Ψ_1$	$Ψ_2$	$Ψ_3$	$Ψ_4$	$Ψ_5$	$Ψ_6$	$Ψ_7$	$Ψ_8$
x_1	p	p	p	p	a	a	p	a
x_2	p	p	p	p	a	a	p	a
x_3	a	p	p	a	a	a	a	a
x_4	p	p	p	p	p	p	p	p

Figure 2. Example for Examining Gene Profiles.

Gene profile operations are used for analysing biological phenomena of interes as gene conservation or gain, for a specific genome (e.g., Ψ) in the context of o genomes (e.g., $\Psi_1,\ldots, \Psi_K$). For the example shown in Figure 2, gene x_4 of Ψ is conserv across Ψ_1 to Ψ_8, while gene x_3 of Ψ is gained with respect to Ψ_1 and Ψ_4 to Ψ_8.

The gene profile operations are also key components of the microbial genome functional characterization process, which, as mentioned above, is based on a number of assumptions regarding (phylogenetically) related genomes, whereby pairs of genes with similar functions are often both present or both absent (i.e, have similar profiles), tend to be collocated, and/or are components of conserved collocated genes across such genomes.

THE INTEGRATED MICROBIAL GENOMES SYSTEM

The abstract microbial genome data model discussed in the previous section helped design the IMG data management system.

System Architecture

Microbial genome data analysis involves large amounts of data distributed across diverse, usually heterogeneous, data sources. Effective data analysis requires providing a coherent view of the biological phenomena that may be concealed by the fragmentation or ambiguity of genomic data.

Integration of biological data has been considered extensively over the years because of the continuous proliferation of these sources and the need to access multiple sources inherent to biological data exploration. Data integration can be carried out using data warehouse or federated database approaches. Both approaches are based on a common (global) view of the data and involve transformation of the data from individual data sources to a common view. While data warehouses use extracttransform- load (ETL) tools for assembling and then regularly updating data in a centralized system, database federations extract and assemble data dynamically from individual data sources through data adapters.

Data warehouse and analytical processing methodologies have proven to be successful in building academic systems (e.g., scc,), as well as commercial systems. While the data federation methodology is more appealing and has been the subject of extensive research, the data warehouse approach has proven to be better suited for dealing with inherently imprecise biological data that require substantial manual data creation. Consequently, a data warehouse approach has been adopted for IMG, as reflected in the system architecture shown in Figure 3. A biological data management system often involves a mix of commercial off-the-shelve (COTS), open source, and custom tools,

… and methods need to be adapted in order to address application- … For IMG, proven open source tools developed under the Apache … provide the components needed for the Web server, a high … MS is employed for the data warehouse, while custom tools … er to address problems specific to the microbial genome data … A variety of open source bioinformatics tools have also been … ost prominent among these tools is BLAST which is widely used to identify … logous (similar) genes in different organisms. IMG has a multi-tier architecture, as shown in Figure 3. Gene predictions (called gene models) are validated and corrected manually by expert scientists using custom tools. An ETL tool kit is employed for extracting, cleaning, integrating and loading data from public data sources into the IMG warehouse. Gene relationships and clusters are computed using custom tools and are then loaded into the IMG warehouse.

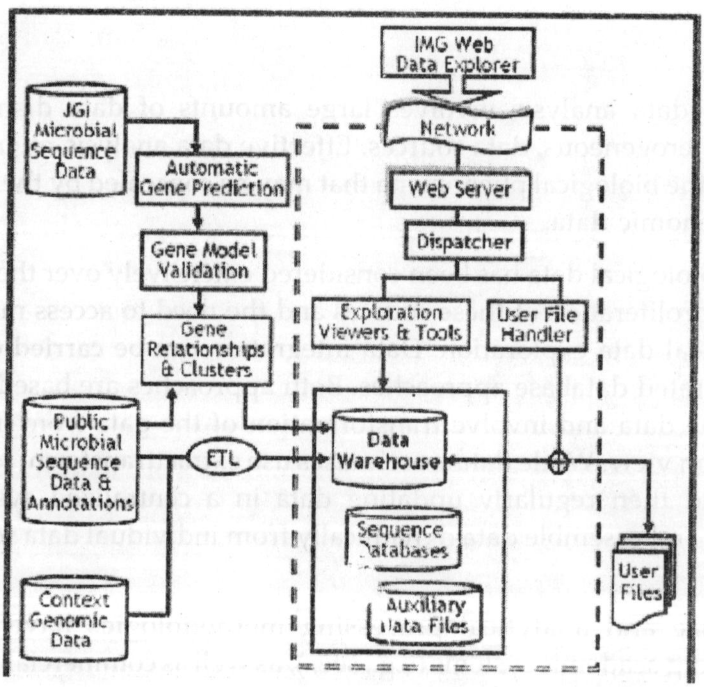

Figure 3. IMG System Architecture

The IMG back-end consists of the IMG warehouse, sequence database for similarity (BLAST) searches, and various auxiliary data files containing scaffold DNA sequences, pathway map images, and cached data for improving performance, such as precomputed statistics and homolog results.The IMG user interface follows t interface paradigm of transparent operations, whereby simplicity is preferre flexibility. For example, preformatted (canned) queries support various Web fo while dynamic construction of queries is not supported. A client web browser is us for accessing IMG: the Exploration Viewers and Tools component handles the dat exploration operations and provides support for running application-specific bioinformatics tools (e.g., BLAST), while the User File Handler component handles files consisting of genes and organisms of interest to users. These files can be generated using IMG's data export capabilities or can be created locally, and allow users to save the results of their analysis.

Web based technologies have limitations in terms of user interactivity and the amount of data that can be efficiently transferred to web pages, which are reflected in certain restrictions on analysis workflows. Such limitations can be addressed by client applications. For example, a separate Java client application that is an extension of the analytical tool prototype mentioned in section 3.2 allows JGI scientists to analyze and curate (edit) IMG data. This tool has substantially more power than IMG's Web based interface, but requires users to tolerate higher complexity. Developing an equally powerful, but less complex interface remains a challenge that needs to be addressed for future versions of IMG.

Data Structure

The structure for the IMG data warehouse was defined after analyzing data content requirements and the characteristics of various public data sources considered for IMG. An outline of the IMG data warehouse and its main data sources are shown in figure 4. Genes are represented in IMG using several classes of related objects: in addition to class Gene that represents curated or predicted hypothetical genes, non-coding RNA genes, other related gene features such as mRNA transcripts and proteins are represented by classes Transcript and Protein respectively, while class Feature represents additional sub-sequence features such as promoters. Gene similarity relationships are represented by classes Ortholog and Paralog.

Genomes are represented in IMG with their taxonomic lineage (domain, phylum, order, class, family, genus, species, strain) using class Taxon, while chromosomes and plasmids are represented by class Chromosome. Gene functions are represented by several classes in IMG: for example, class GO Ontology represents the vocabulary of terms used to describe gene functions following and class Enzyme further characterizes the gene product in terms of molecular function. Finally, pathways are represented in IMG by several classes, including KEGG Pathways, Reaction and Compound, which represent pathways, reactions and compounds, respectively, with class Image ROI modelling the association of reactions with enzymes.

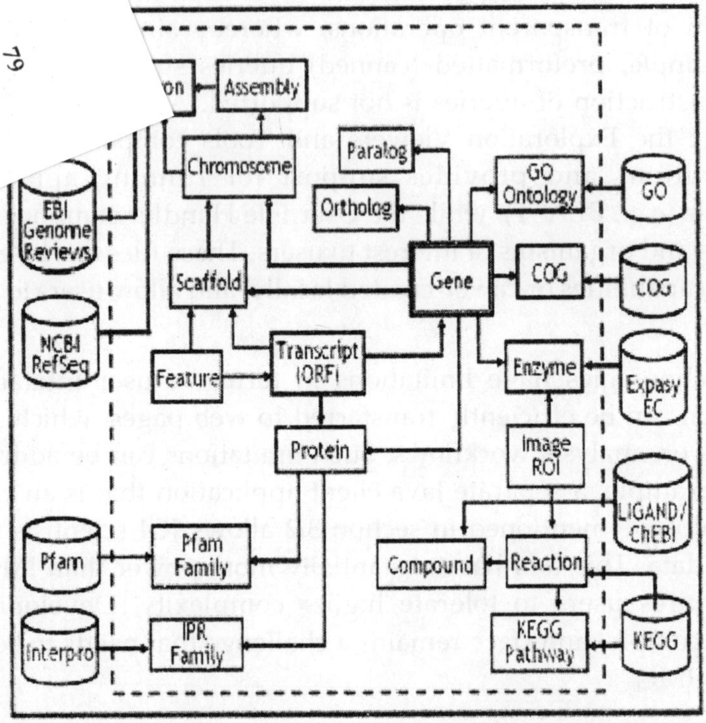

Figure 4. IMG Data Model Outline and Data Sources.

It is important to note that a data model such as that underlying IMG needs to be extensible, whereby data structure changes reflect the evolution of the biological application domain or respond to new system requirements. Keeping such changes documented as well as transparent to users, poses an additional challenge that needs to be addressed in the development of biological data management systems. In IMG, this problem has been addressed by using the OPM toolkit that has been used in developing several biological database, such as GDB. OPM tools allow describing database structures in terms of classes of objects and provide support for dealing effectively with rapid schema evolution.

REFERENCES

Environmental Defense Fund. *Genetically Engineered Foods: Who's Minding the Store?* New York. NY: Environmental Defense Fund. 1995.

Fowle, John R., III. (ed.). *Application of Biotechnology: Environmental and Policy Issues.* Boulder. Colorado: Westview Press for the American Association of Advancement of Science. 1987.

Peters, Pamela. *Biotechnology: A Guide to Genetic Engineering.* Dubuque. IA: Wm. C. Brown Publishers. 1993.

5

Fermentation Technology

Fermentation word has been derived from the word "fevere" which is the process of deriving energy from the oxidation of organic compounds, such as carbohydrates, using an endogenous electron acceptor, which is usually an organic compound. This is in contrast to cellular respiration, where electrons are donated to an exogenous electron acceptor, such as oxygen, via an electron transport chain. Fermentation does not necessarily have to be carried out in an anaerobic environment, however. For example, even in the presence of abundant oxygen, yeast cells greatly prefer fermentation to oxidative phosphorylation, as long as sugars are readily available for consumption.

Sugars are the most common substrate of fermentation, and typical examples of fermentation products are ethanol, lactic acid, and hydrogen. However, more exotic compounds can be produced by fermentation, such as butyric acid and acetone. Yeast carries out fermentation in the production of ethanol in beers, wines and other alcoholic drinks, along with the production of large quantities of carbon dioxide. Fermentation occurs in mammalian muscle during periods of intense exercise where oxygen supply becomes limited, resulting in the creation of lactic acid.

ENERGY SOURCE IN ANAEROBIC CONDITIONS

Fermentation products contain chemical energy (they are not fully oxidized) but are considered waste products, since they cannot be metabolized further without the use of oxygen (or other more highly-oxidized electron acceptors). A consequence is that the production of adenosine triphosphate (ATP) by fermentation is less efficient than oxidative phosphorylation, whereby pyruvate is fully oxidized to carbon dioxide.

Ethanol fermentation (performed by yeast and some types of bacteria) breaks the pyruvate down into ethanol and carbon dioxide. It is important in bread-making, brewing, and wine-making. Usually only one of the products is desired; in bread-making, the alcohol is baked out, and, in alcohol production, the carbon dioxide is released into the atmosphere or used for carbonating the beverage. When the ferment has a high concentration of pectin, minute quantities of methanol can be produced.

Lactic acid fermentation breaks down the pyruvate into lactic acid. It occurs in the muscles of animals when they need energy faster than the blood can supply oxygen. It also occurs in some bacteria and some fungi. It is this type of bacteria that converts lactose into lactic acid in yogurt, giving it its sour taste. These lactic acid bacteria can be classed as homofermentative, where the end product is mostly lactate, or heterofermentative, where some lactate is further metabolized and results in carbon dioxide, acetate or other metabolic products.

Hydrogen gas is produced in many types of fermentation (mixed acid fermentation, butyric acid fermentation, caproate fermentation, butanol fermentation, glyoxylate fermentation), as a way to regenerate NAD^+ from NADH. Electrons are transferred to ferredoxin, which in turn is oxidized by hydrogenase, producing H_2. Hydrogen gas is a substrate for methanogens and sulfate reducers, which keep the concentration of hydrogen sufficiently low to allow the production of such an energy-rich compound.

Fermentation has always been an important part of our lives: foods can be spoiled by microbial fermentations, foods can be made by microbial fermentations, and muscle cells use fermentation to provide us with quick responses. Fermentation could be called the staff of life because it gives us the basic food, bread. But how fermentation actually works was not understood until the work of Louis Pasteur in the latter part of the nineteenth century and the research which followed.

Fermentation is the process that produces alcoholic beverages or acidic dairy products. For a cell, fermentation is a way of getting energy without using oxygen. In general, fermentation involves the breaking down of complex organic substances into simpler ones. The microbial or animal cell obtains energy through glycolysis, splitting a sugar molecule and removing electrons from the molecule. The electrons are then passed to an organic molecule such as pyruvic acid. This results in the formation of a waste product that is excreted from the cell. Waste products formed in this way include ethyl alcohol, butyl alcohol, lactic acid, and acetone—the substances vital to our utilization of fermentation.

During lactic acid fermentation, the electrons released during glycolysis are passed to pyruvic acid to form two molecules of lactic acid. Lactic acid fermentation is carried out by many bacteria, most notably by the lactic acid bacteria used in the production of yogurt, cheese, sauerkraut, and pickles. Some animal cells such as muscle cells can also use fermentation for a quick burst of energy.

Alcohol fermentation also begins with glycolysis to produce two molecules of pyruvic acid, two molecules of ATP, and four electrons. Each pyruvic acid is modified to acetaldehyde and CO_2. Two molecules of ethyl alcohol are formed when each acetaldehyde molecule accepts two electrons. Alcohol fermentation is carried out by many bacteria and yeasts.

FERMENTATION IN INDUSTRY

In industry, as well as other areas, the uses of fermentation progressed rapidly after Pasteur's discoveries. Between 1900 and 1930, ethyl alcohol and butyl alcohol were the most important industrial fermentations in the world. But by the 1960s, chemical synthesis of alcohols and other solvents were less expensive and interest in fermentations waned. Questions can be raised about chemical synthesis, however. Chemical manufacture of organic molecules such as alcohols and acetone rely on starting materials made from petroleum. Petroleum is a nonrenewable resource; dependence on such resources could be considered short-sighted. Additionally, the use of petroleum has concomitant environmental and political problems.

Interest in microbial fermentations is experiencing a renaissance. In 1995, J. W. Frost and K. M. Draths wrote that "chemistry is moving into a new era" in which renewable resources and microbial biocatalysts will be prominent. Plant starch, cellulose from agricultural waste, and whey from cheese manufacture are abundant and renewable sources of fermentable carbohydrates. Additionally these materials, not utilized, represent solid waste that must be buried in dumps or treated with waste water.

Microbial fermentations have other benefits. For one, they don't use toxic reagents or require the addition of intermediate reagents. Microbiologists are now looking for naturally occurring microbes that produce desired chemicals. In addition, they are now capable of engineering microbes to enhance production of these chemicals. In recent years, microbial fermentations have been revolutionized by the application of genetically-engineered organisms. Many fermentations use bacteria but a growing number involve culturing mammalian cells.

FUNCTIONING OF FERMENTATION

In the pharmaceutical and biotechnology industries, fermentation is any large-scale cultivation of microbes or other single cells, occurring with or without air. In the teaching lab or at the research bench, fermentation is often demonstrated in a test tube, flask, or bottle-in volumes from a few milliliters to two liters. At the production and manufacturing level, large vessels called fermenters or bioreactors are used. A bioreactor may hold several liters to several thousand liters. Bioreactors are equipped with aeration devices as well as nutrients, stirrers, and pH and temperature controls.

At Genentech, Inc., for example, in order to get a product from fermentation, fermentation scientists develop media and test growth conditions. Then, a scale-up must be done to reproduce the process at a large volume. During production, technicians monitor temperature, pH, and growth in the bioreactors to ensure that conditions are optimum for cell growth and product. Bioreactors are used to make products such as insulin and human growth hormone from genetically engineered microorganisms as well as products from naturally-occurring cells, such as the food additive xanthan.

The products being developed by the biotechnology industry have enormous implications for our future health and well-being. All of the exciting discoveries in current biotechnical research and its applications will, of course, have repercussions within human history. Science and politics have always interacted, in both direct and indirect ways.

The uses for rubber were limited until 1898 when John Dunlop used vulcanized (heated or fireproofed) rubber to make automobile tires. The rest, as they say, is history: By 1918, there were more than nine million cars in the United States and the United States was using 50 percent of the world's rubber production. Already, by around 1900, the growing demand for rubber and the desire by countries to be self-sufficient motivated scientists to develop synthetic rubber. The greatest stimulus for development of synthetic rubber, however, was the blockade of Germany during World War I. Faced with a cutoff of its supply of natural rubber, Germany succeeded in manufacturing synthetic rubber by polymerizing butadiene, which is obtained from petroleum or alcohol.

In 1904, Chaim Weizmann was a chemistry professor at Manchester University in England trying to make synthetic rubber. He was looking for a microbe that would produce the necessary butyl alcohol. Weizmann was a Russian-born Jew who was active in the Zionist movement which advocated the creation of a homeland for Jews in Palestine. During his stay in England, he became a leader of the international Zionist movement.

By 1914, Weizmann had isolated Clostridium acetobutylicum, a bacterium which used inexpensive starch to produce a high yield of butyl alcohol and acetone. However, World War I broke out in August of 1914 and diverted attention away from synthetic rubber and toward gunpowder (cordite). As it turns out, the solvent for making nitrocellulose and thus cordite was acetone. Weizmann was instrumental in making available a source for the creation of this acetone.

Acetone had previously been made from calcium acetate imported from Germany. Since importation of the German calcium acetate was not possible and the United States did not have a large supply, Weizmann was recruited by Winston Churchill and the British government to set up his microbial fermentation for the production of acetone from corn at the Nicholson Distillery in London. The grain supply was unreliable, however, because of the German blockade and it was necessary to look for a different fermentable carbohydrate. Food was being rationed so a substrate that could not be used for human food was needed. In 1916, Weizmann even tried to use horse chestnuts collected by children, but the supply was insufficient for a large-scale fermentation. The British turned to other parts of the British Empire and to their allies for a fermentable carbohydrate. Consequently, in 1916, the Weizmann process was moved to a distillery in Toronto (Canada) and another was built in India. In 1917, a plant was set up to ferment corn in Indiana (U.S.).

After the war, when British Prime Minister Lloyd George asked what honors Weizmann might want for his considerable contributions, Weizmann answered, "There is only one thing I want. A national home for my people." Lord Balfour then gave Weizmann 15 minutes to explain why that national homeland should be Palestine. Weizmann was an eloquent spokesman and convincingly stated his case. The result was the Balfour Declaration, which affirmed Britain's commitment to the establishment of a Jewish homeland.

Weizmann went on to make significant contributions to both microbiology and politics. In 1920, he began a long tenure as President of the World Zionist Organization. In the years that followed, he campaigned with great zeal. In 1948, when the United States was going to reverse its decision to support the independent state of Israel, Weizmann used his considerable negotiating skills to convince President Truman that the United States should affirm their support for the new country, leading to the founding of Israel. In 1949, he was elected the first president of Israel.

From microbiologist to President, Weizmann illustrates not only the persistence necessary in both research and politics, but the strange and interesting ways research and politics interact. What further developments will the products of biotechnical research inspire?

Fermentation is supported by a rich and dense collection of microbes. Each milliliter of rumen content contains roughly 10 to 50 billion bacteria, 1 million protozoa and variable numbers of yeasts and fungi. The micrograph below, of sheep rumenal fluid, shows a Gulliver like ciliated protozoon in the midst of thousands of bacteria (the small specks).

The environment of the rumen and large intestine is anaerobic and, as expected, almost all these microbes are anaerobes or facultative anaerobes. Fermentative microbes interact and support one another in a complex food web, with the waste products of some species serving as nutrients for other species.

Fermentative bacteria representing many genera provide a comprehensive battery of digestive capabilities. These organisms are often classified by their substrate preferences or the end products they produce. Although there is some specialization, many bacteria utilize multiple substrates. Some of the major groups, each of which contain multiple genera and species, are:

- Cellulolytic (digest cellulose)
- Hemicellulolytic (digest hemicellulose)
- Amylolytic (digest starch)
- Proteolytic (digest proteins)
- Sugar utilizing (utilize monosaccharides and disaccharides)
- Acid utilizing (utilize such substrates as lactic, succinic and malic acids)
- Ammonia producers
- Vitamin synthesizers
- Methane producers

Protozoa, predominantly ciliates, appear to contribute substantially to the fermentation process. Several experiments have demonstrated that lambs and calves deprived of their ruminal protozoa show depressed growth rates and are relative "poor-doers" compared to controls with both bacteria and protozoa. In general, protozoa utilize the same set of substrates as bacteria and, as with bacteria, different populations of protozoa show distinctive substrate preferences. Many utilize simple sugars and some store ingested carbohydrate as glycogen.

An interesting feature of some protozoa is their inability to regulate glycogen synthesis: when soluble carbohydrates are in abundance, they continue to store glycogen until they burst. An additional feature of protozoa is that many species consume bacteria, which is thought to perhaps play a role in limiting bacterial overgrowth.

The distribution of microbial species varies with diet. Some of this appears to reflect substrate availability; for example, populations of cellulolytic bugs are depressed in animals fed diets rich in grain. Environmental conditions in the fermentation vat also can have profound effects on the microbial flora. Rumen fluid normally has a pH between 6 and 7, but may fall if large amounts of soluble carbohydrate are consumed. If pH drops to about 5.5, protozoal populations become markedly depressed due to acid intolerance. More drastic lowering of rumen pH, as can occur with grain engorgement, can destroy many species and have serious consequences to the animal.

BASIC FERMENTATION CHEMISTRY

The forestomach of ruminants and large intestine of caudal fermenters are magnificent, continuous flow fermentation systems containing enormous numbers of microbes. What do these microbes and the process of fermentation provide the herbivore? This section will concentrate on fermentation per se, but the microbes that digest cellulose and other substrates also provide at least three other major services:

- Synthesis of high quality protein in the form of microbial bodies. Caudal fermentors cannot take advantage of this service, but in ruminants, bacteria and protozoa are constantly flowing into the abomasum and small intestine, where they are digested and absorbed. All vertebrates require certain amino acids which their cells cannot synthesize (the "essential amino acids"). Fermentative microbes can synthesize all the amino acids and thereby provide them to their host.
- Synthesis of protein from non-protein nitrogen sources. Fermentative microbes can, for example, utilize urea to synthesize protein. In some situations, ruminants are fed urea as a inexpensive dietary supplement. They also secrete urea formed during protein metabolism into saliva, which flows into the rumen and serves as another nitrogen source for the microbes.
- Synthesis of B vitamins. Mammals can synthesize only two of the B vitamins and require dietary sources of the others. Fermentative microbes are able to synthesize all the B vitamins, and deficiency states are rarely encountered.

Substrates for Fermentation

With few exceptions, all dietary carbohydrates and proteins can serve as substrates for microbial fermentation. Nonetheless, the crucial advantage of being a herbivore is the ability to efficiently extract energy from cellulose and other components of plant cell walls.

Cellulose fibers account for 40-50% of the total dry weight of stems, leaves and roots. These fibers are embedded in a matrix of hemicelluloses and phenolic polymers (lignin-carbohydrate complexes) that are covalently crosslinked. Cellulose itself is a linear polymer of glucose molecules linked to one another by beta[1-4] glycosidic bonds and herein lies the problem for the vertebrate digestive system.

As far as is known, no enzyme able to hydrolyze beta[1-4] glycosidic bonds has evolved in vertebrates. However, a variety of such beta-glucanases are synthesized by microbes. Thus, the diverse population of bacteria and protozoa in the rumen or hindgut produce all the enzymes necessary to digest cellulose and hemicellulose. The glucose released in this process is then taken up and metabolized by the microbes, and the waste products of microbial metabolism are passed on to the host animal. Sugars derived from digestion of soluble carbohydrates such as starch are processed similarly.

Products of Fermentation

Fermentation occurs under anaerobic conditions. As a consequence, sugars are metabolized predominantly to volatile fatty acids (VFAs). Additional major products include lactic acid, carbon dioxide and methane.

The principle VFAs are acetic, proprionic and butyric acids, which collectively provide for the majority of a herbivore's energy needs. The ratio of these VFA's vary with diet, although the majority product is always acetate. On a diet high in fiber, the molar ratio of acetic to proprionic to butyric acids is roughly 70:20:10.

Proteins are also important substrates for fermentation. In caudal fermenters, much of the dietary protein is digested and absorbed prior to the large gut, but in ruminants, all dietary protein enters the rumen. The bulk of this protein is digested by microbial proteases and peptidases. The resulting peptides and amino acids are taken up by microbes and used in several ways, including microbial protein synthesis. However, a large quantity of amino acids ingested by fermentative microbes are deaminated and enter some of the same pathways used for carbohydrate metabolism. The net result is that much of dietary protein is metabolized to VFAs.

Clearly, from the standpoint of the host animal, VFAs are the important product of fermentation. These small lipids are used for many purposes, but the paramount importance of VFAs to herbivores is that they are absorbed and serve as the animal's major fuel for energy production, serving much the same function that glucose does in you.

INDUSTRIAL FERMENTATION

Fermentation has many important uses in industry. Though the word fermentation can have stricter definitions, when speaking of it in industrial fermentation it more loosely refers to the breakdown of organic substances and re-assembly into other substances. Somewhat paradoxically, fermenter culture in industrial capacity often refers to highly oxygenated and aerobic growth conditions, whereas fermentation in the biochemical context is a strictly anaerobic process. A very old method is ABE fermentation.

Ancient fermented food processes, such as making bread, wine, cheese, curds, idli, dosa, etc., can be dated to more than 6,000 years ago. They were developed long before man had any knowledge of the existence of the microorganisms involved. Also, fermentation is a powerful economic incentive for semi-industrialized countries, in their willingness to produce bio-ethanol.

There are 5 major groups of commercially important fermentation:

— Microbial cells or biomass as the product, e.g. single cell protein, bakers yeast lactobacillus, etc.

- Microbial enzymes: catalase, amylase, protease, pectinase, glucose isomerase, cellulase, hemicellulase, lipase, lactase, streptokinase, etc.

Microbial metabolites:

- Primary metabolites – ethanol, citric acid, glutamic acid, lysine, vitamins, polysaccharides etc.
- Secondary metabolites: all antibiotic fermentation
- Recombinant products: insulin, HBV, interferon, GCSF, streptokinase
- Biotransformations: phenyl acetyl carbinol, steroid biotransformation, etc.

Growth media are required for industrial fermentation, since any microbe requires water, oxygen, an energy source, a carbon source, a nitrogen source and micronutrients for growth.

Trace elements: Fe, Zn, Cu, Mn, Mo, Co

Antifoaming agents : Esters, fatty acids, fats, silicones, sulphonates, polypropylene glycol

Buffers: Calcium carbonate, phosphates

Growth factors: Some microorganisms cannot synthesize the required cell components themselves and need to be supplemented, e.g. with thiamine, biotin, calcium pentothenate

Precursors: Directly incorporated into the desired product: Phenyl ethylamine into Benzyl penicillin, Phenyl acetic acid into Penicillin G

Inhibitors: To get the specific products: e.g. sodium barbital for rifamycin

Inducers: The majority of the enzymes used in industrial fermentation are inducible and are synthesized in response of inducers: e.g. starch for amylases, maltose for pollulanase, pectin for pectinase,olive oil and tween are also used at times.

Chelators: Chelators are the chemicals used to avoid the precipitation of metal ions. Chelators like EDTA, citric acid, polyphosphates are used in low concentrations.

Sewage Disposal

In the process of sewage disposal, sewage is digested by enzymes secreted by bacteria. Solid organic matters are broken down into harmless, soluble substances and carbon dioxide. Liquids that result are disinfected to remove pathogens before being discharged into rivers or the sea or can be used as liquid fertilisers. Digested solids, known also as sludge, is dried and used as fertilisers. Gaseous by-products such as methane, can be utilised as biogas to fuel generators. One advantage of bacterial digestion is that it reduces the bulk and odour of sewage, thus reducing space needed for dumping, on the other hand, a major disadvantage of bacterial digestion in sewage disposal is that it is a very slow process.

Phases of Microbial Growth

When a particular organism is introduced into a selected growth medium, the medium is inoculated with the particular organism. Growth of the inoculum does not occur immediately, but takes a little while. This is the period of adaptation, called the lag phase. Following the lag phase, the rate of growth of the organism steadily increases, for a certain period—this period is the log or exponential phase. After a certain time of exponential phase, the rate of growth slows down, due to the continuously falling concentrations of nutrients and/or a continuously increasing (accumulating) concentrations of toxic substances. This phase, where the increase of the rate of growth is checked, is the deceleration phase. After the deceleration phase, growth ceases and the culture enters a stationary phase or a steady state. The biomass remains constant, except when certain accumulated chemicals in the culture lyse the cells (chemolysis). Unless other micro-organisms contaminate the culture, the chemical constitution remains unchanged. Mutation of the organism in the culture can also be a source of contamination, called internal contamination.

FOOD FERMENTATION

Fermentation in food processing typically refers to the conversion of sugar to alcohol using yeast under anaerobic conditions. A more general definition of fermentation is the chemical conversion of carbohydrates into alcohols or acids. When fermentation stops prior to complete conversion of sugar to alcohol, a stuck fermentation is said to have occurred. The science of fermentation is known as zymology.

Fermentation usually implies that the action of the microorganisms is desirable, and the process is used to produce alcoholic beverages such as wine, beer, and cider. Fermentation is also employed in preservation to create lactic acid in sour foods such as pickled cucumbers, kimchi and yogurt.

History of Wine and Beer

Since fruits ferment naturally, fermentation precedes human history. Since prehistoric times, however, humans have been controlling the fermentation process. The earliest evidence of winemaking dates from eight thousand years ago, in Georgia, in the Caucasus area. Seven-thousand-year-old jars of wine have been excavated in the Zagros Mountains in Iran, which are now on display at the University of Pennsylvania. There is strong evidence that people were fermenting beverages in Babylon circa 5000 BC, ancient Egypt circa 3150 BC, pre-Hispanic Mexico circa 2000 BC, and Sudan circa 1500 BC. There is also evidence of leavened bread in ancient Egypt circa 1500 BC and of milk fermentation in Babylon circa 3000 BC.

French chemist Louis Pasteur was the first known zymologist, when in 1854 he connected yeast to fermentation. Pasteur originally defined fermentation as "respiration without air". Pasteur performed careful research and concluded: "I am of the opinion that alcoholic fermentation never occurs without simultaneous organization, development and multiplication of cells.... If asked, in what consists the chemical act whereby the sugar is decomposed... I am completely ignorant of it." The German Eduard Buchner, winner of the 1907 Nobel Prize in chemistry, later determined that fermentation was actually caused by a yeast secretion that he termed zymase.

Wine Fermentation

The process of fermentation in wine is the catalyst function that turns grape juice into an alcoholic beverage. During fermentation yeast interact with sugars in the juice to create ethanol, commonly known as ethyl alcohol, and carbon dioxide (as a by-product). In winemaking the temperature and speed of fermentation is an important consideration as well as the levels of oxygen present in the must at the start of the fermentation. The risk of stuck fermentation and the development of several wine faults can also occur during this stage which can last anywhere from 5 to 14 days for primary fermentation and potentially another 5 to 10 days for a secondary fermentation. Fermentation may be done in stainless steel tanks, which is common with many white wines like Riesling, in an open wooden vat, inside a wine barrel and inside the wine bottle itself as in the production of many sparkling wines.

The natural occurrence of fermentation means it was probably first observed long ago by humans. The earliest uses of the word "Fermentation" in relation to winemaking was in reference to the apparent "boiling" within the must that came from the anaerobic reaction of the yeast to the sugars in the grape juice and the release of carbon dioxide. The Latin fervere means, literally, to boil. In the mid-19th century, Louis Pasteur noted the connection between yeast and the process of the fermentation in which the yeast act as catalyst and mediator through a series of a reaction that convert sugar into alcohol. The discovery of the Embden–Meyerhof–Parnas pathway by Gustav Embden, Otto Fritz Meyerhof and Jakub Karol Parnas in the early 20th century contributed more to the understanding of the complex chemical processes involved the conversion of sugar to alcohol.

In winemaking there are distinctions made between ambient yeasts which are naturally present in wine cellars, vineyards and on the grapes themselves (sometimes known as a grape's "bloom" or "blush") and cultured yeast which are specifically isolated and inoculated for use in winemaking. The most common genera of wild yeasts found in winemaking include Candida, Klöckera/Hanseniaspora, Metschnikowiaceae, Pichia and Zygosaccharomyces. Wild yeasts can produce high-quality, unique-flavored wines; however, they are often unpredictable and may introduce less desirable traits to the wine, and can even contribute to spoilage. Traditional wine makers, particularly in Europe, advocate use of ambient yeast as a characteristic of the region's terroir; nevertheless, many winemakers prefer to control fermentation with predictable cultured yeast. The cultured yeasts most commonly used in winemaking belong to the Saccharomyces cerevisiae (also known as "sugar yeast") species. Within this species are several hundred different strains of yeast that be used during fermentation to affect the heat or vigor of the process and enhance or suppress certain flavor characteristics of the varietal. The use of different strains of yeasts are a major contributor to the diversity of wine, even among the same grape variety.

The addition of cultured yeast normally occurs with the yeast first in a dried or "inactive" state and is reactivated in warm water or diluted grape juice prior to being added to the must. To thrive and be active in fermentation, the yeast needs access to a continuous supply of carbon, nitrogen, sulfur, phosphorus as well as access to various vitamins and minerals. These components are naturally present in the grape must but their amount may be corrected by adding nutrient packets to the wine, in order to foster a more encouraging environment for the yeast. Oxygen is needed as well but in wine making the risk of oxidation and the lack of alcohol production from oxygenated yeast requires the exposure of oxygen to be kept at a minimum.

Upon the introduction of active yeasts to the grape must, phosphates are attached to the sugar and the six-carbon sugar molecules begin to be split into three-carbon pieces and go through a series of rearrangement reactions. During this process the carboxylic carbon atom is released in the form of carbon dioxide with the remaining components becoming acetaldehyde. The absence of oxygen in this anaerobic process allows the acetaldehyde to be eventually converted, by reduction, to ethanol. During the conversion of acetaldehyde a small amount is converted, by oxidation, to acetic acid which, in excess, can contribute to the wine fault known as volatile acidity (vinegar taint). After the yeast has exhausted its life cycle they fall to the bottom of the fermentation tank as sediment known as lees.

The metabolism of amino acids and breakdown of sugars by yeasts has the affect of creating other biochemical compounds that can contribute to the flavor and aroma of wine. These compounds can be considered "volatile" like aldehydes, ethyl acetate, ester, fatty acids, fusel oils, hydrogen sulfide, ketones and mercaptans) or "non-volatile" like glycerol, acetic acid and succinic acid. Yeast also has the effect during fermentation of releasing glycoside hydrolase which can hydrolyse the flavor precursors of aliphatics (a flavor component that reacts with oak), benzene derivities, monoterpenes (responsible for floral aromas from grapes like Muscat and Traminer), norisoprenoids (responsible for some of the spice notes in Chardonnay), and phenols. Some strains of yeasts can generate volatile thiols which contribute to the fruity aromas in many wines such as the gooseberry scent commonly associates with Sauvignon blanc. Brettanomyces yeasts are responsible for the "barnyard aroma" characteristic in some red wines like Burgundy Pinot noir.

Winemaking considerations

During fermentation there are several factors that winemakers take into consideration. The most notable is that of the internal temperature of the must. The biochemical process of fermentation itself creates a lot of residual heat which can take the must out of the ideal temperature range for the wine. Typically white wine is fermented between 64-68 °F (18-20°C) though a wine maker may choose to use a higher temperature to bring out some of the complexity of the wine. Red wine is typically fermented at higher temperatures up to 85 °F (29°C). Fermentation at higher temperatures may have adverse effect on the wine in stunning the yeast to inactivity and even "boiling off" some of the flavors of the wines. Some winemakers may ferment their red wines at cooler temperatures more typical of white wines in order to bring out more fruit flavors.

To control the heat generated during fermentation the winemaker has to choose a suitable vessel size or to use cooling devices of various sorts from the ancient Bordeaux traditions of placing the fermentation vat on top of blocks of ice to today's modern use of sophisticated fermentation tanks with built in cooling rings.

A risk factor involved with fermentation is the development of chemical residue and spoilage which can be corrected with the addition of sulfur dioxide (SO_2), although excess SO_2 can lead to a wine fault. A winemaker who wishes to make a wine with high levels of residual sugar (like a dessert wine) may stop fermentation early either by dropping the temperature of the must to stun the yeast or by adding a high level of alcohol (like brandy) to the must to kill off the yeast and create a fortified wine.

Other types of fermentation

In winemaking there are different processes that fall under the title of "Fermentation" but might not follow the same procedure commonly associated with wine fermentation.

Bottle fermentation: Bottle fermentation is a method of sparkling wine production originating in the Champagne region where after the cuvee has gone through a primary yeast fermentation the wine is then bottled and goes through a secondary fermentation where sugar and additional yeast known as liqueur de tirage is added to the wine. This secondary fermentation is what creates the carbon dioxide bubbles that sparkling wine is known for.

Carbonic maceration: The process of carbonic maceration is also known as whole grape fermentation where instead of yeast being added to grape must fermentation is encouraged to take place inside the individual grape berries. This method is common in the creation of Beaujolais wine and involves whole clusters of grapes being stored in a closed container with the oxygen in the container being replaced with carbon dioxide. Unlike normal fermentation where yeast converts sugar into alcohol, carbonic maceration works by enzymes within the grape breaking down the cellular matter to form ethanol and other chemical properties. The resulting wines are typically soft and fruity.

Malolactic fermentation: Instead of yeast, bacteria plays a fundamental role in malolactic fermentation which is essentially the conversion of malic acid into lactic acid. This has the benefit of reducing some of the tartness and making the resulting wine taste softer. Depending on the style of wine that the winemaker is trying to produce, malolactic fermentation may take place at the same time as the yeast fermentation.

Uses

The primary benefit of fermentation is the conversion of sugars and other carbohydrates, e.g., converting juice into wine, grains into beer, carbohydrates into carbon dioxide to leaven bread, and sugars in vegetables into preservative organic acids.

Food fermentation has been said to serve five main purposes:

- enrichment of the diet through development of a diversity of flavors, aromas, and textures in food substrates.
- preservation of substantial amounts of food through lactic acid, alcohol, acetic acid and alkaline fermentations.
- biological enrichment of food substrates with protein, essential amino acids, essential fatty acids, and vitamins.
- detoxification during food-fermentation processing.
- a decrease in cooking times and fuel requirements.

Fermentation has some uses exclusive to foods. Fermentation can produce important nutrients or eliminate antinutrients. Food can be preserved by fermentation, since fermentation uses up food energy and can make conditions unsuitable for undesirable microorganisms. For example, in pickling the acid produced by the dominant bacteria inhibit the growth of all other microorganisms. Depending on the type of fermentation, some products (e.g., fusel alcohol) can be harmful to people's health.

Stuck Fermentation

A stuck fermentation is where a fermentation has stopped before completion; i.e., before the anticipated percentage of sugars has been converted by yeast into alcohol or carbohydrates into carbon dioxide. Typically, a stuck fermentation may be caused by: 1) insufficient or incomplete nutrients required to allow the yeast to complete fermentation; 2) low temperatures, or temperature changes which have caused the yeast to stop working early; or 3) a percentage of alcohol which has grown too high for the particular yeast chosen for the fermentation.

Corrections to stuck fermentations may include:

1) repitching a different yeast
2) incorporation of nutrients in conjunction with the repitched yeast;
3) restoration of accommodative temperatures for the given yeast.

Risks of Consuming Fermented Foods

Alaska, despite its small population, has witnessed a steady increase of cases of botulism since 1985. It has more cases of botulism than anywhere else in the United States of America. This is caused by the traditional Eskimo practice of allowing animal products such as whole fish, fish heads, walrus, sea lion and whale flippers, beaver tails, seal oil, birds, etc., to ferment for an extended period of time before being consumed. The risk is exacerbated when a plastic container is used for this purpose instead of the old-fashioned method, a grass-lined hole, as the botulinum bacteria thrive in the anaerobic conditions created by the plastic.

Diversity of Fermented Foods

Numerous fermented foods are consumed around the world. Each nation has its own types of fermented food, representing the staple diet and the raw ingredients available in that particular place. Although the products are well know to the individual, they may not be associated with fermentation. Indeed, it is likely that the methods of producing many of the worlds fermented foods are unknown and came about by chance. Some of the more obvious fermented fruit and vegetable products are the alcoholic beverages - beers and wines. However, several more fermented fruit and vegetable products arise from lactic acid fermentation and are extremely important in meeting the nutritional requirements of a large proportion of the worlds population.

Organisms Responsible for Food Fermentations

The most common groups of micro-organisms involved in food fermentations are:

— Bacteria

— Yeasts

— Moulds

Bacteria

Several bacterial families are present in foods, the majority of which are concerned with food spoilage. As a result, the important role of bacteria in the fermentation of foods is often overlooked. The most important bacteria in desirable food fermentations are the *lactobacillaceae* which have the ability to produce lactic acid from carbohydrates.

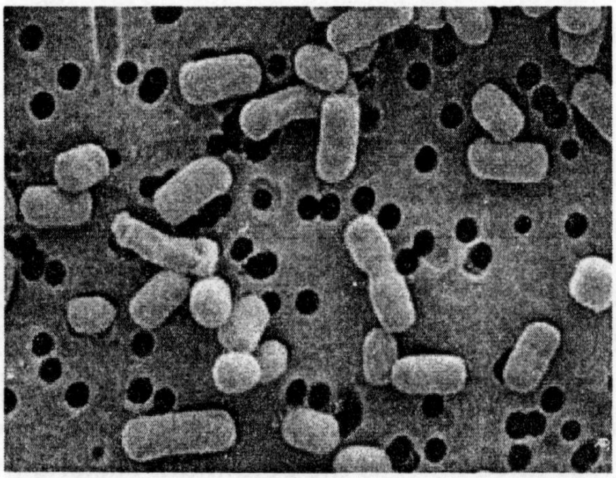

Lactobacillaceae

Other important bacteria, especially in the fermentation of fruits and vegetables, are the acetic acid producing *acetobacter* species.

Yeasts

Yeasts and yeast-like fungi are widely distributed in nature. They are present in orchards and vineyards, in the air, the soil and in the intestinal tract of animals. Like bacteria and moulds, yeasts can have beneficial and non-beneficial effects in foods. The most beneficial yeasts in terms of desirable food fermentation are from the *Saccharomyces* family, especially *S. cerevisiae*. Yeasts are unicellular organisms that reproduce asexually by budding. In general, yeasts are larger than most bacteria. Yeasts play an important role in the food industry as they produce enzymes that favour desirable chemical reactions such as the leavening of bread and the production of alcohol and invert sugar.

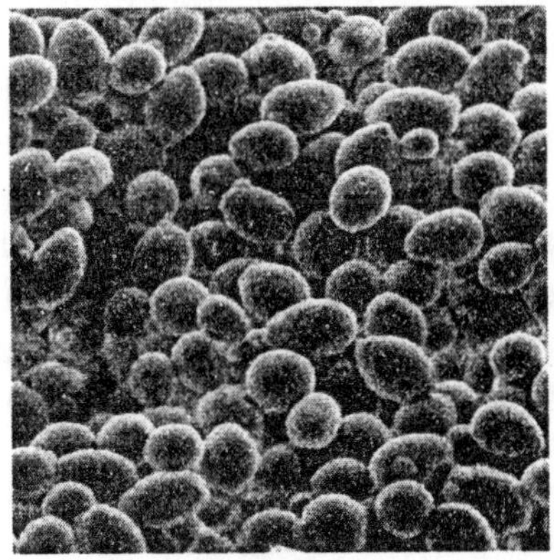

Yeasts

Moulds

Moulds are also important organisms in the food industry, both as spoilers and preservers of foods. Certain moulds produce undesirable toxins and contribute to the spoilage of foods. The *Aspergillus* species are often responsible for undesirable changes in foods. These moulds are frequently found in foods and can tolerate high concentrations of salt and sugar. However, others impart characteristic flavours to foods and others produce enzymes, such as amylase for bread making. Moulds from the genus *Penicillium* are associated with the ripening and flavour of cheeses. Moulds are aerobic and therefore require oxygen for growth. They also have the greatest array of enzymes, and can colonise and grow on most types of food. Moulds do not play a significant role in the desirable fermentation of fruit and vegetable products.

Moulds

When micro-organisms metabolise and grow they release by-products. In food fermentations the by-products play a beneficial role in preserving and changing the texture and flavour of the food substrate. For example, acetic acid is the by-product of the fermentations of some fruits. This acid not only affects the flavour of the final product, but more importantly has a preservative effect on the food. For food fermentations, micro-organisms are often classified according to these by-products. The fermentation of milk to yoghurt involves a specific group of bacteria called the lactic acid bacteria (*Lactobacillus* species). This is a general name attributed to those bacteria which produce lactic acid as they grow. Acidic foods are less susceptible to spoilage than neutral or alkaline foods and hence the acid helps to preserve the product. Fermentations also result in a change in texture. In the case of milk, the acid causes the precipitation of milk protein to a solid curd.

Enzymes

The changes that occur during fermentation of foods are the result of enzymic activity. Enzymes are complex proteins produced by living cells to carry out specific biochemical reactions. They are known as catalysts since their role is to initiate and control reactions, rather than being used in a reaction. Because they are proteinaceous in nature, they are sensitive to fluctuations in temperature, pH, moisture content, ionic strength and concentrations of substrate and inhibitors. Each enzyme has requirements at which it will operate most efficiently. Extremes of temperature and pH will denature the protein and destroy enzyme activity. Because they are so sensitive, enzymic reactions can easily be controlled by slight adjustments to temperature, pH or other reaction conditions. In the food industry, enzymes have several roles - the liquefaction and saccharification of starch, the conversion of sugars and the modification of proteins. Microbial enzymes play a role in the fermentation of fruits and vegetables.

Nearly all food fermentations are the result of more than one micro-organism, either working together or in a sequence. For example, vinegar production is a joint effort between yeast and acetic acid forming bacteria. The yeast convert sugars to alcohol, which is the substrate required by the *acetobacter* to produce acetic acid. Bacteria from different species and the various micro-organisms - yeast and moulds -all have their own preferences for growing conditions, which are set within narrow limits. There are very few pure culture fermentations. An organism that initiates fermentation will grow there until it's by-products inhibit further growth and activity. During this initial growth period, other organisms develop which are ready to take over when the conditions become intolerable for the former ones.

In general, growth will be initiated by bacteria, followed by yeasts and then moulds. There are definite reasons for this type of sequence. The smaller micro-organisms are the ones that multiply and take up nutrients from the surrounding area most rapidly. Bacteria are the smallest of micro-organisms, followed by yeasts and moulds. The smaller bacteria, such as *Leuconostoc* and *Streptococcus* grow and ferment more rapidly than their close relations and are therefore often the first species to colonise a substrate.

Desirable Fermentation

It is essential with any fermentation to ensure that only the desired bacteria, yeasts or moulds start to multiply and grow on the substrate. This has the effect of suppressing other micro-organisms which may be either pathogenic and cause food poisoning or will generally spoil the fermentation process, resulting in an end-product which is neither expected or desired. An everyday example used to illustrate this point is the differences in spoilage between pasteurised and unpasteurised milk. Unpasteurised milk will spoil naturally to produce a sour tasting product which can be used in baking to improve the texture of certain breads. Pasteurised milk, however, spoils (non-desirable fermentation) to produce an unpleasant product which has to be disposed of. The reason for this difference is that pasteurisation (despite being a very important process to destroy pathogenic micro-organisms) changes the micro-organism environment and if pasteurised milk is kept unrefrigerated for some time, undesirable micro-organisms start to grow and multiply before the desirable ones. In the case of unpasteurised milk, the non-pathogenic lactic acid bacteria start to grow and multiply at a greater rate that any pathogenic bacteria. Not only do the larger numbers of lactic acid bacteria compete more successfully for the available nutrients, but as they grow they produce lactic acid which increases the acidity of the substrate and further suppresses the bacteria which cannot tolerate an acid environment.

Most food spoilage organisms cannot survive in either alcoholic or acidic environments. Therefore, the production of both these end products can prevent a food from spoilage and extend the shelf life. Alcoholic and acidic fermentations generally offer cost effective methods of preserving food for people in developing countries, where more sophisticated means of preservation are unaffordable and therefore not an option.

The principles of microbial action are identical both in the use of micro-organisms in food preservation, such as through desirable fermentations, and also as agents of destruction via food spoilage. The type of organisms present and the environmental conditions will determine the nature of the reaction and the ultimate products. By manipulating the external reaction conditions, microbial reactions can be controlled to produce desirable results. There are several means of altering the reaction environment to encourage the growth of desirable organisms.

Manipulation of Microbial Activity

There are six major factors that influence the growth and activity of micro-organisms in foods. These are moisture, oxygen concentration, temperature, nutrients, pH and inhibitors. The food supply available to the micro-organisms depends on the composition of the food on which they grow. All micro-organisms differ in their ability to metabolise proteins, carbohydrates and fats. Obviously, by manipulating any of these six factors, the activity of micro-organisms within foods can be controlled.

Moisture

Water is essential for the growth and metabolism of all cells. If it is reduced or removed, cellular activity is decreased. For example, the removal of water from cells by drying or the change in state of water (from liquid to solid) affected by freezing, reduces the availability of water to cells (including microbial cells) for metabolic activity. The form in which water exists within the food is important as far as microbial activity is concerned.

There are two types of water - free and bound. Bound water is present within the tissue and is vital to all the physiological processes within the cell. Free water exists in and around the tissues and can be removed from cells without seriously interfering with the vital processes. Free water is essential for the survival and activity of micro-organisms. Therefore, by removing free water, the level of microbial activity can be controlled. The amount of water available for micro-organisms is referred to as the water activity (aw). Pure water has a water activity of 1.0. Bacteria require more water than yeasts, which require more water than moulds to carry out their metabolic activities. Almost all microbial activity is inhibited below aw of 0.6.

Most fungi are inhibited below aw of 0.7, most yeasts are inhibited below aw of 0.8 and most bacteria below aw 0.9. Naturally, there are exceptions to these guidelines and several species of micro-organism can exist outside the stated range. The water activity of foods can be changed by altering the amount of free water available. There are several ways to achieve this – drying to remove water; freezing to change the state of water from liquid to solid; increasing or decreasing the concentration of solutes by adding salt or sugar or other hydrophylic compounds. Addition of salt or sugar to a food will bind free water and so decrease the aw. Alternatively, decreasing the concentration will increase the amount of free water and in turn the aw. Manipulation of the aw in this manner can be used to encourage the growth of favourable micro-organisms and discourage the growth of spoilage ones.

Oxidation-Reduction potential

Oxygen is essential to carry out metabolic activities that support all forms of life. Free atmospheric oxygen is utilised by some groups of micro-organisms, while others are able to metabolise the oxygen which is bound to other compounds such as carbohydrates. This bound oxygen is in a reduced form.

Micro-organisms can be broadly classified into two groups - aerobic and anaerobic. Aerobes grow in the presence of atmospheric oxygen while anaerobes grow in the absence of atmospheric oxygen. In the middle of these two extremes are the facultative anaerobes which can adapt to the prevailing conditions and grow in either the absence or presence of atmospheric oxygen. Microaerophilic organisms grow in the presence of reduced amounts of atmospheric oxygen. Thus, controlling the availability of free oxygen is one means of controlling microbial activity within a food. In aerobic fermentations, the amount of oxygen present is one of the limiting factors. It determines the type and amount of biological product obtained, the amount of substrate consumed and the energy released from the reaction.

Moulds do not grow well in anaerobic conditions, therefore they are not important in terms of food spoilage or beneficial fermentation, in conditions of low oxygen availability.

Temperature

Temperature affects the growth and activity of all living cells. At high temperatures, organisms are destroyed, while at low temperatures, their rate of activity is decreased or suspended. Micro-organisms can be classified into three distinct categories according to their temperature preference.

Nutritional requirements

The majority of organisms are dependent on nutrients for both energy and growth. Organisms vary in their specificity towards different substrates and usually only colonise foods which contain the substrates they require. Sources of energy vary from simple sugars to complex carbohydrates and proteins. The energy requirements of micro-organisms are very high. Limiting the amount of substrate available can check their growth.

Hydrogen ion concentration

The pH of a substrate is a measure of the hydrogen ion concentration. A food with a pH of 4.6 or less is termed a high acid or acid food and will not permit the growth of bacterial spores. Foods with a pH above 4.6. are termed low acid and will not inhibit the growth of bacterial spores. By acidifying foods and achieving a final pH of less than 4.6, most foods are resistant to bacterial spoilage.

The optimum pH for most micro-organisms is near the neutral point (pH 7.0). Certain bacteria are acid tolerant and will survive at reduced pH levels. Notable acid-tolerant bacteria include *the Lactobacillus* and *Streptococcus* species, which play a role in the fermentation of dairy and vegetable products. Moulds and yeasts are usually acid tolerant and are therefore associated with spoilage of acidic foods.

Micro-organisms vary in their optimal pH requirements for growth. Most bacteria favour conditions with a near neutral pH (7). Yeasts can grow in a pH range of 4 to 4.5 and moulds can grow from pH 2 to 8.5, but favour an acid pH. The varied pH requirements of different groups of micro-organisms is used to good effect in fermented foods where successions of micro-organisms take over from each other as the pH of the environment changes. For instance, some groups of micro-organisms ferment sugars so that the pH becomes too low for the survival of those microbes. The acidophilic micro-organisms then take over and continue the reaction. The affinity for different pH can also be used to good effect to occlude spoilage organisms.

Inhibitors

Many chemical compounds can inhibit the growth and activity of micro-organisms. They do so by preventing metabolism, denaturation of the protein or by causing physical damage to the cell. The production of substrates as part of the metabolic reaction also acts to inhibit microbial action.

Controlled Fermentation

Controlled fermentations are used to produce a range of fermented foods, including sauerkraut, pickles, olives, vinegar, dairy and other products. Controlled fermentation is a form of food preservation since it generally results in a reduction of acidity of the food, thus preventing the growth of spoilage micro-organisms. The two most common acids produced are lactic acid and acetic acid, although small amounts of other acids such as propionic, fumaric and malic acid are also formed during fermentation.

It is highly probable that the first controlled food fermentations came into existence through trial and error and a need to preserve foods for consumption later in the season. It is possible that the initial attempts at preservation involved the addition of salt or seawater. During the removal of the salt prior to consumption, the foods would pass through stages favourable to acid fermentation. Although the process worked, it is likely that the causative agents were unknown. Today, there are numerous examples of controlled fermentation for the preservation and processing of foods. However, only a few of these have been studied in any detail - these include sauerkraut, pickles, kimchi, beer, wine and vinegar production. Although the general principles and processes for many of the fermented fruit and vegetable products are the same -relying mainly on lactic acid and acetic acid forming bacteria, yeasts and moulds, the reactions have not been quantified for each product. The reactions are usually very complex and involve a series of micro-organisms, either working together or in succession to achieve the final product.

Yeast Fermentation

A yeast is a unicellular fungus which reproduces asexually by budding or division, especially the genus *Saccharomyces* which is important in food fermentations. Yeasts and yeast-like fungi are widely distributed in nature. They are present in orchards and vineyards, in the air, the soil and the intestinal tract of animals. Like bacteria and moulds, they can have beneficial and non-beneficial effects in foods. Most yeasts are larger than most bacteria. The most well known examples of yeast fermentation are in the production of alcoholic drinks and the leavening of bread. For their participation in these two processes, yeasts are of major importance in the food industry.

Some yeasts are chromogenic and produce a variety of pigments, including green, yellow and black. Others are capable of synthesising essential B group vitamins.

Although there is a large diversity of yeasts and yeast-like fungi, (about 500 species), only a few are commonly associated with the production of fermented foods. They are all either ascomycetous yeasts or members of the genus *Candida*. Varieties of the *Saccharomyces cervisiae* genus are the most common yeasts in fermented foods and beverages based on fruit and vegetables. All strains of this genus ferment glucose and many ferment other plant derived carbohydrates such as sucrose, maltose and raffinose. In the tropics, *Saccharomyces pombe* is the dominant yeast in the production of traditional fermented beverages, especially those derived from maize and millet.

Conditions necessary for fermentation

Most yeasts require an abundance of oxygen for growth, therefore by controlling the supply of oxygen, their growth can be checked. In addition to oxygen, they require a basic substrate such as sugar. Some yeasts can ferment sugars to alcohol and carbon dioxide in the absence of air but require oxygen for growth. They produce ethyl alcohol and carbon dioxide from simple sugars such as glucose and fructose.

In conditions of excess oxygen (and in the presence of acetobacter) the alcohol can be oxidised to form acetic acid. This is undesirable if the end product is a fruit alcohol, but is a technique employed for the production of fruit vinegars.

Yeasts are active in a very broad temperature range - from 0 to 50° C, with an optimum temperature range of 20° to 30° C.

The optimum pH for most micro-organisms is near the neutral point (pH 7.0). Moulds and yeasts are usually acid tolerant and are therefore associated with the spoilage of acidic foods. Yeasts can grow in a pH range of 4 to 4.5 and moulds can grow from pH 2 to 8.5, but favour an acid pH.

In terms of water requirements, yeasts are intermediate between bacteria and moulds. Bacteria have the highest demands for water, while moulds have the least need. Normal yeasts require a minimum water activity of 0.85 or a relative humidity of 88%.

Yeasts are fairly tolerant of high concentrations of sugar and grow well in solutions containing 40% sugar. At concentrations higher than this, only a certain group of yeasts – the osmophilic type – can survive. There are only a few yeasts that can tolerate sugar concentrations of 65-70% and these grow very slowly in these conditions. Some yeasts – for example the *Debaromyces* - can tolerate high salt concentrations. Another group which can tolerate high salt concentrations and low water activity is *Zygosaccharomyces rouxii,* which is associated with fermentations in which salting is an integral part of the process.

Production of fruit alcohol

Alcohol and acids are two primary products of fermentation, both used to good effect in the preservation of foods. Several alcohol-fermented foods are preceded by an acid fermentation and in the presence of oxygen and *acetobacter*, alcohol can be fermented to produce acetic acid. Most food spoilage organisms cannot survive in either alcoholic or acidic environments. Therefore, the production of both these end products can prevent a food from undergoing spoilage and extend its shelf life.

Primitive wines and beers have been produced, with the aid of yeasts, for thousands of years, although it was not until about four hundred years ago that micro-organisms associated with the fermentation were observed and identified. It was not until the 1850's that Louis Pasteur demonstrated unequivocally the involvement of yeasts in the production of wines and beers. Since then, the knowledge of yeasts and the conditions necessary for fermentation of wine and beer has increased to the point where pure culture fermentations are now used to ensure consistent product quality. Originally, alcoholic fermentations would have been spontaneous events that resulted from the activity of micro-organisms naturally present. These non-scientific methods are still used today for the home preparation of many of the worlds traditional beers and wines.

Alcoholic drinks fall into two broad categories: wines and beers. Wines are made from the juice of fruits and beers from cereal grains. The principal carbohydrates in fruit juices are soluble sugars; the principal carbohydrate in grains is starch, an insoluble polysaccharide. The yeasts that bring about alcoholic fermentation can attack soluble sugars but do not produce starch-splitting enzymes. Wines can therefore be made by the direct fermentation of the raw material, while the production of beer requires the hydrolysis of starch to yield sugars fermentable by yeast, as a preliminary step.

Raw fruit juice is usually a strongly acidic solution, containing from 10 to 25 percent soluble sugars. Its acidity and high sugar concentration make it an unfavourable medium for the growth of bacteria but highly suitable for yeasts and moulds. Raw fruit juice naturally contains many yeasts, moulds, and bacteria, derived from the surface of the fruit. Normally the yeast used in alcoholic fermentation is a strain of the species *Saccharomyces cerevisiae*.

The fermentation may be allowed to proceed spontaneously, or can be "started" by inoculation with a must that has been previously successfully fermented by *S. cerevisiae* var. *ellipsoideus*. Many modern wineries eliminate the original microbial population of the must by pasteurisation or by treatment with sulphur dioxide. The must is then inoculated with a *starter culture* derived from a pure culture of a suitable strain of wine yeast. This procedure eliminates many of the uncertainties and difficulties of older methods. At the start of the fermentation, the must is aerated slightly to build up a large and vigorous yeast population; once fermentation sets in, the rapid production of carbon dioxide maintains anaerobic conditions, which prevent the growth of undesirable aerobic organisms, such as bacteria and moulds. The temperature of fermentation is usually from 25 to 30oC, and the duration of the fermentation process may extend from a few days to two weeks. As soon as the desired degree of sugar disappearance and alcohol production has been attained, the microbiological phase of wine making is over. Thereafter, the quality and stability of the wine depend very largely on preventing further microbial activity, both during the "aging" in wooden casks and after bottling.

At all stages during its manufacture, fruit juice alcohol is subject to spoilage by undesirable microorganisms. Pasteur, whose descriptions of the organisms responsible and recommendations for overcoming them are still valid today, first scientifically explored the problem of the "diseases" of wines. The most serious aerobic spoilage processes are brought about by film-forming yeasts and acetic acid bacteria, both of which grow at the expense of the alcohol, converting it to acetic acid or to carbon dioxide and water. The chief danger from these organisms arises when access of air is not carefully regulated during aging. Much more serious are the diseases caused by fermentative bacteria, particularly rod-shaped lactic acid bacteria, which utilise any residual sugar and impart a mousy taste to the wine. Such wines are known as turned wines. Since oxygen is unnecessary for the growth of lactic acid bacteria, wine spoilage of this kind can occur even after bottling. These risks of spoilage can be minimised by pasteurisation after bottling.

Grape wine

Grape wine is perhaps the most common fruit juice alcohol. Because of the commercialisation of the product for industry, the process has received most research attention and is documented in detail.

The production of grape wine involves the following basic steps: crushing the grapes to extract the juice; alcoholic fermentation; maltolactic fermentation if desired; bulk storage and maturation of the wine in a cellar; clarification and packaging. Although the process is fairly simple, quality control demands that the fermentation is carried out under controlled conditions to ensure a high quality product.

The distinctive flavour of grape wine originates from the grapes as raw material and subsequent processing operations. The grapes contribute trace elements of many volatile substances which give the final product the distinctive fruity character. In addition, they contribute non-volatile compounds (tartaric and malic acids) which impact on flavour and tannins which give bitterness and astringency. The latter are more prominent in red wines as the tannin components are located in the grape skins.

Although yeasts are the principal organisms involved, filamentous fungi, lactic acid bacteria, acetic acid bacteria and other bacterial groups all play a role in the production of alcoholic fruit products. Normal grapes harbour a diverse micro-flora, of which the principal yeasts (*Saccharomyces cerevisiae*) involved in desirable fermentation are in the minority. Lactic acid bacteria and acetic acid bacteria are also present. The proportions of each and total numbers present are dependent upon a number of external environmental factors including the temperature, humidity, stage of maturity, damage at harvest and application of fungicides. It is essential to ensure proliferation of the desired species at the expense of the non-desired ones. This is achieved through ensuring fermentation conditions are such to encourage *Saccharomyces* species.

The fermentation may be initiated using a starter culture of *Saccharomyces cerevisiae* – in which case the juice is inoculated with populations of yeast of 106 to 107 cfu/ml juice. This approach produces a wine of generally expected taste and quality. If the fermentation is allowed to proceed naturally, utilising the yeasts present on the surface of the fruits, the end result is less controllable, but produces wines with a range of flavour characteristics. It is likely that natural fermentations are practiced widely around the world, especially for home production of wine.

During alcoholic fermentation, yeasts are the prominent species. The composition of fruit juice – its acid and sugar level and low pH favour the growth of yeasts and production of ethanol that restricts the growth of bacteria and fungi.

In natural fermentations, there is a progressive pattern of yeast growth. Several species of yeast, including *Kloeckera, Hanseniaspora, Candida* and *Metschnikowia,* are active for the first two to three days of fermentation. The build up of end products (ethanol) is toxic to these yeasts and they die off, leaving *Saccharomyces cerevisiae* to continue the fermentation to the end. *S. cerevisiae* can tolerate much higher levels of ethanol (up to 15% v/v or more) than the other species who only tolerate up to 5 or 8% alcohol. Because of its tolerance of alcohol, *S. cerevisiae* dominates wine fermentation and is the species that has been commercialised for starter cultures.

Traditionally, fermentation was carried out in large wooden barrels or concrete tanks. Modern wineries now use stainless steel tanks as these are more hygienic and provide better temperature control. White wines are fermented at 10 to 18º C for about seven to fourteen days. The low temperature and slow fermentation favours the retention of volatile compounds. Red wines are fermented at 20 to 30ºC for about seven days. This higher temperature is necessary to extract the pigment from the grape skins.

Factors affecting wine fermentation.

There are several variables which can affect the fermentation process and final quality of wine. Factors which are most important to control are:

- the clarification and pre-treatment of juice
- chemical composition of the juice
- temperature of the fermentation
- the influences of other micro-organisms.

Clarification and pre-treatment of juice: Excessive clarification removes many of the natural yeasts and flora. This is beneficial if a tightly controlled induced fermentation is desired, but less so if the fermentation is a natural one. Long periods of settling out however, encourage the growth of natural flora, which can contribute to the fermentation.

Chemical composition of juice: The main consituents of grape juice are glucose (75 to 150 g/l), fructose (75 to 150 g/l), tartaric acid (2 to 10 g/l), malic acid (1 to 8 g/l) and free amino acids (0.2 to 2.5 g/l). The main reaction is the fermentation of glucose and fructose to ethanol and carbon dioxide. However, the presence of nitrogenous and sulphurous products also contributes to the fermentation. The addition of sulphur dioxide to the juice delays the growth of yeast, but does not necessarily inhibit growth of the non-Saccharomyces strains. Fruits generally contain sufficient substrates - soluble sugars - for the yeast to ferment and convert into an acceptable concentration of alcohol. Sugar can be added to fruit juices with a low sugar content, to increase the amount of fermentable substrate.

Temperature: Temperature has an impact on the growth and activity of different strains of yeast. At temperatures of 10 to 15° C, the non-Saccharomyces species have an increased tolerance to alcohol and therefore have the potential to contribute to the fermentation.

Influence of other micro-organisms: Other micro-organisms have the potential to influence wine production at all stages of the process. Prior to harvest, yeasts grow on the surface of grapes. Fungicides are used in an attempt to control their growth, but these disturb the natural balance of flora, thus making it difficult to carry out a 'natural' fermentation. Overuse of fungicides can lead to the development of resistant strains of yeast which have the potential to produce toxins which destroy the desirable yeast species. These yeasts are known as 'killer' strains. Other microbes have further chances to influence the fermentation during the clarification process, after fermentation and during maturation and bottling when *acetobacter* species can oxidise the alcohol and produce acetic acid.

About two to three weeks after the alcoholic fermentation is finished wines often undergo a malo-lactic fermentation. This occurs naturally and lasts for about four weeks. It is a lactic acid fermentation, initiated by lactic acid bacteria resident in the wine. Inoculating the fermented wine with cultures of *Leuconostoc oenos* can start the process if it is desired. The main reaction of these bacteria is the decarboxylation of L-malic acid to L-lactic acid, which decreases the acidity of the wine and increases its pH by about 0.3 to 0.5 units.

Wines produced from grapes grown in colder climates tend to have a higher concentration of malic acid and a lower pH (3.0 to 3.5) and the taste benefits from this slight decrease in acidity. The benefits of this process are that it imparts a more mellow flavour to the wine. The growth of malo-lactic bacteria also contributes to the taste of the wine. Wines that have undergone a malo-lactic fermentation appear to be less susceptible to any further damage from other bacteria. This could be because *L. oenos* has used up all available substrate, or it may have secreted bacteriocins which prevent the growth of other species.

Although the malo-lactic fermentation seems to be a useful process, not all wines benefit from it. Wines produced from grapes in warmer climates tend to be less acidic (pH > 3.5) and a further reduction in acidity may have adverse effects on the quality of the wine. Decreasing the acidity also increases the pH to values which can allow spoilage organisms to multiply. It is difficult to prevent the malo-lactic fermentation from taking place naturally, especially later on after the wine has been bottled. In low acid wines, the acidity may be adjusted after this fermentation has taken place. The malo-lactic fermentation can be prevented by controlling several factors: the wine pH (< 3.2); ethanol content (> 14%) and levels of sulphur dioxide (>50 mg/l). The bacteriocin nisin can also be used to control the growth of malo-lactic bacteria. However the subtle blend of aromas and flavours that contribute to the final taste may be lost by such stringent control.

The conversion of malic acid to lactic acid is one of the main reactions carried out by wine lactic acid bacteria. *L. oenos* needs to be present in significant numbers (greater than 106 cfu/ml) for the reaction to take place at a suitable pace. The bacteria use residual pentose and hexose sugars in the wine as a substrate for growth. The main reaction is the deacidification (or decarboxylation) of malic acid. In addition to this, the by-products of the reaction impart flavours and aromas to the wine.

During storage, wines are prone to non-desirable microbial changes. Yeasts, lactic acid bacteria, acetic acid bacteria and fungi can all spoil or taint wines after the fermentation process is completed. The changes that occur are increased acidification through the formation of acetic and other acids from alcohol; increased carbonation through a secondary fermentation of residual sugars and flavour changes through the metabolism of numerous compounds.

Bacterial Fermentation

Bacteria are "a large group of unicellular or multi-cellular organisms lacking chlorophyll, with a simple nucleus, multiplying rapidly by simple fission, some species developing a highly resistant resting (spore) phase; some species reproduce sexually, and some are motile. In shape they are spherical, rodlike, spiral, or filamentous. They occur in air, water, soil, rotting organic material, animals and plants. Saprophytic forms are more numerous than parasites. A few forms are autotrophic".

There are several bacterial families present in foods, the majority of which are concerned with food spoilage. The important role of bacteria in the fermentation of foods is often overlooked.

Lactic Acid Bacteria

The lactic acid bacteria are a group of Gram positive bacteria, non-respiring, non-spore forming, cocci or rods, which produce lactic acid as the major end product of the fermentation of carbohydrates. They are the most important bacteria in desirable food fermentations, being responsible for the fermentation of sour dough bread, sorghum beer, all fermented milks, cassava (to produce *gari* and *fufu*) and most "pickled" (fermented) vegetables. Historically, bacteria from the genera *Lactobacillus, Leuconostoc, Pediococcus* and *Streptococcus* are the main species involved. Several more have been identified, but play a minor role in lactic fermentations. Lactic acid bacteria were recently reviewed by Axelsson.

Lactic acid bacteria carry out their reactions - the conversion of carbohydrate to lactic acid plus carbon dioxide and other organic acids - without the need for oxygen. They are described as microaerophilic as they do not utilise oxygen. Because of this, the changes that they effect do not cause drastic changes in the composition of the food. Some of the family are homofermentative, that is they only produce lactic acid, while others are heterofermentative and produce lactic acid plus other volatile compounds and small amounts of alcohol. *Lactobacillus acidophilus, L. bulgaricus, L. plantarum, L. caret, L. pentoaceticus, L brevis* and *L. thermophilus* are examples of lactic acid-producing bacteria involved in food fermentations. All species of lactic acid bacteria have their own particular reactions and niches, but overall, *L. plantarum* – a homofermenter -produces high acidity in all vegetable fermentations and plays the major role.

All lactic acid producers are non-motile gram positive rods that need complex carbohydrate substrates as a source of energy. The lactic acid they produce is effective in inhibiting the growth of other bacteria that may decompose or spoil the food. Because the whole group are referred to as 'lactic acid bacteria' it might appear that the reactions they carry out are very simple, with the production of one substrate. This is far from the truth. The lactic acid bacteria are a diverse group of organisms with a diverse metabolic capacity. This diversity makes them very adaptable to a range of conditions and is largely responsible for their success in acid food fermentations.

Despite their complexity, the whole basis of lactic acid fermentation centres on the ability of lactic acid bacteria to produce acid, which then inhibits the growth of other non-desirable organisms. All lactic acid producers are micro-aerophilic, that is they require small amounts of oxygen to function. Species of the genera *Streptococcus* and *Leuconostoc* produce the least acid. Next are the heterofermentative species of *Lactobacillus* which produce intermediate amounts of acid, followed by the *Pediococcus* and lastly the homofermenters of the *Lactobacillus* species, which produce the most acid. Homofermenters, convert sugars primarily to lactic acid, while heterofermenters produce about 50% lactic acid plus 25% acetic acid and ethyl alcohol and 25% carbon dioxide. These other compounds are important as they impart particular tastes and aromas to the final product. The heterofermentative lactobacilli produce mannitol and some species also produce dextran.

Leuconostoc mesenteroides is a bacterium associated with the sauerkraut and pickle fermentations. This organism initiates the desirable lactic acid fermentation in these products. It differs from other lactic acid species in that it can tolerate fairly high concentrations of salt and sugar (up to 50% sugar). *L. mesenteroides* initiates growth in vegetables more rapidly over a range of temperatures and salt concentrations than any other lactic acid bacteria. It produces carbon dioxide and acids which rapidly lower the pH and inhibit the development of undesirable micro-organisms. The carbon dioxide produced replaces the oxygen, making the environment anaerobic and suitable for the growth of subsequent species of lactobacillus. Removal of oxygen also helps to preserve the colour of vegetables and stabilises any ascorbic acid that is present.

Organisms from the gram positive *Propionibacteriaceae* family are responsible for the flavour and texture of some fermented foods, especially Swiss cheese, where they are responsible for the formation of 'eyes' or holes in the cheese. These bacteria break down lactic acid into acetic and propionic acids and carbon dioxide.

Several other bacteria, for *instance Leuconostoc citrovorum L. Dextranicum, Streptococcus lactis, S. Cremis, & liquefaciens* and *Brevibacterium* species are important in the fermentation of dairy products. They are not discussed in detail in this manuscript.

Lactic acid fermentation

The lactic acid bacteria belong to two main groups – the homofermenters and the heterofermenters. The pathways of lactic acid production differ for the two. Homofermenters produce mainly lactic acid, via the glycolytic (Embden–Meyerhof) pathway). Heterofermenters produce lactic acid plus appreciable amounts of ethanol, acetate and carbon dioxide, via the 6-phosphoglucanate/phosphoketolase pathway. The glycolytic pathway is used by all lactic acid bacteria except leuconostocs, group III lactobacilli, oenococci and weissellas. Normal conditions required for this pathway are excess sugar and limited oxygen. Axelsson gives an in-depth account of the biochemical pathways for both homo- and hetero-fermenters.

Acetic acid bacteria

A second group of bacteria of importance in food fermentations are the acetic acid producers from the *Acetobacter* species. *Acetobacter* are important in the production of vinegar (acetic acid) from fruit juices and alcohols. The same reaction also occurs in wines, oxygen permitting, where the *acetobacter* can cause undesirable changes – the oxidation of alcohol to acetic acid. This produces a vinegary off-taste in the wine.

The most desirable action of acetic acid bacteria is in the production of vinegar. The vinegar process is essentially a two stage process, where yeasts convert sugars into alcohol, followed by *acetobacter*, which oxidise alcohol to acetic acid.

Bacteria of alkaline fermentations

A third group of bacteria are those which bring about alkaline fermentations - the *Bacillus* species. Of note are *Bacillus subtilis, B. licheniformis* and *B. pumilius. Bacillus subtilis* is the dominant species, causing the hydrolysis of protein to amino acids and peptides and releasing ammonia, which increases the alkalinity and makes the substrate unsuitable for the growth of spoilage organisms. Alkaline fermentations are more common with protein rich foods such as soybeans and other legumes, although there are a few examples utilising plant seeds. For example water melon seeds (*Ogiri* in Nigeria) and sesame seeds (*Ogiri-saro* in Sierra Leone) and others where coconut and leaf proteins are the substrates (Indonesian *semayi* and Sudanese *kawal* respectively).

Although the range of products of alkaline fermentation does not match those brought about by acid fermentations, they are important in that they provide protein rich, low cost condiments from leaves, seeds and beans, which contribute to the diet of millions of people in Africa and Asia. Steinkraus presents a comprehensive review of the acid, alkaline and alcoholic fermentations from around the world, which the reader is referred to for further information.

Conditions required for bacterial fermentations

Micro-organisms vary in their optimal pH requirements for growth. Most bacteria favour conditions with a near neutral pH (7). The varied pH requirements of different groups of micro-organisms is used to good effect in fermented foods where successions of micro-organisms take over from each other as the pH of the environment changes. Certain bacteria are acid tolerant and will survive at reduced pH levels. Notable acid-tolerant bacteria include *the Lactobacillus* and *Streptococcus* species, which play a role in the fermentation of dairy and vegetable products.

Oxygen requirements vary from species to species. The lactic acid bacteria are described as microaerophilic as they carry out their reactions with very little oxygen. The acetic acid bacteria however, require oxygen to oxidise alcohol to acetic acid. In vinegar production, oxygen has to be made available for the production of acetic acid, whereas with wine it is essential to exclude oxygen to prevent oxidation of the alcohol and spoilage of the wine.

Temperature

Different bacteria can tolerate different temperatures, which provides enormous scope for a range of fermentations. While most bacteria have a temperature optimum of between 20 to 30ºC, there are some (the thermophiles) which prefer higher temperatures (50 to 55ºC) and those with colder temperature optima (15 to 20ºC). Most lactic acid bacteria work best at temperatures of 18 to 22ºC. The *Leuconostoc* species which initiate fermentation have an optimum of 18 tc 22ºC. Temperatures above 22ºC, favour the *lactobacillus* species.

Salt concentration

Lactic acid bacteria tolerate high salt concentrations. The salt tolerance gives them an advantage over other less tolerant species and allows the lactic acid fermenters to begin metabolism, which produces acid, which further inhibits the growth of non-desirable organisms. Leuconostoc is noted for its high salt tolerance and for this reason, initiates the majority of lactic acid fermentations.

Water activity

In general, bacteria require a fairly high water activity (0.9 or higher) to survive. There are a few species which can tolerate water activities lower than this, but usually the yeasts and fungi will predominate on foods with a lower water activity.

Hydrogen ion concentration (pH)

The optimum pH for most bacteria is near the neutral point (pH 7.0). Certain bacteria are acid tolerant and will survive at reduced pH levels. Notable acid-tolerant bacteria include *the Lactobacillus* and *Streptococcus* species, which play a role in the fermentation of dairy and vegetable products.

Oxygen availability

Some of the fermentative bacteria are anaerobes, while others require oxygen for their metabolic activities. Some, lactobacilli in particular, are microaerophilic. That is they grow in the presence of reduced amounts of atmospheric oxygen. In aerobic fermentations, the amount of oxygen present is one of the limiting factors. It determines the type and amount of biological product obtained, the amount of substrate consumed and the energy released from the reaction. Acetobacter require oxygen for the oxidation of alcohol to acetic acid.

Nutrients

All bacteria require a source of nutrients for metabolism. The fermentative bacteria require carbohydrates – either simple sugars such as glucose and fructose or complex carbohydrates such as starch or cellulose. The energy requirements of micro-organisms are very high. Limiting the amount of substrate available can check their growth.

Principles of lactic acid fermentation

Sauerkraut is one example of an acid fermentation of vegetables. The name sauerkraut literally translates as acid cabbage. The 'sauerkraut process' can be applied to any other suitable type of vegetable product. Because of the importance of this product in the German diet, the process has received substantial research in order to commercialise and standardise production. As a result, the process and the contributing micro-organisms are known intimately. Other less well known fermented fruits and vegetables have received less research attention, therefore little is known of the exact process. It is safe to assume however that the acid fermentation of vegetables is based on this process.

Lactic acid fermentations are carried out under three basic types of condition:– dry salted, brined and non-salted. Salting provides a suitable environment for lactic acid bacteria to grow which impart the acid flavour to the vegetable.

Dry salted fermented vegetables

With dry salting, the vegetable is treated with dry salt. The salt extracts the juice from the vegetable and creates the brine. The vegetable is prepared, washed in potable cold water and drained. For every 100 kg of vegetables 3 kg of salt is needed. The vegetables are placed in a layer of about 2.5cm depth in the fermenting container (a barrel or keg). Salt is sprinkled over the vegetables. Another layer of vegetables is added and more salt added. This is repeated until the container is three quarters full. A cloth is placed above the vegetables and a weight added to compress the vegetables and assist the formation of a brine which takes about 24 hours. As soon as the brine is formed, fermentation starts and bubbles of carbon dioxide begin to appear. Fermentation takes between one and four weeks depending on the ambient temperature. Fermentation is complete when no more bubbles appear, after which time the pickle can be packaged in a variety of mixtures. These can be vinegar and spices or oil and spices.

The 'sauerkraut' process.

Lactic acid bacteria are the primary group of organisms involved in sauerkraut fermentation. They can be divided into three groups according to their types and end products:

Leuconostoc mesenteroides	an acid and gas producing coccus
Lactobacillus plantarum	and bacilli that produce acid and a small amount of gas
L. Cucumeris	
Lactobacillus pentoaceticus	acid and gas producing bacilli (*L. Brevis*)

In addition to the desirable bacteria there are a range of undesirable micro-organisms present on cabbage (and other vegetable material) which can interfere with the sauerkraut process if allowed to multiply unchecked. The quality of the final product depends largely on how well the undesirable organisms are controlled during the fermentation process. Some of the typical spoilage organisms utilise the protein as an energy source, producing unpleasant odours and flavours.

Shredded cabbage or other suitable vegetables are placed in a jar and salt is added. Mechanical pressure is applied to the cabbage to expel the juice, which contains fermentable sugars and other nutrients suitable for microbial activity. The first micro-organisms to start acting are the gas-producing cocci *(L. Mesenteroides).* These microbes produce acids. When the acidity reaches 0.25 to 0.3% (calculated as lactic acid), these bacteria slow down and begin to die off, although their enzymes continue to function. The activity initiated by the *L. mesenteroides* is continued by the lactobacilli (*L. plantarum and L. cucumeris)* until an acidity level of 1.5 to 2% is attained. The high salt concentration and low temperature inhibit these bacteria to some extent. Finally, *L. pentoaceticus* continues the fermentation, bringing the acidity to 2 to 2.5% thus completing the fermentation.

The end products of a normal kraut fermentation are lactic acid along with smaller amounts of acetic and propionic acids, a mixture of gases of which carbon dioxide is the principal gas, small amounts of alcohol and a mixture of aromatic esters. The acids, in combination with alcohol form esters, which contribute to the characteristic flavour of sauerkraut. The acidity helps to control the growth of spoilage and putrefactive organisms and contributes to the extended shelf life of the product. Changes in the sequence of desirable bacteria, or indeed the presence of undesirable bacteria, alter the taste and quality of the product.

The optimum temperature for sauerkraut fermentation is around 21ºC. A variation of just a few degrees from this temperature alters the activity of the microbial process and affects the quality of the final product. Therefore, temperature control is one of the most important factors in the sauerkraut process. A temperature of 18º to 22º C is most desirable for initiating fermentation since this is the optimum temperature range for the growth and metabolism of *L. mesenteroides*. Temperatures above 22ºC favour the growth of *Lactobacillus* species.

Salt plays an important role in initiating the sauerkraut process and affects the quality of the final product. The addition of too much salt may inhibit the desirable bacteria, although it may contribute to the firmness of the kraut. The principle function of salt is to withdraw juice from the cabbage (or other vegetable), thus making a more favourable environment for development of the desired bacteria.

Generally, salt is added to a final concentration of 2.0 to 2.5%. At this concentration, lactobacilli are slightly inhibited, but cocci are not affected. Unfortunately, this concentration of salt has a greater inhibitory effect against the desirable organisms than against those responsible for spoilage. The spoilage organisms can tolerate salt concentrations up to between 5 and 7%, therefore it is the acidic environment created by the lactobacilli that keep the spoilage bacteria at bay, rather than the addition of salt.

In the manufacture of sauerkraut, dry salt is added at the rate if 1 to 1.5 kg per 50kg cabbage (2 to 3%). The use of salt brines is not recommended in sauerkraut making, but is common in vegetables that have a low water content. It is essential to use pure salt since salts with added alkali may neutralise the acid.

In order to produce sauerkraut of consistent quality, starter cultures (similar to those used in the dairy industry) have been recommended. Not only do starter cultures ensure consistency between batches, they speed up the fermentation process as there is no time lag while the relevant microflora colonise the sample. Because the starter cultures used are acidic, they also inhibit the undesirable micro-organisms. It is possible to add starters traditionally used for milk fermentation, such as *Streptococcus lactis*, without adverse effect on final quality. Because these organisms only survive for a short time (long enough to initiate the acidification process) in the kraut medium, they do not disturb the natural sequence of micro-organisms.

On the other hand, if *Leuconostoc mesenteroides* is added in the early stages, it gives a good flavour to the final product, but alters the sequence of subsequent bacterial growth and results in a product that is incompletely fermented. If gas producing rods (for example *L pentoaceticus)* are added to the sauerkraut, this disturbs the balance between acetic and lactic acids - more acetic acid and less lactic acid are produced than normal - and the fermentation never reaches completion. If lactic acid, non-gas producing *rods (L. Cucumeris)* are used as a starter, again the kraut is not completely fermented and the resulting product is bitter and more susceptible to spoilage by yeasts.

It is possible to use the juice from a previous kraut fermentation as a starter culture for subsequent fermentations. The efficacy of using old juice depends largely on the types of organisms present in the juice and its acidity. If the starter juice has an acidity of 0.3% or more, it results in a poor quality kraut. This is because the cocci which would normally initiate fermentation are suppressed by the high acidity, leaving the bacilli with sole responsibility for fermentation. If the starter juice has an acidity of 0.25% or less, the kraut produced is normal, but there do not appear to be any beneficial effects of adding this juice. Often, the use of old juice produces a sauerkraut which has a softer texture than normal.

The majority of spoilage in sauerkraut is due to aerobic soil micro-organisms which break down the protein and produce undesirable flavour and texture changes. The growth of these aerobes can easily be inhibited by a normal fermentation.

Soft kraut can result from many conditions such as large amounts of air, poor salting procedure and varying temperatures. Whenever the normal sequence of bacterial growth is altered or disturbed, it usually results in a soft product. It is the lactobacilli, which seem to have a greater ability than the cocci to break down cabbage tissues, which are responsible for the softening. High temperatures and a reduced salt content favour the growth of lactobacilli, which are sensitive to higher concentrations of salt. The usual concentration of salt used in sauerkraut production slightly inhibits the lactobacilli, but has no effect on the cocci. If the salt content is too low initially, the lactobacilli grow too rapidly at the beginning and upset the normal sequence of fermentation.

Another problem encountered is the production of dark coloured sauerkraut. This is caused by spoilage organisms during the fermentation process. Several conditions favour the growth of spoilage organisms. For example, an uneven distribution of salt tends to inhibit the desirable organisms while at the same time allowing the undesirable salt tolerant organisms to flourish. An insufficient level of juice to cover the kraut during the fermentation allows undesirable aerobic bacteria and yeasts to grow on the surface of the kraut, causing off flavours and discoloration. If the fermentation temperature is too high, this also encourages the growth of undesirable microflora, which results in a darkened colour.

Pink kraut is a spoilage problem. It is caused by a group of yeasts which produce an intense red pigment in the juice and on the surface of the cabbage. It is caused by an uneven distribution of or an excessive concentration of salt, both of which allow the yeast to multiply. If conditions are optimal for normal fermentation, these spoilage yeasts are suppressed.

Brine salted fermented vegetables

Brine is used for vegetables which inherently contain less moisture. A brine solution is prepared by dissolving salt in water (a 15 to 20% salt solution). Fermentation takes place well in a brine of about 20 salometer. As a general guide, a fresh egg floats in a 10% brine solution (Kordylas, 1990). Properly brined vegetables will keep well in vinegar for a long time. The duration of brining is important for the overall keeping qualities. The vegetable is immersed in the brine and allowed to ferment. The strong brine solution draws sugar and water out of the vegetable, which decreases the salt concentration. It is crucial that the salt concentration does not fall below 12%, otherwise conditions do not allow for fermentation. To achieve this, extra salt is added periodically to the brine mixture.

Once the vegetables have been brined and the container sealed, there is a rapid development of micro-organisms in the brine. The natural controls which affect the microbial populations of the fermenting vegetables include the concentration of salt and temperature of the brine, the availability of fermentable materials and the numbers and types of micro-organisms present at the start of fermentation. The rapidity of the fermentation is correlated with the concentration of salt in the brine and its temperature.

Most vegetables can be fermented at 12.5o to 20o salometer salt. If so, the microbial sequence of lactic acid bacteria generally follows the classical sauerkraut fermentation described by Pederson. At higher salt levels of up to about 40o salometer, the sequence is skewed towards the development of a homofermentation, dominated by *Lactobacillus plantarum*. At the highest concentrations of salt (about 60o salometer) the lactic fermentation ceases to function and if any acid is detected during brine storage it is acetic acid, presumably produced by acid-forming yeasts which are still active at this concentration of salt.

Pickled cucumbers are another fermented product that has been studied in detail and the process is known. The fermentation process is very similar to the sauerkraut process, only brine is used instead of dry salt. The washed cucumbers are placed in large tanks and salt brine (15 to 20%) is added. The cucumbers are submerged in the brine, ensuring that none float on the surface - this is essential to prevent spoilage. The strong brine draws the sugar and water out of the cucumbers, which simultaneously reduces the salinity of the solution. In order to maintain a salt solution so that fermentation can take place, more salt has to be added to the brine solution. If the concentration of salt falls below 12%, it will result in spoilage of the pickles through putrefaction and softening.

A few days after the cucumbers have been placed in the brine, the fermentation process begins. The process generates heat which causes the brine to boil rapidly. Acids are also produced as a result of the fermentation.

During fermentation, visible changes take place which are important in judging the progress of the process. The colour of the cucumber surface changes from bright green to a dark olive green as acids interact with the chlorophyll. The interior of the cucumber changes from white to a waxy translucent shade as air is forced out of the cells. The specific gravity of the cucumbers also increases as a result of the gradual absorption of salt and they begin to sink in the brine rather than floating on the surface.

As with the sauerkraut process, the gram positive coccus - *Leuconostoc mesenteroides* predominates in the first stages of pickle fermentation. This species is more resistant to temperature changes and tolerates higher salt concentration than the subsequent species. As fermentation proceeds and the acidity increases, lactobacilli start to take over from the cocci. The active stage of fermentation continues for between 10 to 30 days, depending upon the temperature of the fermentation. The optimum temperature for *L. Cucumeris* is 29 to 32ºC. During the fermentative period, the acidity increases to about 2% and the strong acid producing types of bacteria reach their maximum growth. If sugar or acetic acid is added to the fermenting mixture during this time it increases the production of acid.

The production of excessive amounts of acid during the fermentation, results in shrivelling of the pickles, possibly due to over-activity of the *L. mesenteroides* species. If the brine is stirred, it may introduce air, which makes conditions more favourable for the growth of spoilage bacteria. In general, if the pickles are well covered with brine, the salt concentration is maintained and the temperature is at an optimum, it should be quite simple to produce good quality pickles.

Non salted, lactic acid fermented vegetables

Some vegetables are fermented by lactic acid bacteria, without the prior addition of salt or brine. Examples of non-salted products include *gundruk* (consumed in Nepal), *sinki* and other wilted fermented leaves. The detoxification of cassava through fermentation includes an acid fermentation, during which time the cyanogenic glycosides are hydrolysed to liberate the toxic cyanide gas.

The fermentation process relies on the rapid colonisation of the food by lactic acid producing bacteria, which lower the pH and make the environment unsuitable for the growth of spoilage organisms. Oxygen is also excluded as the *Lactobacilli* favour an anaerobic atmosphere. Restriction of oxygen ensures that yeasts do not grow.

For the production of *sinki*, fresh radish roots are harvested, washed and wilted by sun-drying for one to two days. They are then shredded, re-washed and packed tightly into an earthenware or glass jar, which is sealed and left to ferment. The optimum fermentation time is twelve days at 30ºC. *Sinki* fermentation is initiated by *L. fermentum* and *L. brevis*, followed by *L. plantarum*. During fermentation the pH drops from 6.7 to 3.3. After fermentation, the radish substrate is sun-dried to a moisture level of about 21%. For consumption, *sinki* is rinsed in water for two minutes, squeezed to remove the excess water and fried with salt, tomato, onion and green chilli. The fried mixture is then boiled in rice water and served hot as soup along with the main meal.

South Pacific pit fermentations are an ancient method of preserving starchy vegetables without the addition of salt. The raw materials undergo an acid fermentation within the pit, to produce a paste with good keeping qualities. Pit fermentations are also used in other parts of the world – for example in Ethiopia, where the false banana (*Ensete ventricosum*) is fermented in a pit to produce a pulp known as *kocho*. Foods preserved in pits can last for years without deterioration, therefore pits provide a good, reliable cheap means of storage.

Root crops and bananas are peeled before being placed in the pit, while breadfruit are scraped and pierced. Food is left to ferment for three to six weeks, after which time it becomes soft, has a strong odour and a paste-like consistency. During fermentation, carbon dioxide builds up in the pit, creating an anaerobic atmosphere. As a result of bacterial activity, the temperature rises much higher than the ambient temperature. The pH of the fruit within the pit decreases from 6.7 to 3.7 within about four weeks. Inoculation of the fruit in the pit with lactic acid bacteria greatly speeds up the process. The fermented paste can be left in the pit and removed as required. Usually, it is removed and replaced with a second batch of fresh food to ferment. The fermented food is washed and fibrous material removed. It is then dried in the sun for several hours to remove the volatile odours, and pounded into a paste. Grated coconut or coconut cream and sugar may be added and the mixture is wrapped in banana leaves and either baked or boiled.

Principles of Acetic Acid Fermentation

The main desirable fermentation carried out by acetic acid bacteria is the production of vinegar. Vinegar, literally translated as sour wine, is one of the oldest products of fermentation used by man. It can be made from almost any fermentable carbohydrate source, for example fruits, vegetables, syrups and wine. Whatever the raw material used, the fermentation process follows a definite sequence. The basic requirement for vinegar production is a raw material that will undergo an alcoholic fermentation. Apples, pears, grapes, honey, syrups, cereals, hydrolysed starches, beer and wine are all ideal substrates for the production of vinegar. The best raw materials are cider and wine, which are widely used in Europe and the United States. To produce a high quality product it is essential that the raw material is mature, clean and in good condition.

Microbes involved in Vinegar Process

The production of vinegar depends on a mixed fermentation, which involves both yeasts and bacteria. The fermentation is usually initiated by yeasts which break down glucose into ethyl alcohol with the liberation of carbon dioxide gas.

The yeasts and bacteria exist together in a form known as commensalism. The acetobacter are dependent upon the yeasts to produce an easily oxidisable substance (ethyl alcohol). It is not possible to produce vinegar by the action of one type of micro-organism alone. For a good fermentation, it is essential to have an alcohol concentration of 10 to 13%. If the alcohol content is much higher, the alcohol is incompletely oxidised to acetic acid. If it is lower than 13%, there is a loss of vinegar because the esters and acetic acid are oxidised. In addition to acetic acid, other organic acids are formed during the fermentation which become esterified and contribute to the characteristic odour, flavour and colour of the vinegar.

Acetaldehyde is an intermediate product in the transformation of the reducing sugar in fruit juice to acetic acid or vinegar. Oxygen is required for the conversion of acetaldehyde to acetic acid. In general, the yield of acetic acid from glucose is approximately 60%. That is three parts of glucose yield two parts acetic acid.

Micro-organisms involved in the fermentation of vinegar

The organisms involved in vinegar production usually grow at the top of the substrate, forming a jelly like mass. This mass is known as 'mother of vinegar'. The mother is composed of both *acetobacter* and yeasts, which work together. The principal bacteria are *Acetobacter acetic, A. Xylinum* and *A. Ascendens*. The main yeasts are *Saccharomyces ellipsoideus* and S *cerevisiae.* It is important to maintain an acidic environment to suppress the growth of undesirable organisms and to encourage the presence of desirable acetic acid producing bacteria. It is common practice to add 10 to 25% by volume of strong vinegar to the alcoholic substrate in order to attain a desirable fermentation.

The alcoholic fermentation of sugars should be completed before the solution is acidified because any remaining sugar will not be converted to alcohol after the acetic acid is added. Incomplete fermentation of the juice results in a "weak" product. The acetic acid strength of good vinegar should be approximately 6%.

Fermentation methods

Small scale production

Vinegar can be made at home at the small scale by introducing oxygen into barrels of wine or cider and allowing fermentation to occur spontaneously. This process is not very rigorously controlled and often results in a poor quality product.

The Orleans process

The Orleans process is one of the oldest and well known methods for the production of vinegar. It is a slow, continuous process, which originated in France. High grade vinegar is used as a starter culture, to which wine is added at weekly intervals. The vinegar is fermented in large (200 litre) capacity barrels. Approximately 65 to 70 litres of high grade vinegar is added to the barrel along with 15 litres of wine. After one week, a further 10 to 15 litres of wine are added and this is repeated at weekly intervals. After about four weeks, vinegar can be withdrawn from the barrel (10 to 15 litres per week) as more wine is added to replace the vinegar.

One of the problems encountered with this method is that of how to add more liquid to the barrel without disturbing the floating bacterial mat. This can be overcome by using a glass tube which reaches to the bottom of the barrel. Additional liquid is poured in through the tube and therefore does not disturb the bacteria. Wood shavings are sometimes added to the fermenting barrel to help support the bacterial mat.

Quick vinegar method

Because the Orleans process is slow, other methods have been adapted to try and speed up the process. The German method is one such method. It uses a generator, which is an upright tank filled with beechwood shavings and fitted with devices which allow the alcoholic solution to trickle down through the shavings in which the acetic acid bacteria are living. The tank is not allowed to fill as that would exclude oxygen which is necessary for the fermentation. Near the bottom of the generator are holes which allow air to be drawn in. the air rises through the generator and is used by the acetic acid bacteria to oxidise the alcohol. This oxidisation also releases considerable amounts of heat which must be controlled to avoid causing damage to the bacteria.

Problems in vinegar production

Many of the problems of vinegar production are concerned with the presence of nematodes, mites, flies and other insects. These pests can be controlled by adherence to good hygiene and pasteurisation of the vinegar. Problems associated with the fermentative process include the presence of a whitish film on the surface of the vinegar. This is sometimes called *Mycoderma vini* and is composed of yeast-like organisms, which grow aerobically and oxidise the carbon containing compounds to carbon dioxide and water. They also alter the flavour and alcohol content of the vinegar. This problem can however be controlled by adding one part vinegar to three parts of the alcoholic solution or by storing the alcoholic liquid in filled closed containers.

REFERENCES

Beuchat, L.R. and A.D. Hocking, "Some considerations when analysing foods for the presence of xerophilic fungi", *J. Food Prot.*, 1990.

D'Aoust, J.-Y., "Effective enrichment-plating conditions for detection of Salmonella in foods", *J. Food Prot.*, 1984.

Environmental Defense Fund. *Genetically Engineered Foods: Who's Minding the Store?* New York. NY: Environmental Defense Fund. 1995.

Jarvis, B. and A.P. Williams, *Methods for detecting fungi in foods and beverages*, In: Beuchat, L.R. (ed.), Food and beverage mycology. Second edition, Avi Publishing Co., Westport, Connecticut, 1987.

Organisation for Economic Cooperation and Development (OECD). *Biotechnology, Agriculture and Food.* Paris. OECD. 1992.

6

Microbial Biodegradation

Interest in the microbial biodegradation of pollutants has intensified in recent years as humanity strives to find sustainable ways to cleanup contaminated environments. These bioremediation and biotransformation methods endeavour to harness the astonishing, naturally occurring, ability of microbial xenobiotic metabolism to degrade, transform or accumulate a huge range of compounds including hydrocarbons (e.g. oil), polychlorinated biphenyls (PCBs), polyaromatic hydrocarbons (PAHs), pharmaceutical substances, radionuclides and metals. Major methodological breakthroughs in recent years have enabled detailed genomic, metagenomic, proteomic, bioinformatic and other high-throughput analyses of environmentally relevant microorganisms providing unprecedented insights into key biodegradative pathways and the ability of organisms to adapt to changing environmental conditions.

The elimination of a wide range of pollutants and wastes from the environment is an absolute requirement to promote a sustainable development of our society with low environmental impact. Biological processes play a major role in the removal of contaminants and they take advantage of the astonishing catabolic versatility of microorganisms to degrade/convert such compounds. New methodological breakthroughs in sequencing, genomics, proteomics, bioinformatics and imaging are producing vast amounts of information. In the field of Environmental Microbiology, genome-based global studies open a new era providing unprecedented in silico views of metabolic and regulatory networks, as well as clues to the evolution of degradation pathways and to the molecular adaptation strategies to changing environmental conditions. Functional genomic and metagenomic approaches are increasing our understanding of the relative importance of different pathways and regulatory networks to carbon flux in particular environments and for particular compounds and they will certainly accelerate the development of bioremediation technologies and biotransformation processes.

AEROBIC BIODEGRADATION OF POLLUTANTS

The burgeoning amount of bacterial genomic data provides unparalleled opportunities for understanding the genetic and molecular bases of the degradation of organic pollutants. Aromatic compounds are among the most recalcitrant of these pollutants and lessons can be learned from the recent genomic studies of Burkholderia xenovorans LB400 and Rhodococcus sp. strain RHA1, two of the largest bacterial genomes completely sequenced to date. These studies have helped expand our understanding of bacterial catabolism, non-catabolic physiological adaptation to organic compounds, and the evolution of large bacterial genomes. First, the metabolic pathways from phylogenetically diverse isolates are very similar with respect to overall organization. Thus, as originally noted in pseudomonads, a large number of "peripheral aromatic" pathways funnel a range of natural and xenobiotic compounds into a restricted number of "central aromatic" pathways.

Nevertheless, these pathways are genetically organized in genus-specific fashions, as exemplified by the b-ketoadipate and Paa pathways. Comparative genomic studies further reveal that some pathways are more widespread than initially thought. Thus, the Box and Paa pathways illustrate the prevalence of non-oxygenolytic ring-cleavage strategies in aerobic aromatic degradation processes. Functional genomic studies have been useful in establishing that even organisms harboring high numbers of homologous enzymes seem to contain few examples of true redundancy. For example, the multiplicity of ring-cleaving dioxygenases in certain rhodococcal isolates may be attributed to the cryptic aromatic catabolism of different terpenoids and steroids. Finally, analyses have indicated that recent genetic flux appears to have played a more significant role in the evolution of some large genomes, such as LB400's, than others. However, the emerging trend is that the large gene repertoires of potent pollutant degraders such as LB400 and RHA1 have evolved principally through more ancient processes. That this is true in such phylogenetically diverse species is remarkable and further suggests the ancient origin of this catabolic capacity.

ANAEROBIC BIODEGRADATION OF POLLUTANTS

Anaerobic microbial mineralization of recalcitrant organic pollutants is of great environmental significance and involves intriguing novel biochemical reactions. In particular, hydrocarbons and halogenated compounds have long been doubted to be degradable in the absence of oxygen, but the isolation of hitherto unknown anaerobic hydrocarbon-degrading and reductively dehalogenating bacteria during the last decades provided ultimate proof for these processes in nature. Many novel biochemical reactions were discovered enabling the respective metabolic pathways, but progress in the molecular understanding of these bacteria was rather slow, since genetic systems are not readily applicable for most of them. However, with the increasing application of genomics in the field of environmental microbiology, a new and promising perspective is now at hand to obtain molecular insights into these new metabolic properties. Several complete genome sequences were determined during the last few years from bacteria capable of anaerobic organic pollutant degradation.

The ~4.7 Mb genome of the facultative denitrifying Aromatoleum aromaticum strain EbN1 was the first to be determined for an anaerobic hydrocarbon degrader (using toluene or ethylbenzene as substrates). The genome sequence revealed about two dozen gene clusters (including several paralogs) coding for a complex catabolic network for anaerobic and aerobic degradation of aromatic compounds. The genome sequence forms the basis for current detailed studies on regulation of pathways and enzyme structures.

Further genomes of anaerobic hydrocarbon degrading bacteria were recently completed for the iron-reducing species Geobacter metallireducens and the perchlorate-reducing Dechloromonas aromatica, but these are not yet evaluated in formal publications. Complete genomes were also determined for bacteria capable of anaerobic degradation of halogenated hydrocarbons by halorespiration: the ~1.4 Mb genomes of Dehalococcoides ethenogenes strain 195 and Dehalococcoides sp. strain CBDB1 and the ~5.7 Mb genome of Desulfitobacterium hafniense strain Y51. Characteristic for all these bacteria is the presence of multiple paralogous genes for reductive dehalogenases, implicating a wider dehalogenating spectrum of the organisms than previously known. Moreover, genome sequences provided unprecedented insights into the evolution of reductive dehalogenation and differing strategies for niche adaptation.

BIOAVAILABILITY

Bioavailability, or the amount of a substance that is physiochemically accessible to microorganisms is a key factor in the efficient biodegradation of pollutants. Chemotaxis, or the directed movement of motile organisms towards or away from chemicals in the environment is an important physiological response that may contribute to effective catabolism of molecules in the environment. In addition, mechanisms for the intracellular accumulation of aromatic molecules via various transport mechanisms are also important.

OIL BIODEGRADATION

Petroleum oil contains aromatic compounds that are toxic for most life forms. Episodic and chronic pollution of the environment by oil causes major ecological perturbations. Marine environments are especially vulnerable since oil spills of coastal regions and the open sea are poorly containable and mitigation is difficult. In addition to pollution through human activities, millions of tons of petroleum enter the marine environment every year from natural seepages. Despite its toxicity, a considerable fraction of petroleum oil entering marine systems is eliminated by the hydrocarbon-degrading activities of microbial communities, in particular by a remarkable recently discovered group of specialists, the so-called hydrocarbonoclastic bacteria (HCB) . Alcanivorax borkumensis was the first HCB to have its genome sequenced.

ANALYSIS OF WASTE BIOTREATMENT

Sustainable development requires the promotion of environmental management and a constant search for new technologies to treat vast quantities of wastes generated by increasing anthropogenic activities. Biotreatment, the processing of wastes using living organisms, is an environmentally friendly, relatively simple and cost-effective alternative to physico-chemical clean-up options. Confined environments, such as bioreactors], have been engineered to overcome the physical, chemical and biological limiting factors of biotreatment processes in highly controlled systems. The great versatility in the design of confined environments allows the treatment of a wide range of wastes under optimized conditions. To perform a correct assessment, it is necessary to consider various microorganisms having a variety of genomes and expressed transcripts and proteins. A great number of analyses are often required. Using traditional genomic techniques, such assessments are limited and time-consuming. However, several high-throughput techniques originally developed for medical studies can be applied to assess biotreatment in confined environments.

METABOLIC ENGINEERING AND BIOCATALYTIC APPLICATIONS

The study of the fate of persistent organic chemicals in the environment has revealed a large reservoir of enzymatic reactions with a large potential in preparative organic synthesis, which has already been exploited for a number of oxygenases on pilot and even on industrial scale. Novel catalysts can be obtained from metagenomic libraries and DNA sequence based approaches. Our increasing capabilities in adapting the catalysts to specific reactions and process requirements by rational and random mutagenesis broadens the scope for application in the fine chemical industry, but also in the field of biodegradation. In many cases, these catalysts need to be exploited in whole cell bioconversions or in fermentations, calling for system-wide approaches to understanding strain physiology and metabolism and rational approaches to the engineering of whole cells as they are increasingly put forward in the area of systems biotechnology and synthetic biology.

BIOREMEDIATION TECHNOLOGY

Bioremediation can be defined as any process that uses microorganisms, fungi, green plants or their enzymes to return the environment altered by contaminants to its original condition. Bioremediation may be employed to attack specific soil contaminants, such as degradation of chlorinated hydrocarbons by bacteria. An example of a more general approach is the cleanup of oil spills by the addition of nitrate and/or sulfate fertilisers to facilitate the decomposition of crude oil by indigenous or exogenous bacteria. Naturally-occurring bioremediation and phytoremediation have been used for centuries. Bioremediation technologies can be generally classified as in situ or ex situ. In situ bioremediation involves treating the contaminated material at the site while ex situ involves the removal of the contaminated material to be treated elsewhere. Some examples of bioremediation technologies are bioventing, landfarming, bioreactor, composting, bioaugmentation, rhizofiltration, and biostimulation. Not all contaminants, however, are easily treated by bioremediation using microorganisms.

For example, heavy metals such as cadmium and lead are not readily absorbed or captured by organisms. The assimilation of metals such as mercury into the food chain may worsen matters. Phytoremediation is useful in these circumstances, as many plants are able to bioaccumulate these toxins in their above-ground parts, which are then harvested for removal. The heavy metals in the harvested biomass may be further concentrated by incineration. The use of genetic engineering to create organisms specifically designed for bioremediation has great potential. The bacterium Deinococcus radiodurans has been modified to consume and digest toluene and ionic mercury from highly radioactive nuclear waste.

There are a number of cost/efficiency advantages to bioremediation, which can be employed in areas that are inaccessible without excavation. For example, hydrocarbon spills or certain chlorinated solvents may contaminate groundwater, and introducing the appropriate electron acceptor or electron donor amendment, as appropriate, may significantly reduce contaminant concentrations after a lag time allowing for acclimation. This is typically much less expensive than excavation followed by disposal elsewhere, incineration or other ex situ treatment strategies, and reduces or eliminates the need for "pump and treat", a common practice at sites where hydrocarbons have contaminated groundwater.

Biodegradation is the process by which organic substances are broken down by other living organisms. The term is often used in relation to ecology, waste management, environmental remediation and to plastic materials, due to their long life span. Organic material can be degraded aerobically, with oxygen, or anaerobically, without oxygen. A term related to biodegradation is biomineralisation, in which organic matter is converted to into minerals.Biodegradable matter is generally organic material such as plant & animal matter and other substances originating from living organisms. Biodegradable waste in landfill degrades in the absence of oxygen through the process of anaerobic digestion. The byproducts of this anaerobic biodegradation are biogas and lignin and cellulose fibres which cannot be broken down by anaerobes. Engineered landfills are designed with liners to prevent toxic leachate seeping into the surrounding soil and groundwater. Paper and other materials that normally degrade in a few years degrade more slowly over longer periods of time. Biogas contains methane which has approximately 21 times the global warming potential of carbon dioxide. In modern landfills this biogas can be collected and used for power generation.

Biodegradation can be measured in a number of ways. The activity of aerobic microbes can be measured by the amount of oxygen they consume or the amount of carbon dioxide they produce. Biodegradation by anaerobic microbes can be measured by the amount of methane they produce. Natural communities of microorganisms present in the subsurface have an amasing physiological versatility. Microorganisms can carry out biodegradation in many different types of habitats and environments, both under aerobic and anaerobic conditions. Communities of bacteria and fungi can degrade a multitude of synthetic compounds and probably every natural product.

In situ bioremediation is the application of biological treatment to the cleanup of hazardous chemicals present in the subsurface. The optimisation and control of microbial transformations of organic contaminants require the integration of many scientific and engineering disciplines. Hazardous compounds persist in the subsurface because environmental conditions are not appropriate for the microbial activity that results in biochemical degradation. The optimisation of environmental conditions is achieved by understanding the biological principles under which these compounds are degraded, and the effect of environmental conditions on both the responsible microorganisms and their metabolic reactions.

Microbial Metabolism

During the process of in situ bioremediation, microorganisms use the organic contaminants for their growth. In addition, compounds providing the major nutrients such as nitrogen, phosphorus, and minor nutrients such as sulfur and trace elements are also required for their growth. In most cases, an organic compound that represents a carbon and energy source is transformed by the metabolic pathways that are characteristic of heterotrophic microorganisms. It should be stressed, however, that an organic compound need not necessarily be a substrate for growth in order for it to be metabolised by microorganisms. Two categories of transformations exist. In the first, biodegradation provides carbon and energy to support growth, and the process, therefore, is growth-linked. In the second, biodegradation is not linked to multiplication, but to obtaining the carbon for respiration in order for the cells to maintain their viability. This maintenance metabolism may take place only when the organic carbon concentrations are very low. Cometabolic transformations also fall into the second category. It has been observed that the number of microbial cells or the biomass of the species acting on the compound of interest increases as degradation proceeds.

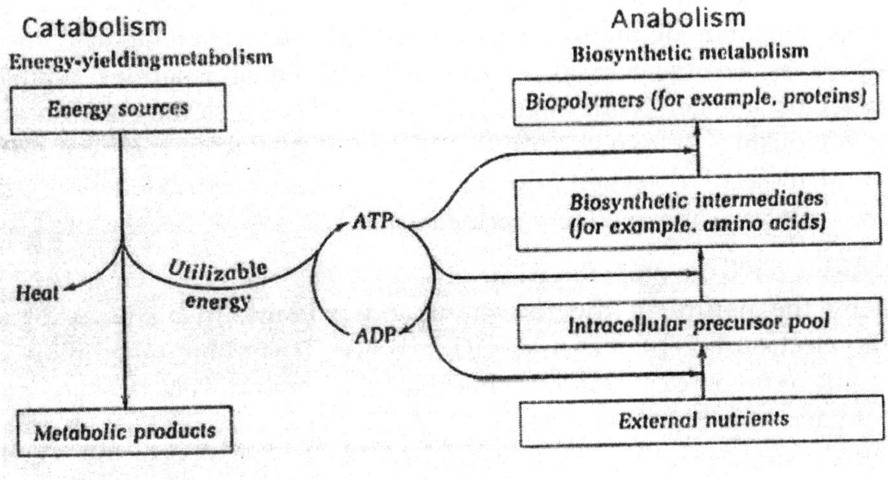

Figure 1. Microbial Metabolism

During a typical growth-linked mineralisation brought about by bacteria, the cells use some of the energy and carbon of the organic substrate to make new cells, and this increasingly larger population causes increasingly rapid mineralisation. Microorganisms need nitrogen, phosphorus, and sulfur, and a variety of trace nutrients other than carbon. These requirements should be satisfied as the responsible species degrade the compound of interest. For heterotrophic microorganisms in most natural systems, usually sufficient amounts of N, P, S, and other trace nutrients are present to satisfy the microbial demand. Because carbon is limiting and because it is the element for which there is intense competition, a species with the unique ability to grow on synthetic molecules has a selective advantage.

Prior to the degradation of many organic compounds, a period is observed in which no degradation of the chemical is evident. This time interval is known as the acclimatisation period or, sometimes, as adaptation or lag period. The length of the acclimatisation period varies and may be less than 1 h or many months. The duration of acclimatisation depends upon the chemical structure, subsurface biogeochemical environmental conditions, and concentration of the compound. Once the indigenous population of microorganisms has become acclimatised to the presence and degradation of a chemical and the activity becomes marked, the microbial community will retain its higher level of activity for some time. Acclimatisation of a microbial population to one substrate frequently results in the simultaneous acclimatisation to some, but not all, structurally related molecules.

The design of bioremediation processes requires determination of the desired degradation reactions to which the target compounds will be subjected. This involves selecting the metabolism mode that will occur in the process. The metabolism modes are broadly classified as aerobic and anaerobic. Aerobic transformations occur in the presence of molecular oxygen, with molecular oxygen serving as the electron acceptor. This form of metabolism is known as aerobic respiration. Anaerobic reactions occur only in the absence of molecular oxygen and the reactions are subdivided into anaerobic respiration, fermentation, and methane fermentation.

Microorganisms have developed a wide variety of respiration systems. These can be characterised by the nature of the reductant and oxidant. In all cases of aerobic respiration, the electron acceptor is molecular oxygen. Anaerobic respiration uses an oxidised inorganic or organic compound other than oxygen as the electron acceptor. The respiration of organic substrates by bacteria is, in most cases, very similar. The substrates are oxidised to CO and H_2O.

Fermentation is the simplest of the three principal modes of energy yielding metabolism. During fermentation, organic compounds serve as both electron donors and electron acceptors. Fermentation can proceed only under strictly anaerobic conditions. The process maintains a strict oxidation-reduction balance. The average oxidation level of the end products is identical to that of the substrate fermented. Thus the substrate yields a mixture of end products, some more oxidised than the substrate and others more reduced. The end products depend on the type of microorganisms but usually include a number of acids, alcohols, ketones, and gases such as CO_2 and CH_4

The metabolism modes that utilise nitrate as an electron acceptor (performed by denitrifying and nitrate-reducing organisms), sulfate and thiosulfate as electron acceptors (performed by sulfate-reducing organisms), and CO_2 as an electron acceptor (performed by methanogenic organisms) can be used to biodegrade various organic contaminants. The utilisation of chlorinated organic compounds as electron acceptors during anaerobic respiration is a recent observation. Another important metabolism concept in bioremediation is cometabolism. In a true sense, cometabolism is not metabolism (energy yielding), but fortuitous transformation of a compound. As noted earlier, it was the traditional belief that microorganisms must obtain energy from an organic compound to biodegrade it. The transformation of an organic compound by a microorganism that is unable to use the substrate as a source of energy is termed cometabolism.

Enzymes generated by an organism growing at the expense of one substrate also can transform a different substrate that is not associated with that organism's energy production, carbon assimilation, or any other growth processes. Contaminants that lend themselves to bioremediation by becoming a secondary substrate through cometabolism are only partially transformed. This transformation may or may not result in reducing toxicity. If all toxicity properties of a hazardous compound are removed via biotransformation, this is referred to as detoxification. Detoxification results in inactivation, with the toxicologically active substance being converted to an inactive product. Detoxification does not imply mineralisation and may include several processes such as hydrolysis, hydroxylation, dechlorination, and demethylation. Fortunately, the metabolites or transformation products from cometabolism by one organism can typically be used as an energy source by another.

Since cometabolism generally leads to a slow degradation of the substrate, attention has been given to enhancing its rate. The addition of a number of organic compounds into the contaminated zone may promote the rate of cometabolism, but the responses to such additions are not predictable. Addition of mineralizable compounds that are structurally analogous to the compound whose cometabolism is desired is known as analog enrichment. The microorganism that grows on the mineralisable compound contains enzymes transforming the analogous molecule by cometabolism. Another aspect of microbial metabolism is the recognition of preferential substrate degradation. Preferential degradation results in a sequential attack where the higher energyyielding compounds are degraded first. In a petroleum spill, benzene will be degraded, under aerobic conditions, at a faster rate than naphthalene, and naphthalene will degrade faster than chrysene.

Mycoremediation

Mycoremediation is a form of bioremediation, the process of using fungi to return an environment (usually soil) contaminated by pollutants to a less contaminated state. The term mycoremediation was coined by Paul Stamets and refers specifically to the use of fungal mycelia in bioremediation.

One of the primary roles of fungi in the ecosystem is decomposition, which is performed by the mycelium. The mycelium secretes extracellular enzymes and acids that break down lignin and cellulose, the two main building blocks of plant fiber. These are organic compounds composed of long chains of carbon and hydrogen, structurally similar to many organic pollutants. The key to mycoremediation is determining the right fungal species to target a specific pollutant. Certain strains have been reported to successfully degrade the nerve gases VX and sarin.

In an experiment conducted in conjunction with Thomas, a major contributor in the bioremediation industry, a plot of soil contaminated with diesel oil was inoculated with mycelia of oyster mushrooms; traditional bioremediation techniques (bacteria) were used on control plots. After four weeks, more than 95% of many of the PAH (polycyclic aromatic hydrocarbons) had been reduced to non-toxic components in the mycelial-inoculated plots. It appears that the natural microbial community participates with the fungi to break down contaminants, eventually into carbon dioxide and water. Wood-degrading fungi are particularly effective in breaking down aromatic pollutants (toxic components of petroleum), as well as chlorinated compounds.

The concept of mycoremediation was explored in the 1984 film Nausicaä of the Valley of the Wind, where vast tracts of fungal forest rehabilitate the planet after catastrophic human polluting and apocalypse. Mycofiltration is a similar or same process, using fungal mycelia to filter toxic waste and microorganisms from water in soil.

Bioremediation of a contaminated site typically works in one of two ways. In the case described above, ways are found to enhance the growth of whatever pollution-eating microbes might already be living at the contaminated site. In the second, less common case, specialized microbes are added to degrade the contaminants.

Bioremediation provides a good cleanup strategy for some types of pollution, but as you might expect, it will not work for all. For example, bioremediation may not provide a feasible strategy at sites with high concentrations of chemicals that are toxic to most microorganisms. These chemicals include metals such as cadmium or lead, and salts such as sodium chloride.

Nonetheless, bioremediation provides a technique for cleaning up pollution by enhancing the same biodegradation processes that occur in nature. Depending on the site and its contaminants, bioremediation may be safer and less expensive than alternative solutions such as incineration or landfilling of the contaminated materials. It also has the advantage of treating the contamination in place so that large quantities of soil, sediment or water do not have to be dug up or pumped out of the ground for treatment.

REFERENCES

Allenby, Braden R. "Integrating Environment And Technology: Design For Environment" in Allenby, Braden R., Richards, Deanna, (ed.) *Greening Industrial Ecosystems.* Washington DC. National Academy Press Office. 1994 .

Alexander, M. *Biodegradation and Bioremediation.* Academic Press. New York, 1994.

D. R. Thevenot, K. Toth, R. A. Durst, G. S. Wilson. *Biosens. Bioelectron.* 16, 121–131 (2001); O. A. Sadik and A. Mulchandani. In *Chemical & Biological Sensors: Meeting the Challenges of Environmental Monitoring.* ACS Symposium Series. Vol. 762. pp. 1–7. American Chemical Society. Washington. DC 2000.

Edgington, Stephen M. "Environmental Biotechnology." *Bio/Technology.* December 1994. p. 1338-42.

Ginzburg, Lev. R. *Assessing Ecological Risks of Biotechnology.* Boston: Butterworth-Heinemann. 1991.

7

Bioreactor Technology

Traditionally, microbiologists have played the dominant role in bioreaction development, with assistance from those in multiple disciplines, including biochemists, geneticists and chemical engineers. And while the fermentation process — the precursor to modern bioreactions — has been used since prehistoric days, the major advancements of the last half century have had as much to do with technology as with biology. It is our objective to illustrate the relevance of established chemical engineering practices and processes as they apply to today's bioreaction engineering, as chemical engineers make further inroads into a field once thought to be the sole domain of biology-based scientists.

Biochemical engineering is a branch of chemical engineering or biological engineering that mainly deals with the design and construction of unit processes that involve biological organisms or molecules. Biochemical engineering is often taught as a supplementary option to chemical engineering or biological engineering due to the similarities in both the background subject curriculum and problem-solving techniques used by both professions. Its applications are used in the food, feed, pharmaceutical, biotechnology, and water treatment industries.

A bioreactor may refer to any device or system that supports a biologically active environment.In one case, a bioreactor is a vessel in which is carried out a chemical process which involves organisms or biochemically active substances derived from such organisms. This process can either be aerobic or anaerobic. These bioreactors are commonly cylindrical, ranging in size from liters to cubic meters, and are often made of stainless steel.

A bioreactor may also refer to a device or system meant to grow cells or tissues in the context of cell culture. These devices are being developed for use in tissue engineering.

On the basis of mode of operation, a bioreactor may be classified as batch, fed batch or continuous (e.g. Continuous stirred-tank reactor model). An example of a continuous bioreactor is the chemostat.

Organisms growing in bioreactors may be suspended or immobilized. The simplest, where cells are immobilized, is a Petri dish with agar gel. Large scale immobilized cell bioreactors are:

- moving media
- packed bed
- fibrous bed
- membrane

Long before anyone understood the concept of bioreaction, humans were taking advantage of its results. Bread, cheese, wine and beer were all made possible through what was traditionally known as fermentation—a little-understood process, successful more by chance than design. It was, in fact, the failure and frustration of French vintners who found they were too often producing vinegar not wine, that led the famous French chemist and microbiologist Louis Pasteur to study the fermentation process at their request.

What Pasteur discovered was that fermentation occurred as a result of the biological activity of a microscopic plant called yeast. When unwanted microbes in- filtrated the wine and "fed"on the alcohol produced by the yeast, the microbes left behind distasteful and harmful wastes, which ruined the wine's flavor. Pasteur's work laid the foundation for bioreactors as we know them today, because once the process was identified and understood, it could be controlled. And it is the control of the process that concerns chemical engineers first and foremost. The scope of bioengineering has grown from simple wine-bottle microbiology to the industrialization of not only beer, wine, cheese and milk production, but also the production of biotechnology's newer products -- antibiotics, enzymes, steroidal hormones, vitamins, sugars and organic acids.

BIOREACTORS VS. CHEMICAL REACTORS

By definition, a bioreactor is a system in which a biological conversion is effected. Although this definition can apply to any conversion involving enzymes, microorganisms, and animal or plant cells, for the purposes of this article, we will limit the definition. The bioreactors referred to here include only mechanical vessels in which (a) organisms are cultivated in a controlled manner and/or (b) materials are converted or transformed via specific reactions.

Quite similar to conventional chemical reactors, bioreactors differ in that they are specifically designed to influence metabolic pathways. Traditional chemical reactor models and designs that may be used for bioreaction as well include: continuous stirred-tank reactors, continuous flow stirred-tank reactors, and plug-flow reactors, singularly or in series; ebullized-bed (*i.e.*, "bubbling and boiling") reactors; and fluidized-bed reactors. The term "bioreactor" is often used synonymously with "fermenter;" however, in the strictest definition, a fermenter is a system that provides an anaerobic process for producing alcohol from sugar.

Bioreactors differ from conventional chemical reactors in that they support and control biological entities. As such, bioreactor systems must be designed to provide a higher degree of control over process upsets and contaminations, since the organisms are more sensitive and less stable than chemicals. Biological organisms, by their nature, will mutate, which may alter the biochemistry of the bioreaction or the physical properties of the organism. Analogous to heterogeneous catalysis, deactivation or mortality occur and promoters or coenzymes influence the kinetics of the bioreaction. Although the majority of fundamental bioreactor engineering and design issues are similar, maintaining the desired biological activity and eliminating or minimizing undesired activities often presents a greater challenge than traditional chemical reactors typically require.

Organisms, influenced by their morphology and the bioreaction medium, are shear-sensitive to varying degrees. A number of bacteria, yeast and fungi cultures that can be relatively tolerant of high-shear environments exhibit a robustness in high-energy mixing vessels. Animal, fish, insect and plant cells are delicate and usually require low-shear environments for viability. The viscosities of bioreaction masses may change during growth and production phases, and, often, the medium becomes non-Newtonian as a cycle progressed. Mixing within the bioreactor is integral to efficient heat and mass transfer during the production phases, which places additional constraints on the suitable agitation mechanism and rheology of the bioreaction medium. Other key differences between chemical reactors and bioreactors are selectivity and rate. In bioreactors, higher selectivity — that is, the measure of the system's capability for producing the preferred product (over other outcomes)—is of primary importance. In fact, selectivity is especially important in the production of relatively complex molecules such as antibiotics, steroids, vitamins, proteins and certain sugars and organic acids. Frequently, the activity and desired selectivity occur in a substantially smaller range of conditions than are present in conventional chemical reactors. Further, deactivation of the biomass often poses more severe consequences than a chemical upset.

Rate is of secondary importance. For many biological systems, an incubation period is needed to prepare a culture used to inoculate the bioreactor with the producing microbes or their precursors. Although a bioreaction can be brief, in systems where organism or biomass growth is necessary, the bioreaction can take 10–20 d for completion of the batch. Further, the bioreactor should not be regarded as an isolated unit, but as part of an integrated unit operation with both upstream (preparation) and downstream (recovery) unit operations.

PRODUCTS OF BIOREACTIONS

Bioreaction products are formed by three basic processes:

1. Processes in which the product is produced by the cells is either extracellular, e.g., alcohols or critic acid, or intracellular, e.g., a metabolite or enzyme. Production of cellular products is divided into two types, based on when they are produced within a biological cycle. Primary metabolites are produced during growth and are essential for continuing growth. Secondary metabolites are produced after growth has ceased. Primary metabolites include amino acids, nucleotides, nucleic acids, proteins, lipids and carbohydrates. Examples of primary products for industrial use include ethanol, citric acid, acetone, butanol, lysine, polysaccharides and vitamins.

   Secondary cellular products are formed from the intermediates and products of primary metabolism, and tend to be specific to a species or group of organisms. Not all microorganisms produce secondary metabolites, but they are widespread among the filamentous fungi and plants. Many secondary products have toxic or antibiotic properties and are, as such, the basis of much of the antibiotic industry. Production of enzymes via bioreactions has displaced inef-ficient extraction techniques with mutation and genetic manipulation. Industrial enzymes find their home in baking, brewing, grain processing, dairy making, and in the production of detergents, juices, wines and other products.

2. Processes that produce a cell mass. Bakers' yeast, used in the baking industry, is an example of a produced cell mass. Others include single-cell proteins for food sources.
3. Processes that modify a compound that is added to the fermentation process are referred to as biotransformations. Biotransformations occur using the inherent enzymatic capability of most cells. Cells of all types can be employed to biocatalyze a transformation of certain compounds via dehydration, oxidation, hydroxylation, amination or isomerization. Enzymatic conversions frequently exhibit lower activation energies and higher selectivity than their chemical counterparts. Steroids, antibiotics and prostaglandins can all be produced via biotransformations.

KEY ISSUES IN BIOREACTOR DESIGN AND OPERATION

The goal of an effective bioreactor is to control, contain and positively influence the biological reaction. To accomplish this, the chemical engineer must take into consideration two areas. One is the suitable reactor parameters for the desired biological, chemical and physical (macrokinetic) system. The macrokinetic system includes microbial growth and metabolite production. Microbes can include bacteria, yeast, fungi, and animal, plant, fish and insect cells, as well as other biological materials.

The other area of major importance in bioreactor design involves the bioreaction parameters, including:

- controlled temperature
- optimum pH
- sufficient substrate (usually a carbon source), such as sugars, proteins and fats
- water availability
- salts for nutrition
- vitamins
- oxygen (for aerobic processes)
- gas evolution and
- product and byproduct removal.

In addition to controlling these, the bioreactor must be designed to both promote formation of the optimal morphology of the organism and to eliminate or reduce contamination by unwanted organisms or mutation of the organism. This article will provide a description and overview of a variety of bioreaction systems, including both bioreactors that are production-oriented and those used for environmental control. We will discuss the advantages and disadvantages of the various systems, with a brief mention of the typical applications for each. It is also worth noting here that there is a wide variety of bioreaction systems, and any attempt to categorize them by their various attributes will naturally result in some overlap of system characteristics.

We will *not* include all the various subtopics that may be relevant, since the study of bioreactions is so expansive that it would be impossible to include subtopics here in any detail. Those subtopics not covered include: microbiology, sterilization, rheology, mixing, agitator design, fluidization, heat transfer, mass transfer, surface phenomena and transport enhancements, kinetics, hydrodynamics, scaleup, modeling, instrumentation and process control. For the most part, an abundance of literature exists on these conventional chemical processes, and the treatment is frequently directly applicable to bioreactors with only slight modifications to accommodate for the living organisms involved in the process.

BIOREACTION/FERMENTATION TECHNOLOGIES

Now we will discuss the engineering aspects and applications for a variety of bioreaction/ fermentation technologies, including the challenges of each and the advantages and disadvantages of the respective technologies. The various types of bioreaction systems covered here include batch, continuous, semi-continuous, surface/tray, submerged, air-lift loop and trickle-bed setups. As stated before, there are overlapping characteristics in several of the technologies discussed.

Batch bioreactions

The majority of bioreactions are batch-wise. The first phase of batch bioreaction is commonly sterilization, after which the sterile culture medium is inoculated with microorganisms that have been cultivated to achieve a specific result. During this dynamic reaction period, cells, substrates (including the nutrient salts and vitamins) and concentrations of the products vary with time. Proper mixing keeps the differences in composition and temperature at acceptable levels. To promote aerobic cultivation, the medium is aerated to provide a continuous flow of oxygen. Gaseous byproducts formed, such as CO_2, are removed, and aeration and gas-removal processes take place semicontinuously.

Next, an acid or alkali is added if the pH needs to be controlled. To keep foaming to acceptable levels, antifoaming agents may be added when indicated by a foam sensor. One of the first types of batch systems is the tray fermenter (Figure 1), used in the early days of commercial aerobic bioreactions for products such as citric acid and penicillin. In this system, the trays are loaded with the culture medium and the organisms, and the air-flow produces the bioreaction, during which exhaust gas is discharged. When the bioreaction is complete, end product is removed from the trays. Because this method is inefficient for producing large commercial quantities, it fell quickly to the wayside with the emergence of submerged tank systems, which are designed to handle significantly higher volumes.

Overall, batch bioreaction systems provide a number of advantages, including:

- Reduced risk of contamination or cell mutation, due to a relatively brief growth period.
- Lower capital investment when compared to continuous processes for the same bioreactor volume.
- More flexibility with varying product/biological systems.
- Higher raw material conversion levels, resulting from a controlled growth period.

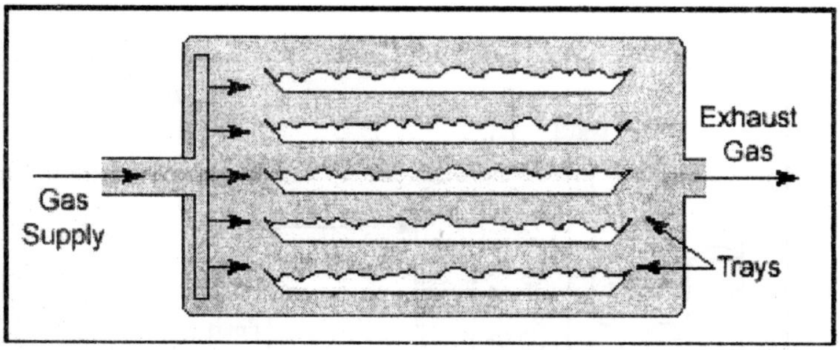

Figure 1. A tray bioreactor is loaded with the culture medium and organisms, then airflow is started to initiate the reaction.

The disadvantages include:

- Lower productivity levels due to time for filling, heating, sterilizing, cooling, emptying and cleaning the reactor.
- Increased focus on instrumentation due to frequent sterilization.
- Greater expense incurred in preparing several subcultures for inoculation.
- Higher costs for labor and/or process control for this non-stationary process.
- Larger industrial hygiene risks due to potential contact with pathogenic microorganisms or toxins.

Common applications for batch bioreactors include:

- Products that must be produced with minimal risk of contamination or organism mutation.
- Operations in which only small amounts of product are produced.
- Processes using one reactor to make various products.
- Processes in which batch or semicontinuous product separation is adequate.

Continuous Bioreactions

The defining characteristic of continuous bioreaction is a perpetual feeding process. A culture medium that is either sterile or comprised of microorganisms is continuously fed into the bioreactor to maintain the steady state. Of course, the product is also drawn continuously from the reactor. The reaction variables and control parameters remain consistent, establishing a time-constant state within the reactor. The result is continuous productivity and output.

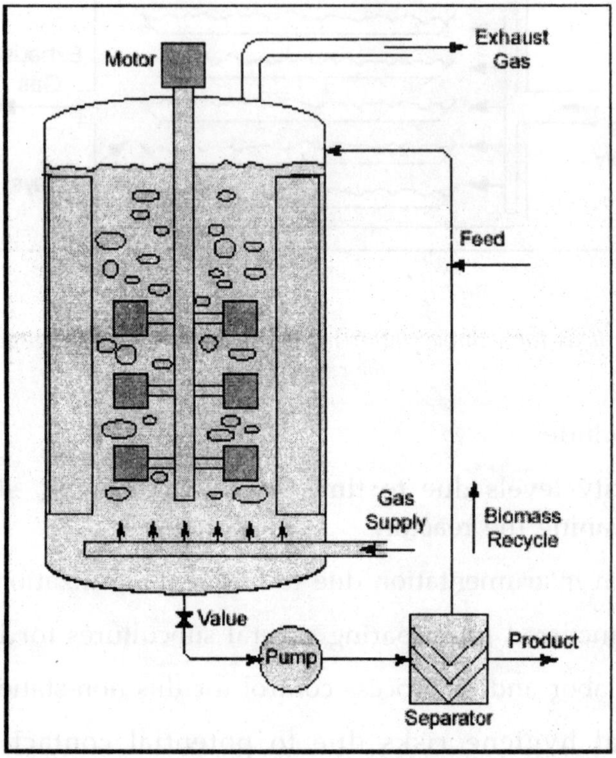

Figure 2. Stirred-tank bioreactor uses baffles and an agitator for optimal mixing, and recycles biomass.

These systems provide a number of advantages, including:

— Increased potential for automating the process.

— Reduced labor expense, due to automation.

— Less non-productive time expended in emptying, filling and sterilizing the reactor.

— Consistent product quality due to invariable operating parameters.

— Decreased toxicity risks to staff, due to automation.

— Reduced stress on instruments due to sterilization.

The disadvantages of continuous bioreactors include:

— Minimal flexibility, since only slight variations in the process are possible (throughput, medium composition, oxygen concentration and temperature).

- Mandatory uniformity of raw material quality is necessary to ensure that the process remains continuous.
- Higher investment costs in control and automation equipment, and increased expenses for continuous sterilization of the medium.
- Greater processing costs with continuous replenishment of non-soluble, solid substrates such as straw.
- Higher risk of contamination and cell mutation, due to the relatively brief cultivation period.

Continuous bioreaction is frequently used for processes with high-volume production; for processes using gas, liquid or soluble solid substrates; and for processes involving microorganisms with high mutation-stability. Typical end products include vinegar, baker's yeast and treated wastewater.

CONTINUOUS VS. BATCH

There are several major advantages to using continuous bioreactions as opposed to the batch mode. First, continuous reactions offer increased opportunities for system investigation and analysis. Because the variables remain unchanged, a benchmark can be determined for the process results, and then the effects of even minor changes to physical or chemical variables can be evaluated. Also, by changing the growth-limiting nutrient, changes in cell composition and metabolic activity can be tracked. The constancy of continuous bioreaction also provides a more accurate picture of kinetic constants, maintenance energy and true growth yields.

Secondly, continuous bioreaction provides a higher degree of control than does batch. Growth rates can be regulated and maintained for extended periods. By varying the dilution rate, biomass concentration can be controlled. Secondary metabolite production can be sustained simultaneously along with growth. In steady-state continuous bioreaction, mixed cultures can be maintained using chemostat cultures — unlike in batch bioreaction, where one organism usually outgrows another. Chemostats are continuous-flow stirred-tank bioreactors (CFSTRs) in an idealized steady-state, *i.e.*, the feed- and outlet-stream compositions and flows are constant, and perfect mixing occurs within the CFSTR vessel. In chemostats, the outlet stream composition is considered to be the same as within the bioreactor. Bioreactors operated as chemostats can be used to enhance the selectivity for thermophiles, osmotolerant strains, or mutant organisms with high growth rates. Also, the medium composition can be optimized for biomass and product formation, using a pulse-and-shift method that injects nutrients directly into the chemostat. As changes are observed, the nutrient is added to the medium supply reservoir and a new steady state is established.

A third advantage is the quality of the product. Because of the steady-state of continuous bioreaction, the results are not only more reliable, but also more easily reproducible. This process also results in higher productivity per unit volume, because time-consuming tasks, such as cleaning and sterilization, are unnecessary. The ability to automate the process also renders it less labor-intensive, and, therefore, more cost-efficient and less sensitive to the impact of human error. Along with the strengths of continuous bioreaction, there are inherent disadvantages that may make this process unsuitable for some types of bioreaction. For example, one challenge lies in controlling the production of some non-growth-related products. For this reason, the continuous process often requires feed-batch culturing, and a continuous nutrient supply. Wall growth and cell aggregation can also cause wash-out or prevent optimum steady-state growth.

Another problem is that the original product strain can be lost over time, if it is overtaken by a faster-growing one. The viscosity and heterogenous nature of the mixture can also make it difficult to maintain filamentous organisms. Long growth periods not only increase the risk of contamination, but also dictate that the bioreactor must be extremely reliable and consistent, incurring a potentially larger initial expenditure in higher-quality equipment.

SEMICONTINUOUS BIOREACTIONS

This hybrid of batch and continuous operations is found in many types of processes. One of the more frequently used is initiating the bioreaction in the batch mode, until the growth-limiting substrate has been consumed. Then, the substrate is fed to the reactor as specified (batch) or is maintained by an extended culture period (continuous). For secondary metabolite production, in which cell growth and product formation often occur in separate phases, the substrate is typically added at a specified rate. Like batch reactors, semicontinuous reactors are non-stationary. These systems provide a number of advantages, including:

- Higher yield, resulting from a well-defined cultivation period during which no cells are added or removed.
- Increased opportunity for optimizing environmental conditions of the microorganisms in regard to the phase of growth or production and age of the culture.
- Nearly stationary operation, important with slightly mutating microorganisms and those at risk for contamination.

The disadvantages include:

— Lower productivity levels due to time-consuming procedures for filling, heating, sterilizing, cooling, emptying and cleaning the reactor.

— Greater expenses in labor and/or dynamic process control for the process.

Semicontinuous bioreactors are typically used when continuous methods are not feasible, for example, those in which slight mutation or contamination of the microorganism occurs. Such bioreactors are also used when batch methods do not offer the desired productivity levels.

SUBMERGED BIOREACTORS

The most common type of aerobic bioreactor in use today is the stirred-tank reactor, which may feature a specific internal configuration designed to provide a specific circulation pattern. Ideal for industrial applications, this unit offers manufacturers both low capital and operating costs. For laboratory experiments with smaller volumes, the mixing vessel is typically made of glass. Stainless steel tank construction is the standard for industrial applications involving larger volumes. The height-to-diameter ratio of the vessel can vary, depending on heat removal requirements. The operating principles of the stirred-tank bioreactor are relatively simple.

The sterile medium and inoculum are introduced into a sterilized tank, and the air supply typically enters at the bottom. For optimal mixing, the tank features not only an agitator system but also baffles, which help prevent a whirlpool effect that could impede proper mixing. In the early stages of the process, warm water may be circulated through the baffles to heat up the system; later, cool water may be circulated inside of them to keep the process from overheating. The number of baffles typically ranges from four to eight.

As the bioreaction progresses, the bubbles produced by the air supply are broken up by the agitator as they travel upward. Many types of agitators are currently used, with the most common one being the four-bladed disk turbine. Newer designs featuring 12 or 18 blades, or concave ones, have also been shown to improve the hydrodynamics. At the top of the tank, exhaust gas is discharged and the product flows back down, where it is drained from the tank. In a continuous flow stirred-tank reactor, the substrate is continuously fed into the system and the product is continually drawn out and separated, with the producing organism recycled back into the tank for reuse. As with conventional chemical reactors, bioreactors can be placed in series or parallel with controlled recycle streams.

AIRLIFT REACTOR SYSTEMS

Also known as a tower reactor, an airlift bioreactor can be described as a bubble column containing a draught tube. Many types of airlift bioreactors are currently in use today. Air is typically fed through a sparger ring into the bottom of a central draught tube that controls the circulation of air and the medium (Figure 3). Air flows up the tube, forming bubbles, and exhaust gas disengages at the top of the column. The degassed liquid then flows downward and the product is drained from the tank. The tube can be designed to serve as an internal heat exchanger, or a heat exchanger can be added to an internal circulation loop.

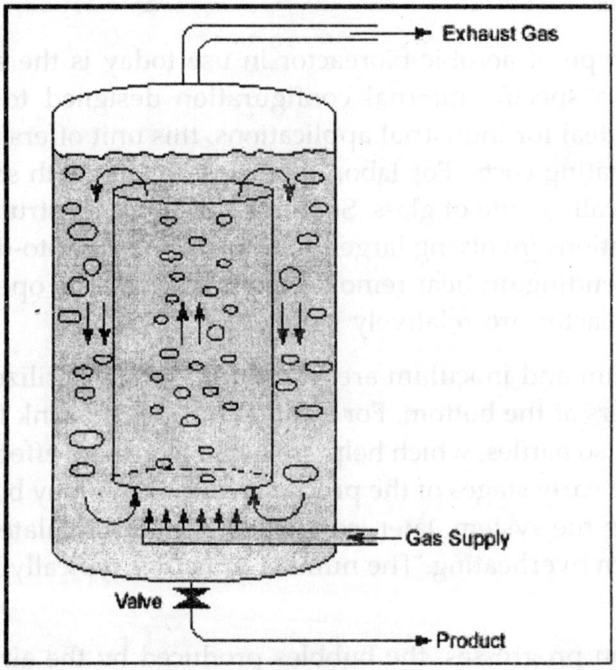

Figure 3. The simple design of a concentric draught tube bioreactor with annular liquid downflow results in less maintenance.

Airlift systems provide some advantages vs. more conventional bioreactors, such as the standard fermenter:

— Simple design with no moving parts or agitator shaft seals, for less maintenance, less risk of defects and easier sterilization.

— Lower shear rate, for greater flexibility — the system can be used for growing both plant and animal cells.

- Efficient gas-phase disengagement.
- Large, specific interfacial contact-area with lowenergy input.
- Well-controlled flow and efficient mixing.
- Well-defined residence time for all phases.
- Increased mass-transfer due to enhanced oxygen solubility achieved in large tanks with greater pressures.
- Large-volume tanks possible, increasing the output.
- Greater heat-removal vs. conventional stirred tanks.

The main disadvantages are:

- Higher initial capital investments due to largescale processes.
- Greater air throughput and higher pressures needed, particularly for large-scale operation.
- Low friction with an optimal hydraulic diameter for the riser and downcomer.
- Lower efficiency of gas compression.
- Inherently impossible to maintain consistent levels of substrate, nutrients and oxygen with the organisms circulating through the bioreactor and conditions changing.
- Inefficient gas/liquid separation when foaming occurs.

However, these disadvantages can and must be minimized in designing airlift systems. For example, if only one location serves as the feed source, the organism would experience continuous cycles of high growth, followed by starvation, resulting in the production of undesirable byproducts, low yields and high mortality rates. A design with multiple feed points eliminates this risk, especially for large-scale operations. The same risks are inherent in a single entry point for oxygen, which must be delivered at various places within the vessel, with the majority of the air entering at the bottom to circulate the fluid through the reactor.

AIRLIFT EXTERNAL-LOOP REACTORS

Another type of airlift system is the airlift external-loop reactor system (AELR), used primarily for batch operation. Figure 4 shows a cut-away view, that is, the vessel and downcomer are actually taller than shown for the particular diameter drawn. A variation of the airlift system, the AELR uses induced circulation to direct air and liquid throughout the vessel. This system consists of a riser and an external downcomer, which are connected at the bottom and the top, respectively. As the injected air at the bottom of the riser creates gas bubbles that begin to rise through the main tank, exhaust gas disengages at the top and the resulting heavier solution descends through the downcomer.

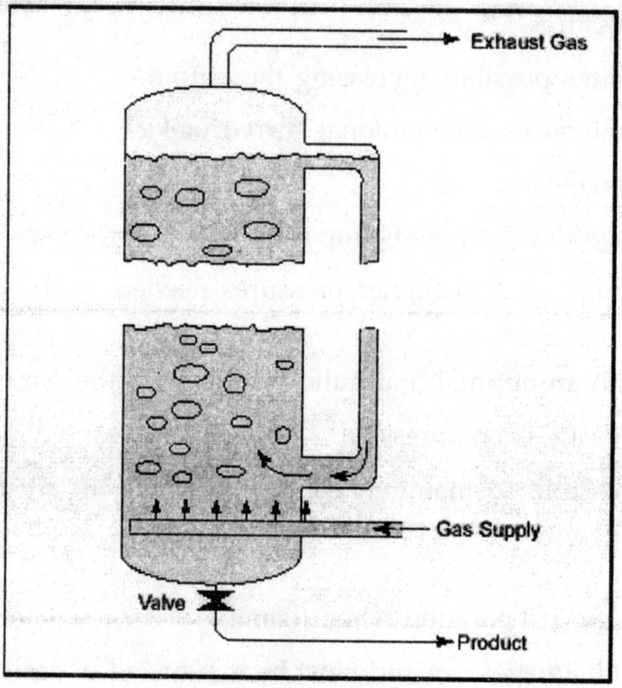

Figure 4. An airlift external-loop reactor has induced circulation that directs air/liquid throughout vessel

The AELR has some advantages over standard airlifts:

- Effective heat-transfer and efficient temperature control.
- Low friction with an optimal hydraulic diameter for both the riser and downcomer.
- Well-defined residence time in the individual section of the AELR.
- Increased opportunity for measurement and control in the riser and the downcomer.
- Independent control of the gas input-rate and liquid velocity by a throttling device between riser and downcomer.

ANAEROBIC BIOREACTIONS

Anaerobic bioreactions are used in applications such as ethanol production, winemaking, beer brewing and wastewater treatment. Technologies are well established in wine, ethanol and beer production, with improvements resulting from product improvements and manufacturing cost reductions. Continuous bioreactors for beer have been commercialized, however, batch fermenters continue to receive capital investments. Waste treatment is a field largely considered to have matured with fully functional technology in place, therefore, not as much attention has been paid to developing newer wastewater treatment processes.

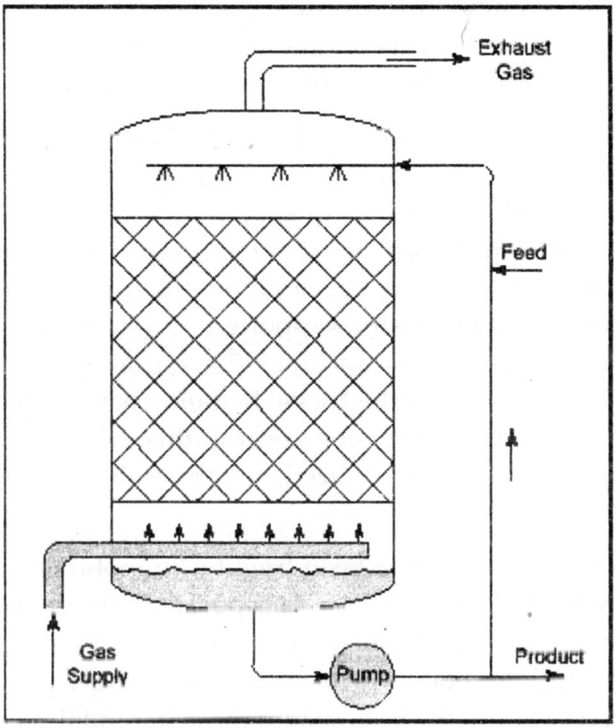

Figure 5. A trickle-bed employs adhered, immobilized enzymes to accomplish a reaction.

TECHNOLOGIES UNDER DEVELOPMENT

A number of new processes are being developed. One involves the use of isolated enzymes rather than whole cells to carry out a chemical change. The advantage is that this process does not require catering to the special requirements of living cells. However, enzymes, too, can undergo changes and, therefore, require determining the optimal conditions to express their catalytic activity. An additional problem is that using isolated enzymes is frequently an expensive undertaking for a single use application.

Consequently, long reaction times may be necessary if cost factors necessitate that expensive enzymes must be used only in low concentrations. There are other disadvantages to their use, too, such as the need to remove the enzyme from the product once the desired bioreaction has taken place. Immobilized enzyme technology is now successfully solving some of these difficulties. With the enzyme immobilized in a bed or tube, the solution of substrate for conversion is then passed through for conversion to product. The product is continuously collected as effluent from the bioreactor. The design and operation of an immobilized system is similar to that of processes employing heterogeneous catalysis. Heterogeneous systems enable product recovery at lower separation costs than do corresponding homogeneous systems.

Gas-liquid-solid contacting bioreactors have been investigated with a number of immobilized enzyme systems. Enzyme immobilization can take a variety of different forms and it has been studied on a range of supports. The method used in a particular application depends on the characteristics of the enzyme, its system, the substrate and the bioreactor fluid. Enzymes may be supported on a mesh-type or conventional mass-transfer structure, encapsulated in a film, supported by gel or silica-derived systems, on macroporous ion-exchange resins, or on other polymeric supports. One system that currently employs this technology is the trickle bed bioreactor. Not unlike certain types of biofilters traditionally used for emission control, this system features a screen onto which the enzyme is adhered and immobilized, and through which the substrate solution is passed for conversion.

Membranes and hollow fibers have been tried for immobilized bioreaction systems. An example is hollow fibers with enzymes incorporated into their walls. The diffusion of the substrate through the tube wall allows contact with the gelled enzyme and conversion into product. Subsequent diffusion of the product provides the separation necessary for its recovery. Under the influence of the differential pressure across the tube wall, the product flows through to the inside of the tube, eventually to be collected at a multitube header. Ethanol production has been achieved in the laboratory with immobilized enzymes in fixed-bed and membrane reactors. Enzymes have already been successfully tested for sugar conversion to ethanol.

Glucose isomerization to fructose is a well-established, high-volume commercial process for the production of high-fructose corn syrup (HFCS). Technologies using immobilized glucose isomerase in fixed-bed and fluidized bioreactor continue to be developed.

BIOREACTOR DESIGN

Bioreactor design is a relatively complex engineering task. Under optimum conditions, the microorganisms or cells are able to perform their desired function with 100 percent rate of success. The bioreactor's environmental conditions like gas (i.e., air, oxygen, nitrogen, carbon dioxide) flow rates, temperature, pH and dissolved oxygen levels, and agitation speed/circulation rate need to be closely monitored and controlled.

Most industrial bioreactor manufacturers use vessels, sensors and a control system networked together.

Fouling can harm the overall sterility and efficiency of the bioreactor, especially the heat exchangers. To avoid it, the bioreactor must be easily cleaned and as smooth as possible (therefore the round shape).

A heat exchanger is needed to maintain the bioprocess at a constant temperature. Biological fermentation is a major source of heat, therefore in most cases bioreactors need refrigeration. They can be refrigerated with an external jacket or, for very large vessels, with internal coils.

In an aerobic process, optimal oxygen transfer is perhaps the most difficult task to accomplish. Oxygen is poorly soluble in water—even less in fermentation broths—and is relatively scarce in air (20.8%). Oxygen transfer is usually helped by agitation, which is also needed to mix nutrients and to keep the fermentation homogeneous. There are, however, limits to the speed of agitation, due both to high power consumption (which is proportional to the cube of the speed of the electric motor) and to the damage to organisms caused by excessive tip speed bio.

PHOTOBIOREACTOR

A photobioreactor (PBR) is a bioreactor which incorporates some type of light source. Virtually any translucent container could be called a PBR, however the term is more commonly used to define a closed system, as opposed to an open tank or pond. Photobioreactors are used to grow phototroph small organisms like cyanobacteria, algae, , or moss plants. These organisms use light through photosynthesis as their energy source and do not require sugars or lipids as energy source. Consequently, risk of contamination with other organisms like bacteria or fungi is lower in photobioreactors when compared to bioreactors for heterotroph organisms.

SEWAGE TREATMENT

Bioreactors are also designed to treat sewage and wastewater. In the most efficient of these systems there is a supply of free-flowing, chemically inert media that acts as a receptacle for the bacteria that breaks down the raw sewage. Examples of these bioreactors often have separate, sequential tanks and a mechanical separator or cyclone to speed the division of water and biosolids. Aerators supply oxygen to the sewage and media further accelerating breakdown. In the process, the liquids Biochemical Oxygen Demand BOD is reduced sufficiently to render the contaminated water fit for reuse. The biosolids can be collected for further processing or dried and used as fertilizer. An extremely simple version of a sewage bioreactor is a septic tank whereby the sewage is left in situ, with or without additional media to house bacteria. In this instance, the biosludge itself is the primary host (activated sludge) for the bacteria. Septic systems are best suited where there is sufficient landmass and the system is not subject to flooding or overly saturated ground and where time and efficiency is not of an essence.

In bioreactors where the goal is to grow cells or tissues for experimental or therapeutic purposes, the design is significantly different from industrial bioreactors. Many cells and tissues, especially mammalian ones, must have a surface or other structural support in order to grow, and agitated environments are often destructive to these cell types and tissues. Higher organisms also need more complex growth medium.

ALGAE BIOREACTOR

An algae bioreaction system that recycles CO_2 from power and manufacturing plant flue gases and converts it to an onsite, continuous supply of biofuel, such as biodiesel, ethanol or methane has been licensed to the Victor Smorgon Group (VSG) by GreenFuel Technologies of Cambridge, Massachusetts. VSG is to have exclusive license to distribute, install and operate GreenFuel's Emissions-to-Biofuels proprietary technology for growing, harvesting, and processing biomass and products derived from algae throughout for Australia and New Zealand.

The technology has been proven in two pilot programs. The second and larger unit was commissioned at a 1,060 MW combined cycle facility in 2005 in the southwest United States. The bioreactor productivities suggest annual yields of 5,000-10,000 gallons of biodiesel and a comparable amount of bioethanol per acre. With low construction, energy and operational costs, the process mitigates CO_2 emissions profitably and is able to produce algae growth rates consistently higher than those ever achieved before. The system does not impact the operations of the power plant, and is designed to be retrofitted to flue stacks with minimal impact on ongoing operations.

Functioning of Algae Bioreactor

CO_2-rich gas streams are introduced to the bioreactor, in which algae are suspended in a media with nutrients added to optimize the growth rate. A portion of the media is withdrawn continuously from the bioreactor and sent to dewatering to harvest the algae. The dewatering operation uses two stages of conventional processing. Primary dewatering increases the algae concentration by a factor of 10-30. Secondary dewatering further increases the algal solids concentration to yield a cake suitable for downstream processing. Water removed from the dewatering steps is returned to the bioreactor, with a small purge stream to prevent precipitation of salts. Make-up water is added to maintain the media volume. A blower pulls the flue gas through the bioreactor. Using an induced draft fan provides several operating advantages, including ensuring minimal disruption to power plant operations, simplifying retrofits to existing facilities.

The downstream unit operations for algal oil extraction and conversion of the dewatered algae into final fuel products, in contrast to the upstream unit operations, are conventional technologies currently practiced on a large scale, e.g. biodiesel is currently produced from vegetable oils via transesterification (several algae species have lipids, starch, and protein compositions similar to soy and canola beans). Consequently the same facilities can be adapted to produce biodiesel from algae and conventional agricultural feeds.

MEMBRANE BIOREACTOR

Membrane bioreactor (MBR) is the combination of a membrane process like microfiltration or ultrafiltration with a suspended growth bioreactor, and is now widely used for municipal and industrial wastewater treatment with plant sizes up to 80,000 population equivalent (i.e. 48 MLD).

When used with domestic wastewater, MBR processes could produce effluent of high quality enough to be discharged to coastal, surface or brackish waterways or to be reclaimed for urban irrigation. Other advantages of MBRs over conventional processes include small footprint, easy retrofit and upgrade of old wastewater treatment plants. Two MBR configurations exist: internal, where the membranes are immersed in and integral to the biological reactor; and external/sidestream, where membranes are a separate unit process requiring an intermediate pumping step.

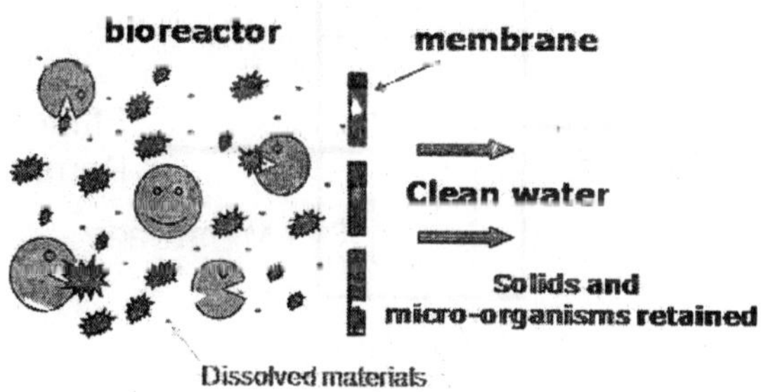

Figure 6. Simple schematic describing the MBR process

Recent technical innovation and significant membrane cost reduction have pushed MBRs to become an established process option to treat wastewaters. As a result, the MBR process has now become an attractive option for the treatment and reuse of industrial and municipal wastewaters, as evidenced by their constantly rising numbers and capacity. The current MBR market has been estimated to value around US$216 million in 2006 and to rise to US$363 million by 2010.

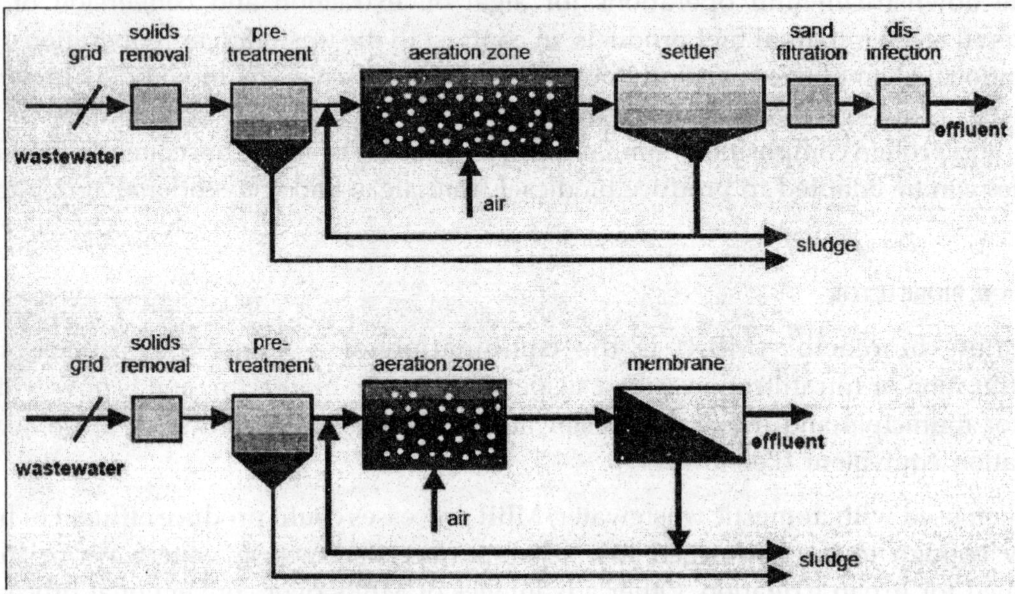

Figure 7. Schematic of conventional activated sludge process (top) and membrane bioreactor (bottom)

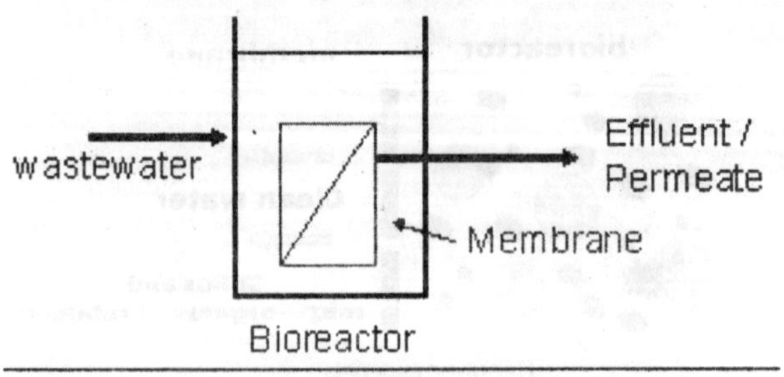

Figure 8. Schematic of a submerged MBR

MBR history and basic operating parameters

The MBR process was introduced by the late 1960s, as soon as commercial scale ultrafiltration (UF) and microfiltration (MF) membranes were available. The original process was introduced by Dorr-Olivier Inc. and combined the use of an activated sludge bioreactor with a crossflow membrane filtration loop. The flat sheet membranes used in this process were polymeric and featured pore sizes ranging from 0.003 to 0.01 μm. Although the idea of replacing the settling tank of the conventional activated sludge process was attractive, it was difficult to justify the use of such a process because of the high cost of membranes, low economic value of the product (tertiary effluent) and the potential rapid loss of performance due to membrane fouling. As a result, the focus was on the attainment of high fluxes, and it was therefore necessary to pump the mixed liquor suspended solids (MLSS) at high crossflow velocity at significant energy penalty (of the order 10 kWh/m^3 product) to reduce fouling. Due to the poor economics of the first generation MBRs, they only found applications in niche areas with special needs like isolated trailer parks or ski resorts for example.

The breakthrough for the MBR came in 1989 with the idea of Yamamoto and co-workers to submerge the membranes in the bioreactor. Until then, MBRs were designed with the separation device located external to the reactor (sidestream MBR) and relied on high transmembrane pressure (TMP) to maintain filtration. With the membrane directly immersed into the bioreactor, submerged MBR systems are usually preferred to sidestream configuration, especially for domestic wastewater treatment. The submerged configuration relies on coarse bubble aeration to produce mixing and limit fouling. The energy demand of the submerged system can be up to 2 orders of magnitude lower than that of the sidestream systems and submerged systems operate at a lower flux, demanding more membrane area. In submerged configurations, aeration is considered as one of the major parameter on process performances both hydraulic and biological. Aeration maintains solids in suspension, scours the membrane surface and provides oxygen to the biomass, leading to a better biodegradability and cell synthesis.

The other key steps in the recent MBR development were the acceptance of modest fluxes (25% or less of those in the first generation), and the idea to use two-phase bubbly flow to control fouling. The lower operating cost obtained with the submerged configuration along with the steady decrease in the membrane cost encouraged an exponential increase in MBR plant installations from the mid 90s. Since then, further improvements in the MBR design and operation have been introduced and incorporated into larger plants. While early MBRs were operated at solid retention times (SRT) as high as 100 days with mixed liquor suspended solids up to 30 g/L, the recent trend is to apply lower solid retention times (around 10-20 days), resulting in more manageable mixed liquor suspended solids (MLSS) levels (10-15 g/L). Thanks to these new operating conditions, the oxygen transfer and the pumping cost in the MBR have tended to decrease and overall maintenance has been simplified. There is now a range of MBR systems commercially available, most of which use submerged membranes although some external modules are available; these external systems also use two-phase flow for fouling control. Typical hydraulic retention times (HRT) range between 3 and 10 hours. In terms of membrane configurations, mainly hollow fibre and flat sheet membranes are applied for MBR applications.

Major Considerations in MBR

Fouling and fouling control

The MBR filtration performance inevitably decreases with filtration time. This is due to the deposition of soluble and particulate materials onto and into the membrane, attributed to the interactions between activated sludge components and the membrane. This major drawback and process limitation has been under investigation since the early MBRs, and remains one of the most challenging issues facing further MBR development.

Illustration of membrane fouling

In recent reviews covering membrane applications to bioreactors, it has been shown that, as with other membrane separation processes, membrane fouling is the most serious problem affecting system performance. Fouling leads to a significant increase in hydraulic resistance, manifested as permeate flux decline or transmembrane pressure (TMP) increase when the process is operated under constant-TMP or constant-flux conditions respectively. Frequent membrane cleaning and replacement is therefore required, increasing significantly the operating costs.Membrane fouling results from interaction between the membrane material and the components of the activated sludge liquor, which include biological flocs formed by a large range of living or dead microorganisms along with soluble and colloidal compounds. The suspended biomass has no fixed composition and varies both with feed water composition and MBR operating conditions employed. Thus though many investigations of membrane fouling have been published, the diverse range of operating conditions and feedwater matrices employed, the different analytical methods used and the limited information reported in most studies on the suspended biomass composition, has made it difficult to establish any generic behaviour pertaining to membrane fouling in MBRs specifically.

Factors influencing fouling (interactions in red)

The air-induced cross flow obtained in submerged MBR can efficiently remove or at least reduce the fouling layer on the membrane surface. A recent review reports the latest findings on applications of aeration in submerged membrane configuration and describes the enhancement of performances offered by gas bubbling. As an optimal air flow-rate has been identified behind which further increases in aeration have no effect on fouling removal, the choice of aeration rate is a key parameter in MBR design. Many other anti-fouling strategies can be applied to MBR applications. They comprise, for example:

- Intermittent permeation, where the filtration is stopped at regular time interval for a couple of minutes before being resumed. Particles deposited on the membrane surface tend to diffuse back to the reactor; this phenomena being increased by the continuous aeration applied during this resting period.
- Membrane backwashing, where permeate water is pumped back to the membrane, and flow through the pores to the feed channel, dislodging internal and external foulants.
- Air backwashing, where pressurized air in the permeate side of the membrane build up and release a significant pressure within a very short period of time. Membrane modules therefore need to be in a pressurised vessel coupled to a vent system.
- Proprietary anti-fouling products, such as Nalco's Membrane Performance Enhancer Technology.

In addition, different types/intensities of chemical cleaning may also be recommended:

- Chemically enhanced backwash (daily);
- Maintenance cleaning with higher chemical concentration (weekly);
- Intensive chemical cleaning (once or twice a year).

Intensive cleaning is also carried out when further filtration cannot be sustained because of an elevated transmembrane pressure (TMP). Each of the four main MBR suppliers (Kubota, Memcor, Mitsubishi and Zenon) have their own chemical cleaning recipes, which differ mainly in terms of concentration and methods. Under normal conditions, the prevalent cleaning agents remain NaOCl (Sodium Hypochlorite) and citric acid. It is common for MBR suppliers to adapt specific protocols for chemical cleanings for individual facilities. Intensive chemical cleaning protocols for four MBR suppliers.

Biological performances/kinetics

COD removal and sludge yield

Simply due to the high number of microorganism in MBRs, the pollutants uptake rate can be increased. This leads to better degradation in a given time span or to smaller required reactor volumes. In comparison to the conventional activated sludge process which typically achieves 95%, COD removal can be increased to 96-99% in MBRs. COD and BOD5 removal are found to increase with MLSS concentration. Above 15g/L COD removal becomes almost independent of biomass concentration at >96%. Arbitrary high MLSS concentrations are not employed, however, as oxygen transfer is impeded due to higher and Non-Newtonian fluid viscosity. Kinetics may also differ due to easier substrate access. In ASP, flocs may reach several 100 μm in size. This means that the substrate can reach the active sites only by diffusion which causes an additional resistance and limits the overall reaction rate (diffusion controlled). Hydrodynamic stress in MBRs reduces floc size (to 3.5 μm in sidestream MBRs) and thereby increases the apparent reaction rate. Like in the conventional ASP, sludge yield is decreased at higher SRT or biomass concentration. Little or no sludge is produced at sludge loading rates of 0.01 kgCOD/(kgMLSS d). Due to the biomass concentration limit imposed, such low loading rates would result in enormous tank sizes or long HRTs in conventional ASP.

Nutrient removal

Nutrient removal is one of the main concerns in modern wastewater treatment especially in areas that are sensitive to eutrophication. Like in the conventional ASP, currently, the most widely applied technology for N-removal from municipal wastewater is nitrification combined with denitrification. Besides phosphorus precipitation, enhanced biological phosphorus removal (EBPR) can be implemented which requires an additional anaerobic process step. Some characteristics of MBR technology render EBPR in combination with post-denitrification an attractive alternative that achieves very low nutrient effluent concentrations.

Anaerobic MBRs

Anaerobic MBRs were introduced in the 1980s in South Africa and currently see a renaissance in research. However, anaerobic processes are normally used when a low cost treatment is required that enables energy recovery but does not achieve advanced treatment (low carbon removal, no nutrients removal). In contrast, membrane-based technologies enable advanced treatment (disinfection), but at high energy cost. Therefore, the combination of both can only be economically viable if a compact process for energy recovery is desired, or when disinfection is required after anaerobic treatment (cases of water reuse with nutrients). If maximal energy recovery is desired, a single anaerobic process will be always superior to a combination with a membrane process.

Mixing/Hydrodynamics

Like in any other reactors, the hydrodynamics (or mixing) within an MBR plays an important role in determining the pollutant removal and fouling control within an MBR. It has a substantial effect on the energy usage and size requirements of an MBR, therefore the whole life cost of an MBR.

The removal of pollutants is greatly influenced by the length of time fluid elements spend in the MBR (i.e. the residence time distribution or RTD). The residence time distribution is a description of the hydrodynamics/mixing in the system and is determined by the design of the MBR (e.g. MBR size, inlet/recycle flowrates, wall/baffle/mixer/aerator positioning, mixing energy input). An example of the effect of mixing is that a continuous stirred-tank reactor will not have as high pollutant conversion per unit volume of reactor as a plug flow reactor.

The control of fouling, as previously mentioned, is primarily undertaken using coarse bubble aeration. The distribution of bubbles around the membranes, the shear at the membrane surface for cake removal and the size of the bubble are greatly influenced by the mixing/hydrodynamics of the system. The mixing within the system can also influence the production of possible foulants. For example, vessels not completely mixed (i.e. plug flow reactors) are more susceptible to the effects of shock loads which may cause cell lysis and release of soluble microbial products.

Example of computational fluid dynamic (CFD) modelling results (streamlines) for a full scale MBR.

Many factors affect the hydrodynamics of wastewater processes and hence MBRs. These range from physical properties (e.g. mixture rheology and gas/liquid/solid density etc) to the fluid boundary conditions (e.g. inlet/outlet/recycle flowrates, baffle/mixer position etc). However, many factors are peculiar to MBRs, these cover the filtration tank design (e.g membrane type, multiple outlets attributed to membranes, membrane packing density, membrane orientation etc) and it's operation (e.g. membrane relaxation, membrane back flush etc).

The mixing modelling and design techniques applied to MBRs are very similar to those used for conventional activated sludge systems. They include the relatively quick and easy compartmental modelling technique which will only derive the RTD of a process (e.g. the MBR) or the process unit (e.g. membrane filtration vessel) and relies on broad assumptions of the mixing properties of each sub unit. Computational fluid dynamics modelling (CFD) on the other hand does not rely on broad assumptions of the mixing characteristics and attempts to predict the hydrodynamics from a fundamental level. It is applicable to all scales of fluid flow and can reveal much information about the mixing in a process, ranging from the RTD to the shear profile on a membrane surface.

Investigations of MBR hydrodynamics have occurred at many different scales, ranging from examination of shear stress at the membrane surface to RTD analysis of the whole MBR. Cui et al investigated the movement of Taylor bubbles through tubular membranes, Prieske et al examined the entire membrane filtration vessel using CFD and velocity measurements, while Brannock et al examined the entire MBR using tracer study experiments and RTD analysis.

MBR suppliers

The design of the reactor (including membrane, baffle and aerator locations) and the mode of operation of the membrane also appear as key parameters in the optimisation of the system. Several immersed MBR designs are currently proposed by the leading membrane suppliers such as GE-Zenon (Canada), X-Flow (The Netherlands), Siemens-Australia (Australia), Mitsubishi and Kubota (Japan). Kubota has the largest installation base of membrane bioreactors worldwide. In each case, the process proposed is very specific. Not only the membrane material and configuration used are different, but the operating conditions, cleaning protocols and reactor designs also change from a company to another. For example, the flat sheet membrane provided by Kubota does not require backwash operation, while hollow fibre membrane type from Zenon and Memcor (USFilter) have been especially designed to hydraulically backwash the membrane on a given frequency (around every 20 min).

REFERENCES

Biotechnology Industry Organization. *Biotechnology in Perspective.* Washington, D.C.: Biotechnology Industry Organization. 1990.

Cox, R.P. and Thomsen, J.K., "Computer-aided identification of lactic acid bacteria using the API 50 CHL system", *Letters in Applied Microbiology,* 1990.

Dekker, J., and G. Comstock. "Ethical and Environmental Considerations in the Release of Herbicide Resistant Crops." *Agriculture and Human Values.* Summer 1992.

Edgington, Stephen M. "Environmental Biotechnology." *Bio/Technology.* December 1994. p. 1338-42.

Jarvis, B. and A.P. Williams, *Methods for detecting fungi in foods and beverages,* In: Beuchat, L.R. (ed.), Food and beverage mycology. Second edition, Avi Publishing Co., Westport, Connecticut, 1987.

Margaret G. Mellon. *Biotechnology and the Environment : A Primer on the Environmental Implications of Genetic Engineering.*

8

Biofilm Technology

Biofilms are a collection of microorganisms surrounded by the slime they secrete, attached to either an inert or living surface. You are already familiar with some biofilms: the plaque on your teeth, the slippery slime on river stones, and the gel-like film on the inside of a vase which held flowers for a week. Biofilm exists wherever surfaces contact water. More than 99 percent of all bacteria live in biofilm communities. Some are beneficial. Sewage treatment plants, for instance, rely on biofilms to remove contaminants from water. But biofilms can also cause problems by corroding pipes, clogging water filters, causing rejection of medical implants, and harboring bacteria that contaminate drinking water

Microbiologists have traditionally focused on free-floating bacteria growing in laboratory cultures; yet they have recently come to realize that in the natural world most bacteria aggregate as biofilms, a form in which they behave very differently. As a result, biofilms are now one of the hottest topics in microbiology. As in any water system, 99 percent of the bacteria in an automated watering system is likely to be in biofilms attached to internal surfaces. Biofilms are the source of much of the free-floating bacteria in drinking water, some of which can cause infection and disease in laboratory animals. One common biofilm bacteria, Pseudomonas aeruginosa, is a secondary pathogen which can infect animals with suppressed immune systems. Besides being a reservoir of bacteria which can affect animal health, biofilms can also cause corrosion in stainless steel piping systems. In order to design and operate automated watering systems that deliver the bacterial quality required by our customers, we should understand how biofilms develop, some of the problems they can cause, and how they can be controlled.

Understanding bacteria in biofilms is one step in preparing for the future. We are currently meeting the most demanding microbiological water quality requirements of many of our customers by supplying chlorinated reverse osmosis water and by maintaining water quality through flushing and sanitization. But, what if chlorine use in animal drinking water is prohibited? Or, what if water quality requirements become even more stringent with the use of new specialized animals? Of course, you might just want to learn about biofilms to marvel at the ability of bacteria to adapt to their environment and to evade our attempts to eliminate them.

STEPS IN BIOFILM DEVELOPMENT

The instant a clean pipe is filled with water, a biofilm begins to form. The development of the biofilm occurs in the following steps:

Step 1. Surface Conditioning

The first substances associated with the surface are not bacteria but trace organics. Almost immediately after the clean pipe surface comes into contact with water, an organic layer deposits on the water/solid interface. These organics are said to form a "conditioning layer" which neutralizes excessive surface charge and surface free energy which may prevent a bacteria cell from approaching near enough to initiate attachment. In addition, the adsorbed organic molecules often serve as a nutrient source for bacteria

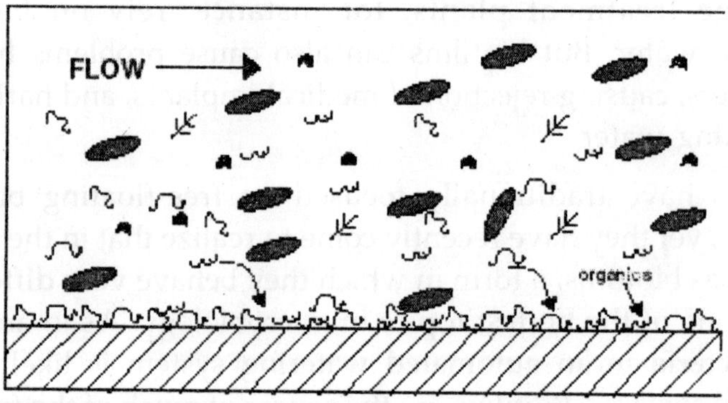

Figure 1. Adsorption of organic molecules on a clean surface forms a conditioning film

Step 2.Adhesion of 'Pioneer' Bacteria

In a pipe of flowing water, some of the planktonic (free-floating) bacteria will approach the pipe wall and become entrained within the boundary layer, the quiescent zone at the pipe wall where flow velocity falls to zero. Some of these cells will strike and adsorb to the surface for some finite time, and then desorb. This is called reversible adsorption. This initial attachment is based on electrostatic attraction and physical forces, not any chemical attachments. Some of the reversibly adsorbed cells begin to make preparations for a lengthy stay by forming structures which may permanently adhere the cell to the surface. These cells become irreversibly adsorbed.

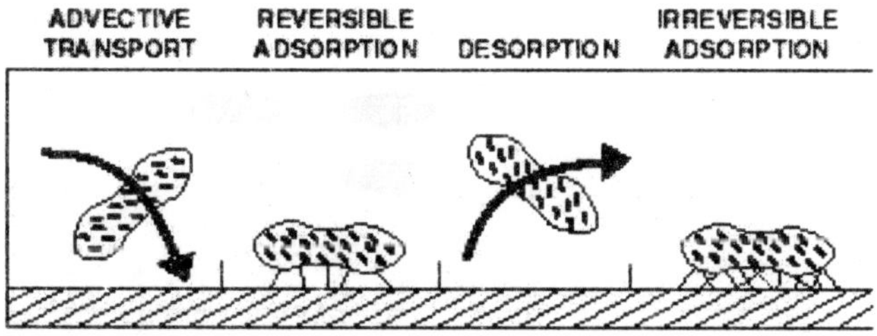

Figure 2. Transport of bacteria cells to the conditioned surface, adsorption, desorption, and irreversible adsorption.

Step 3. Glycocalyx or 'Slime' Formation

Biofilm bacteria excrete extracellular polymeric substances, or sticky polymers, which hold the biofilm together and cement it to the pipe wall. In addition, these polymer strands trap scarce nutrients and protect bacteria from biocides. According to Mittelman, "Attachment is mediated by extracellular polymers that extend outward from the bacterial cell wall (much like the structure of a spider's web). This polymeric material, or glycocalyx, consists of charged and neutral polysaccharides groups that not only facilitate attachment but also act as an ion-exchange system for trapping and concentrating trace nutrients from the overlying water. The glycocalyx also acts as a protective coating for the attached cells which mitigates the effects of biocides and other toxic substances."

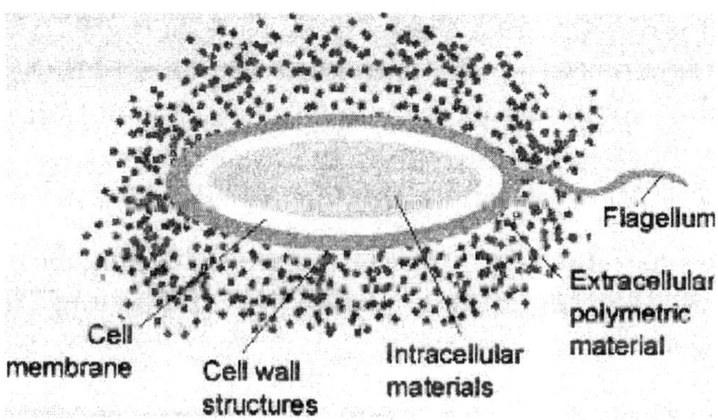

Figure 3. Wild bacteria are "hairy" cells with extracellular polymers which stick to surfaces

As nutrients accumulate, the pioneer cells proceed to reproduce. The daughter cells then produce their own glycocalyx, greatly increasing the volume of ion exchange surface. Pretty soon a thriving colony of bacteria is established.

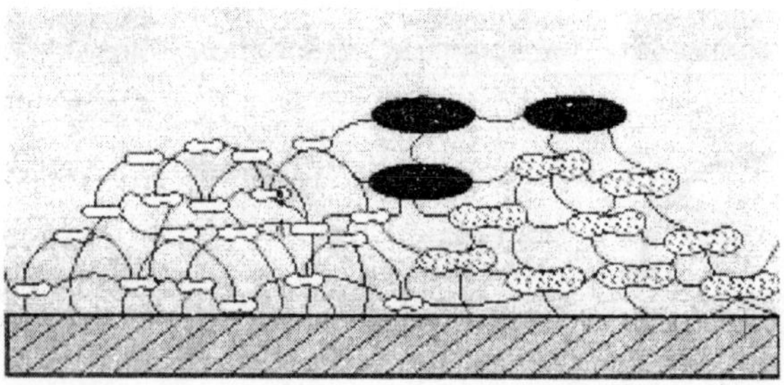

Figure 4. Biofilm is made up microbes and a "spiders web" of extracellular polymers

In a mature biofilm, more of the volume is occupied by the loosely organized glycocalyx matrix (75-95%) than by bacterial cells (5-25%). Because the glycocalyx matrix holds a lot of water, a biofilm-covered surface is gelatinous and slippery.

Step 4. Secondary Colonizers

As well as trapping nutrient molecules, the glycocalyx net also snares other types of microbial cells through physical restraint and electrostatic interaction. These secondary colonizers metabolize wastes from the primary colonizers as well as produce their own waste which other cells then use in turn. According to Borenstein, these "other bacteria and fungi become associated with the surface following colonization by the pioneering species over a matter of days."

Step 5. Fully Functioning Biofilm A Cooperative "Consortia" of Species

The mature, fully functioning biofilm is like a living tissue on the pipe surface. It is a complex, metabolically cooperative community made up of different species each living in a customized microniche. Biofilms are even considered to have primitive circulatory systems. Mature biofilms are imaginatively described in the article "Slime City":

"Different species live cheek-by-jowl in slime cities, helping each other to exploit food supplies and to resist antibiotics through neighborly interactions. Toxic waste produced by one species might be hungrily devoured by its neighbor. And by pooling their biochemical resources to build a communal slime city, several species of bacteria, each armed with different enzymes, can break down food supplies that no single species could digest alone." "The biofilms are permeated at all levels by a network of channels through which water, bacterial garbage, nutrients, enzymes, metabolites and oxygen travel to and fro. Gradients of chemicals and ions between microzones provide the power to shunt the substances around the biofilm."

BIOFILMS GROWTH

A biofilm can spread at its own rate by ordinary cell division and it will also periodically release new 'pioneer' cells to colonize downstream sections of piping. As the film grows to a thickness that allows it to extend through the boundary layer into zones of greater velocity and more turbulent flow, some cells will be sloughed off. According to Mayette, "These later pioneer cells have a somewhat easier time of it than their upstream predecessors since the parent film will release wastes into the stream which may serve as either the initial organic coating for uncolonized pipe sections down stream or as nutrient substances for other cell types."

According to Mittelman, the development of a mature biofilm may take several hours to several weeks, depending on the system. Pseudomonas aeruginosa is a common 'pioneer' bacteria and is used in a lot of biofilm research. In one experiment, researchers found that Pseudomonas cells adhere to stainless steel, even to electropolished surfaces, within 30 seconds of exposure.

The association of bacteria with a surface and the development of a biofilm can be viewed as a survival mechanism. Bacteria benefit by acquiring nutrients and protection from biocides.

REASONS FOR DEVELOPMENT

Potable water, especially high-purity water systems, are nutrient-limited environments, but even nutrient concentrations too low to measure are sufficient to permit microbial growth and reproduction. Bacteria and other organisms capable of growth in nutrient-limited environments are called oligotrophs. Bacteria have evolved the means to find and attach to surfaces in order to increase the chances of encountering nutrients. What advantages are offered by adhesion to surfaces and development of biofilm?

- Trace organics will concentrate on surfaces.
- Extracellular polymers will further concentrate trace nutrients from the bulk water.
- Secondary colonizers utilize the waste products from their neighbors.
- By pooling their biochemical resources, several species of bacteria, each armed with different enzymes, can break down food supplies that no single species could digest alone.

Means to Attach to Surfaces

Motility and Chemotaxis

Motile bacteria can swim along a chemical concentration gradient towards a higher concentration of a nutrient. The movement of organisms in response to a chemical (nutrient) gradient is called chemotaxis. Pseudomonas aeruginosa is one of the motile bacteria which uses a flagellum to move toward higher nutrient concentrations at the pipe wall. In a study on the attachment of Pseudomonas to stainless steel surfaces, researchers put cells in a blender to remove the flagella. They found that the rate of cell attachment decreased at least 90% when flagella were removed.

Hydrophobic Cell Wall

Many organisms faced with the starvation conditions encountered in purified water systems respond by altering their cell wall structure to increase their affinity for surfaces. By altering the protein and lipid composition of the outer membrane, the charge and hydrophobicity can be changed. The cell wall becomes hydrophobic. "Such hydrophobic cells want nothing more than to find their way out of the water column. Once in the boundary layer (the dead zone at the piping wall where flow velocity falls to zero), they are attracted to the pipe surface."

Extracellular Polymer Production

Once at the surface, bacteria cells anchor themselves to the surface with their sticky polymers. Simple shear (flushing) is no longer adequate to remove these cells.

Protection from Disinfectants

Once the microorganisms have attached, they must be capable of withstanding normal disinfection processes. Biofilm bacteria display a resistance to biocides that may be considered stunning.

Biofilm associated bacteria may be 150-3000 times more resistant to free chlorine and 2-100 times more resistant to monochloramine than free-floating bacteria.

Another researcher's work suggests that Pseudomonas has a clever way of eluding its attackers: It secretes a sticky slime that builds up on the pipe interior. A germicide flushed through the water distribution system kills free-floating microbes, but it can't touch bacteria embedded in the slimy biofilm.

When bacteria are in a film, they are very resistant to biocides. *"In fact, they often produce more exopolymers after biocide treatment to protect themselves"*.

Protection from Disinfectants

Protective shield

In order to destroy the cell responsible for forming the biofilm, the disinfectant must first react with the surrounding polysaccharide network. The cells themselves are not actually more resistant, rather they have surrounded themselves with a protective shield. The disinfectant's oxidizing power is used up before it can reach the cell.

Diffusion limitations

When cells are attached to a pipe wall, delivery of the disinfectant is limited by the rate of diffusion of the compound across the boundary layer and through the film. It takes a higher concentration over a longer contact time for the disinfectant to reach the bacteria cells in a biofilm compared to free-floating organisms

- Trace organics will concentrate on surfaces.
- Extracellular polymers will further concentrate trace nutrients from the bulk water.
- Secondary colonizers utilize the waste products from their neighbors.

By pooling their biochemical resources, several species of bacteria, each armed with different enzymes, can break down food supplies that no single species could digest alone.

NEW DISCOVERIES

Recent research from the Center for Biofilm Engineering has dispelled some earlier assumptions about bacteria and biofilms.

Biofilm Structure

In the past, microbiologists assumed that biofilms contained disorderly clumps of bacteria located in no particular structure or pattern. New techniques to magnify biofilms without destroying the gel-like structures have enabled researchers to discover the complex structure of biofilms as if viewing a city from a satellite. This structure is described in the recent article "Slime City":

> "In most cases, the base of the biofilm is a bed of dense, opaque slime 5 to 10 micrometers (197-394 microinch) thick. It is a sticky mix of polysacharides, other polymeric substances and water, all produced by the bacteria. Soaring 100 to 200 micrometers (3940 7870 microinch) upwards are colonies of bacteria shaped like mushrooms or cones. Above street level comes more slime, this time of a more watery makeup and variable consistency with a network of channels through which water, bacterial garbage, nutrients, enzymes, metabolites and oxygen travel."

Some microcolonies are simple conical structures, while others are mushroom shaped. Water currents flow in channels between the colonies carrying nutrients and waste.

Biochemistry of Biofilm Bacteria

Past researchers assumed that biofilm bacteria behaved much like solitary, free-floating microorganisms. Now, they are discovering that while it's true that biofilm bacteria have exactly the same genetic makeup as their free-roving cousins, their biochemistry is very different because they switch to using a different set of genes. For example, the Center for Biofilm Engineering has studied how Pseudomonas aeruginosa forms biofilms. The instant the bacteria dock to glass, they switch on certain genes involved in the synthesis of alginate (an unusually sticky form of slime), switching them off again once the bacteria are engulfed in alginate. Researchers now estimate that as many as 30 to 40 percent of the proteins present in bacterial cell walls differ between sessile and planktonic bacteria (called 'city dwellers' and 'free-rovers'). Some of the targets for antibiotics are not there any more, so bacteria become difficult to kill. This is primarily a problem with biofilms inside humans and animals.

Chemical Signals

Researchers are studying the chemicals (called sigma factors) which signal bacteria to change their biochemistry to life in a biofilm. If they can discover a *"reverse sigma factor"* which would change biofilm bacteria into planktonic free-floaters, it might be possible to dissolve biofilms by "sending *the equivalent of an evacuation signal.*"

Implications for sanitization

Traditional disinfectant testing has been done with single-species free-floating laboratory cultures. The CT constant for a disinfectant is the product of (concentration) x (time) required to kill a particular bacteria. However, CT values shouldn't be extrapolated to bacteria in biofilms. What does this mean for automated drinking water systems? For one thing, it explains how bacteria counts can be measured even when the water contains low levels of chlorine. Typical chlorine levels in tap water are between 0.5-2.0 ppm. This amount of chlorine has been shown to kill free-floating bacteria, but may not be enough to kill biofilm bacteria. Chunks of sloughed off biofilm can contain viable bacteria which show up in plate counts. This is a particular problem with Pseudomonas which is a great slime producer and so is more chlorine resistant. One animal facility determined through their own testing that they need approximately 3 ppm chlorine in RO water to achieve low Pseudomonas counts.

GROWTH FACTORS

Factors that effect biofilm attachment and growth:

Surface Material

The material of the surface has little or no effect on biofilm development. Stainless steel is just as susceptible as plastic pipe.

Surface area

One major factor influencing biofilm development in purified-water systems is surface area. Industrial water systems, unlike most natural environments (lakes and rivers), offer a tremendous amount of surface area for attachment. RO membranes, DI resins, storage tanks, cartridge filters, and piping systems all provide surfaces suitable for bacterial attachment and growth.

Surface smoothness

Although smoother surfaces delay the initial buildup of attached bacteria, smoothness does not appear to significantly effect the total amount of biofilm that will attach to a surface. According to Meltzer, "no surfaces have been found that are exempt from biofouling. Surface structure does appear to influence the rate of fouling, but only initially over the first few hours of exposure. In general, smooth surfaces foul at a slower initial rate than do rough ones, but biofilm formation after a period of days is inevitable." This conclusion is based on research on Pseudomonas attachment to stainless steel.

Flow Velocity

High water flow rates may alter biofilm growth but will not prevent the attachment of bacteria to pipe surfaces. This conclusion is supported by Mittelman, Patterson and Meltzer. High flow will not prevent bacteria attachment or remove existing biofilm for the following reasons.

Low flow in the boundary layer

Regardless of the water velocity, it flows slowest in the layers adjacent to pipe surfaces. Even when water flow in the center of the pipe is turbulent, the flow velocity falls to zero at the pipe wall. The distance out from the pipe wall in which the flow rate is not turbulent is called the boundary layer or laminar sublayer. The thickness of the laminar sublayer was calculated by Pittner for various flow velocities and for 5 size pipes. Pittner calculated that the shear forces within the laminar sublayer are much less than that required to dislodge a bacteria cell.

Strong adhesion by exopolymers.In water systems with continuous high-velocity flow, the bacteria that accumulate in biofilm tend to be filamentous varieties (like Pseudomonas) especially suited for attachment by filaments. The bacteria anchor themselves to the surface with their 'sticky' exopolymers.

Although high flow velocity will not prevent the attachment of bacteria to pipe surfaces, it does have the following effects on the biofilm structure.

Table 1: Laminar Sublayer Thickness (microns)

Pipe Size	Velocity (ft/sec)					
	0.2	1.0	2.0	5.0	8.0	12.0
E.I. RDS						
0.428"ID	*	*	125	55	37	26
1/2" Sch.80	*	*	136	60	40	28
1" Sch.80	*	265	146	65	43	30
2" Sch.80	537	291	158	69	46	32
3" Sch.80	563	305	165	74	48	33
4" Sch.80	582	312	170	75	50	34

* *Flow may or may not be turbulent at these conditions. Current E.I. RDS flush velocity is approximately 2 ft/sec.*

- *Denser biofilm.* According to Mittelman , "at higher flow rates, a denser, somewhat more tenacious biofilm is formed. As a result, these surfaces often appear to be free from foulants, since they are not slimy to the touch."
- *Limited biofilm thickness.* The maximum thickness of the biofilm can be considered to be the thickness of the laminar flow layer. In a constant flow system, "an equilibrium thickness is reached which is dependent on water velocity and nutrients. Growth of the biofilm beyond the laminar layer will result in the release of planktonic 'pioneer' cells that will, conditions permitting, establish the biofilm in another section of pipe." In systems that have fluctuating water flow, such as automated watering systems with periodic flushing, bacteria will be sloughed off during the flush. This results in random 'particle showers' of bacteria which can explain day-to-day fluctuations seen in total bacteria count results.
- *Limited Nutrients.* Like other living creatures, bacteria require certain nutrients for growth and reproduction. Limiting these nutrients will limit bacteria growth, but "nutrient levels in high-purity systems are unequivocally sufficient to permit microbial growth and reproduction to a troublesome extent." Table 2 lists some sources of nutrients in purified water systems.

Substrate nutrition

Most plastics are not biodegradable, but pipe cements and plastisizers that leach from epoxy resins, PVC pipe and polyamide pipe can be organic carbon sources for bacteria. Cellulose-based RO membranes can also be a nutrient source. That is why we must chlorinate RO feedwater. Also, bacteria can obtain trace metal nutrients from stainless steel and other metal components.

Table 2: Nutrients for bacterial growth found in pure water systems

Nutrient	*Sources*
Organic Carbon	Humic and fulvic acids (source water)
	Pipe plasticizers and solvents
	Fiberglass-reinforced plastics (FRPs)
	Pump and gage lubricants
	Microbial byproducts
	Personnel
	Airborne dust
Nitrogen	Humic and fulvic acids (source water)
	Nitrates and nitrites (source water)
	Microbial byproducts
	Airborne dust
Phosphorus	Phosphates (source water)
	Microbial byproducts
	Airborne dust
Sulfur	Sulfates (source waters)
	Sulfuric acid (RO pretreatment)
	Membrane surfactants
	Airborne dust
Trace metals and salts	Source waters
	Process piping
	Fiberglass-reinforced plastics (FRPs)
	Stainless steel system components
	RO pretreatment chemicals
	Personnel
	Airborne dust

Purified Water

Under perfect growth conditions, a bacterial cell divides into two daughter cells once every 20 minutes. This means that a single cell and its descendants will grow exponentially to more than 2 million cells in eight hours or to 4 million pounds of bacteria in 24 hours! Of course, these growth rates are never actually realized (especially in clean drinking water) because they are limited by space and available nutrients.

Can bacteria be "starved to death" or at least inhibited in their growth by depriving them of organic nutrients and oxygen? Unfortunately, even minute amounts of organic matter will support many bacteria. This was explained in the following example by Pittner:

> "If only one part per billion of organic matter in a 1-milliliter water sample were converted to bacterial bodies (assuming the bacteria to be 20% organic matter and the specific gravity of bacteria to be about that of water), approximately 9,500 bacteria, each 1.0 micron in diameter, would be present in a 1-milliliter sample."
>
> or
>
> 1 ppb organic matter 9,500 bacteria/ml

Currently available technology cannot reduce nutrient levels completely, so total control of bacteria is not achievable by simply controlling nutrients. Similarly, "very small quantities of oxygen will adequately support luxurious bacterial growth even if the bacteria do not revert to anaerobic respiration, which most bacteria have the ability to do. For these reasons, a thriving bacterial population can exist even in high purity water systems."

Nutrient-limiting environments can actually promote the attachment of bacteria to surfaces because that is where the trace organics accumulate and extracellular polymers in a biofilm capture trace nutrients. Although we can't completely starve bacteria out, nutrient-poor reverse osmosis water will support less biofilm than regular tap water supplies.

SIZE, PROPORTIONS, AND WATERING SYSTEMS

So far, this chapter has described how a biofilm layer develops on the inside of water pipes and how this layer will reach a certain equilibrium thickness depending on flow velocity and nutrient levels. And it has discussed how surface smoothness isn't a significant factor effecting biofilm attachment. This section will show you how the size and proportion of an individual bacteria cell compares to surface roughness, biofilm thickness, and pipe diameter.

Surface finish and cell size: For many years, the finish provided on stainless steel surfaces was a dairy standard defined by Number or grit such as #4 or 150 grit. Grit finish is used with mechanical polishing and refers to the number of grit lines per inch of abrasive; the higher the number, the smoother to finish. Although the dairy and pharmaceutical industries still use grit coding for finishes, they are moving toward a system where the surface roughness can be more accurately measured.

Surface roughness can be measured by a profilometer, a stylus device used to trace across the surface profile. The results are expressed either as RA, which is the arithmetic average deviation from the center line of the surface, or as RMS, which is the root mean square of the deviations from the center line. RA or RMS values are given in either microns (same as micrometers or μm) or micro-inches (μ"). On Edstrom Industries' drawings, surface roughness is specified as RA in micro-inches. RMS will be approximately 11 percent higher than the RA number for a given surface.

Pharmaceutical Water Stainless Steel Piping Finish: According to Meltzer, "there is no universally accepted standard for surface finishing for stainless steel. Tubing in the US is usually of an interior 150 to 180-grit finish. Of four large pharmaceutical manufacturers in the Chicago area, two rely on 150-grit finishes, one on a 180-grit finish, and one on a 180-grit finish followed by electropolishing. Water-for-Injection is usually flowed through pipe finished to a 240- to 320-grit surface."

Smoothness of Plastic Pipe: According to Gillis, extruded polypropylene and PVDF pipes are as smooth as electropolished stainless steel. There are no irregularities of significant magnitudes in the size range of a bacterial cell.

Finish on Edstrom Industries Fittings and Valves: For machined fittings and drinking valves, the default finish specified in the title block of our drawings is 64 microinch RA. When a smoother finish is needed, for o-ring sealing surfaces for example, a smoother 32 or 16 microinch finish may be specified. If machined parts are electropolished, roughness should be reduced by 30-40%.

Finish on E.I. Stainless Steel RDS Tubing: The stainless steel tubing used in Edstrom room piping and manifolds has a welded seam, but it does not have a defined interior surface smoothness. It is a rolled finish which appears smooth but could have crevices formed by flattened metal during rolling. Assume it is no smoother than 180 grit finish.

Profile height: On most surfaces the total profile height of the surface roughness, or the peak-to-valley height will be approximately four times the RA value. Knowing the measured roughness, an approximate profile of the surface can be drawn (Figure 5).

Typical Size of Biofilm Bacteria Cells. A very common biofilm bacteria is Pseudomonas aeruginosa. Cells of Pseudomonas are rod shaped and approximately 0.3-0.8 microns wide by 1.0 1.2 microns long. This is equivalent to 12-31 microinches wide x 40-47 microinches long.

Comparing Surface Profile to the Size of Bacteria Cells

The roughness profile of various stainless steel finishes used in water systems is shown schematically in Figure 5. Notice that a 32 microinch" RA or 180-grit finish (which is considered sanitary for dairy, food, and pharmaceutical uses) has scratches large enough to harbor bacteria. A 12 microinch" RA (320-grit) finish, which is typical of Water-for-Injection applications, has scratches only as deep as approximately one bacteria cell. A 320 grit surface followed by electropolishing has only minor surface variations relative to cell size.

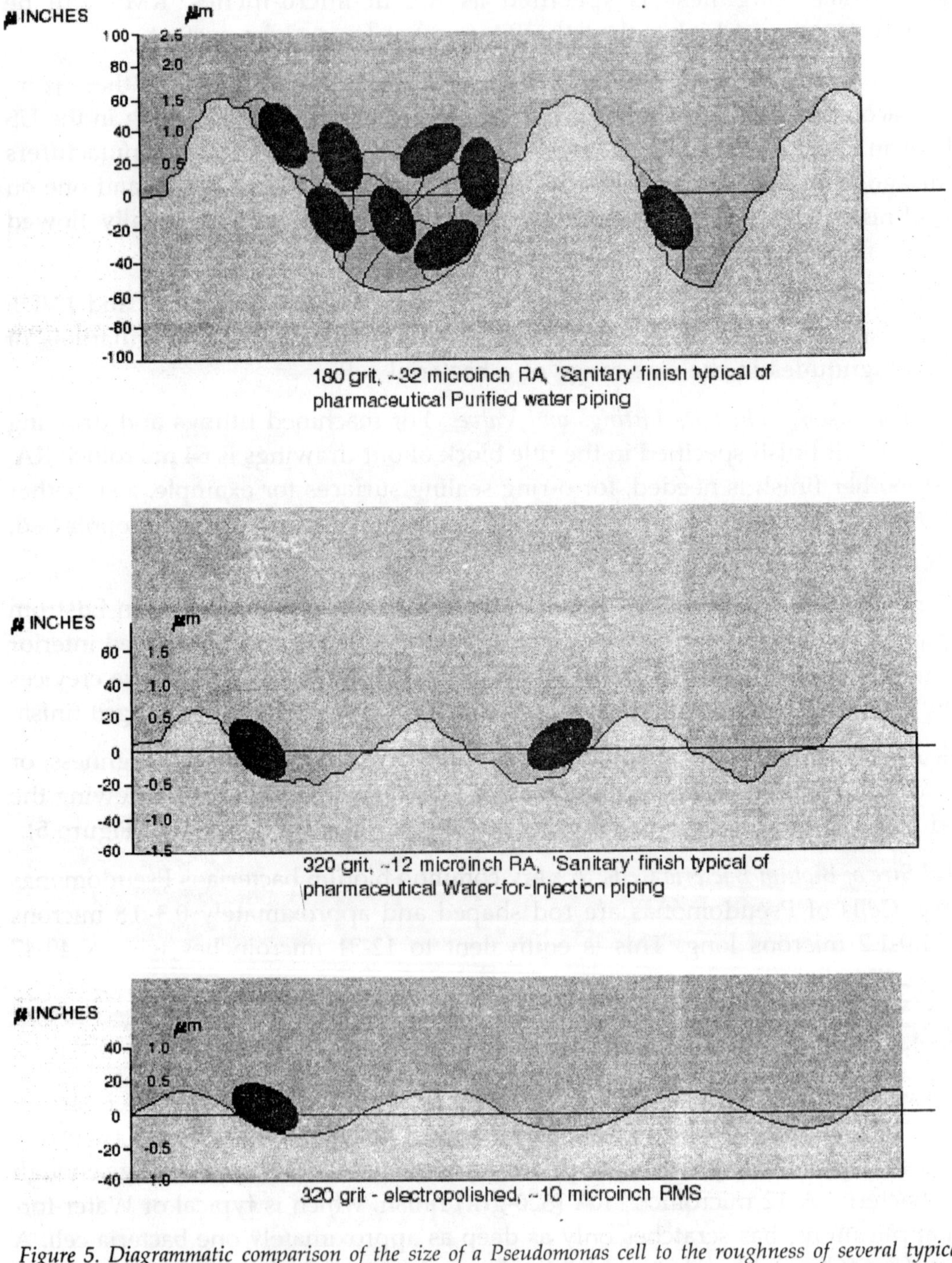

Figure 5. Diagrammatic comparison of the size of a Pseudomonas cell to the roughness of several typical stainless steel surfaces.

Figures 6 and 7 (below) show actual photographs of Pseudomonas cells on a 180 grit stainless steel surface. The relative sizes of bacteria cells and surface scratches shown in these photos are similar to the 180 grit profile of Figure 5.

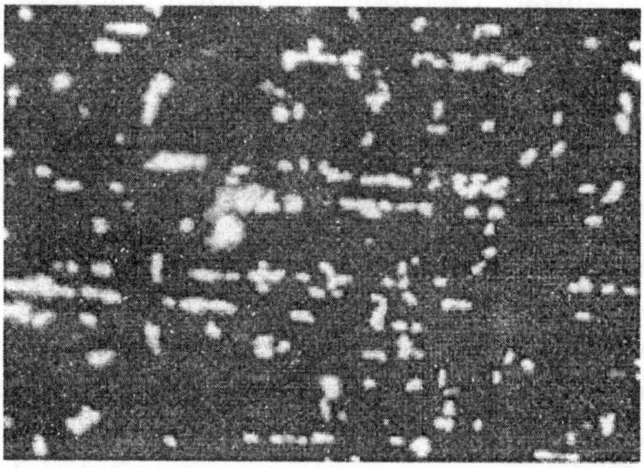

Figure 6. Scanning electron micrograph at x 400 magnification of 180 grit mechanically polished 316L stainless steel surface after 180-minute incubation with Pseudomonas aeruginosa. Notice that cells tended to congregate along polishing marks.

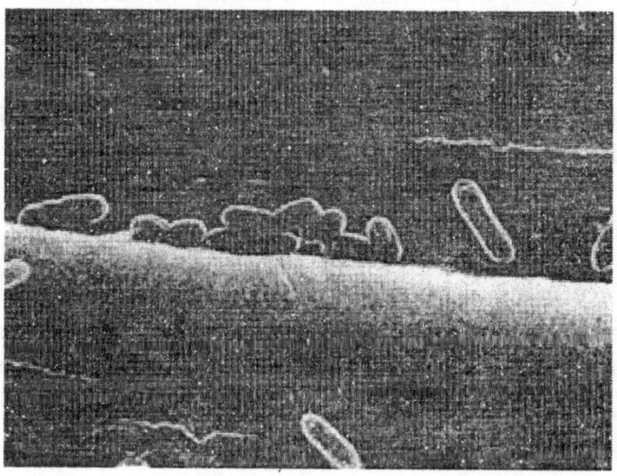

Figure 7. Same 180 grit surface as Figure 6, but at x 5000 magnification. Notice scratches are large enough to harbor bacteria.

FLUSHING

Flushing will limit biofilm thickness in automated watering systems. As discussed earlier, shear forces caused by flushing will slough off biofilm which extends out into the turbulent flow in the center of a pipe. Therefore, the maximum thickness of the biofilm will be approximately the same as the laminar layer for a particular flow rate.

Anaerobic Surface Conditions

Aerobic bacteria near the outer surface of a biofilm consume oxygen. If the biofilm is thick enough, oxygen will be depleted at the pipe surface creating an anaerobic environment. Anaerobic surface conditions are undesirable because there can be more corrosion problems. Could the biofilm in automated watering systems be thick enough to have anaerobic zones? One source indicates that oxygen can be depleted within 30-40 microns of the water/biofilm interface. The depth of the oxygen gradient into the biofilm will vary depending on oxygen content in the bulk water, water temperature, and water flow, but this gives a rough idea of how far oxygen may diffuse.

Aerobic P. aeruginosa biofilms grew to 30-40 μm in depth as monocultures, but increased in depth to 130 μm when the culture was amended with anaerobic bacteria. This indirect evidence suggests that depletion of oxygen - not of nutrients - limited the vertical development of the P. aeruginosa biofilm. If biofilm thickness in an automated watering system is only limited by flushing, it could be 50-125 microns thick and have some anaerobic zones. Of course, crevices such as o-ring pipe joints and threaded fittings can have much deeper biofilms and are most likely to have anaerobic zones.

MICROBIOLOGICALLY INFLUENCED CORROSION

The physical presence of microbial cells on a metal surface, as well as their metabolic activities, can cause Microbiologically Influenced Corrosion (MIC) or biocorrosion. The forms of corrosion caused by bacteria are not unique. Biocorrosion results in pitting, crevice corrosion, selective dealloying, stress corrosion cracking, and under-deposit corrosion. The following mechanisms are some of the causes of biocorrosion.

Oxygen depletion or differential aeration cells

Nonuniform (patchy) colonies of biofilm result in the formation of differential aeration cells where areas under respiring colonies are depleted of oxygen relative to surrounding noncolonized areas. Having different oxygen concentrations at two locations on a metal causes a difference in electrical potential and consequently corrosion currents. Under aerobic conditions, the areas under the respiring colonies become anodic and the surrounding areas become cathodic.

Stainless Steels' Protective Film

Oxygen depletion at the surface of stainless steel can destroy the protective passive film. Remember that stainless steels rely on a stable oxide film to provide corrosion resistance. Corrosion occurs when the oxide film is damaged or oxygen is kept from the metal surface by microorganisms in a biofilm.

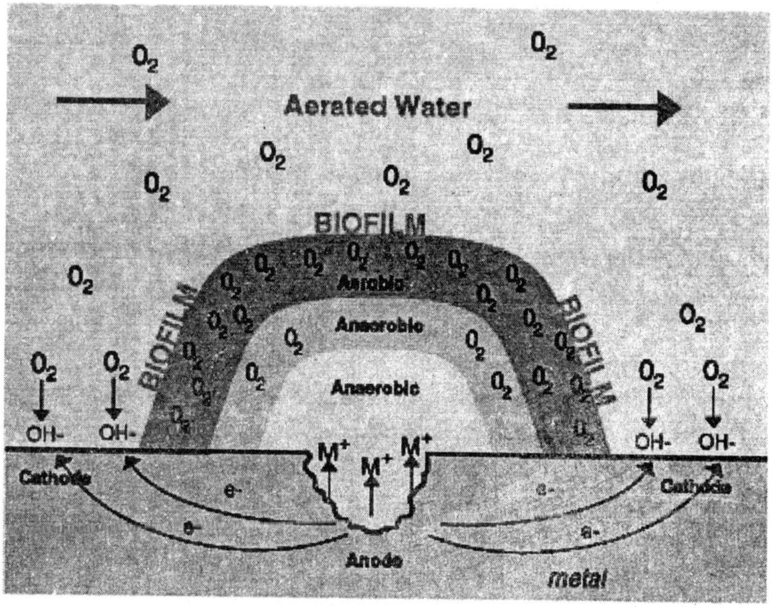

Figure 8. Nonuniform (patchy) colonization by bacteria results in differential aeration cells. This schematic shows pit initiation due to oxygen depletion under a biofilm.

Sulfate-reducing Bacteria

Oxygen depletion at the surface also provides a condition for anaerobic organisms like sulfate-reducing bacteria (SRB) to grow. This group of bacteria are one of the most frequent causes for biocorrosion. They reduce sulfate to hydrogen sulfide which reacts with metals to produce metal sulfides as corrosion products. Aerobic bacteria near the outer surface of the biofilm consume oxygen and create a suitable habitat for the sulfate reducing bacteria at the metal surface. SRBs can grow in water trapped in stagnant areas, such as dead legs of piping. Symptoms of SRB-influenced corrosion are hydrogen sulfide (rotten egg) odor, blackening of waters, and black deposits. The black deposit is primarily iron sulfide.

One way to limit SRB activity is to reduce the concentration of their essential nutrients: phosphorus, nitrogen, and sulfate. Thus, purified (RO or DI) waters would have less problem with SRBs. Also, any practices which minimize biofilm thickness (flushing, sanitizing, eliminating dead-end crevices) will minimize the anaerobic areas in biofilm which SRBs need.

Byproducts of Bacterial Metabolism

Another corrosion mechanism is based on the by-products of bacterial metabolism.

Acid-producing bacteria

Bacteria can produce aggressive metabolites, such as organic or inorganic acids. For example, Thiobacillus thiooxidans produces sulfuric acid and Clostridium aceticum produces acetic acid. Acids produced by bacteria accelerate corrosion by dissolving oxides from the metal surface and accelerating the cathodic reaction rate.

Hydrogen-producing bacteria

Many microorganisms produce hydrogen gas as a product of carbohydrate fermentation. Hydrogen gas can diffuse into metals and cause hydrogen embrittlement.

Iron bacteria

Iron-oxidizing bacteria, such as Gallionella, Sphaerotilus, Leptothrix, and Crenothrix, are aerobic and filamentous bacteria which oxidize iron from a soluble ferrous (Fe^{2+}) form to an insoluble ferric (Fe^{3+}) form. The dissolved ferrous iron could be from either the incoming water supply or the metal surface. The ferric iron these bacteria produce can attract chloride ions and produce ferric chloride deposits which can attack austenitic stainless steel. For iron bacteria on austenitic stainless steel, the deposits are typically brown or red-brown mounds.

SANITIZATION METHODS

Biofilm can be removed and/or destroyed by chemical and physical treatments. Chemical biocides can be divided into two major groups: oxidizing and nonoxidizing. Physical treatments include mechanical scrubbing and hot water.

Table 3. Typical biocide dosage levels.

Biocide	Dosage Level (mg/l)	Contact Time (hours)
Chlorine	50 [illegible]	1-2
Ozone	10-50*	<1
Chlorine dioxide	50-100	1-2
Hydrogen peroxide	10% (v/v)	2-3
Iodine	100-200	1-2
Quaternary ammonium cmpds.	300-1000	2-3
Formaldehyde	1-2% (v/v)	2-3
Anionic & nonionic surfactants	300-500	3-4

* Ozone dosage is 10-50 mg/l, but the residual levels in water were 1-2 mg/l.

Oxidizing biocides: Mittelman says the effectiveness of the oxidizing biocides in purified-water systems on an equal milligram-per-liter-dosage basis decreases in the following order:

ozone → chlorine dioxide → chlorine → iodine → hydrogen peroxide

Chlorine: According to Mittelman "*Chlorine is probably the most effective and least expensive of all oxidizing and nonoxidizing biocides.*" The activity of chlorine against attached biofilms is particularly high; not only are planktonic and biofilm bacteria killed, but chlorine also reacts with and destroys the polysaccharide web and its attachments to the surface. By destroying the extracellular polymers, chlorine breaks up the physical integrity of the biofilm.

Characklis recommends improving a chlorine treatment program by taking the following measures:

1. *Increase the Chlorine Concentration at the Water-Biofilm Interface.* As chlorine diffuses into a biofilm, it is used up in reactions with bacteria cells and extracellular materials. At low chlorine levels, biofilm bacteria can produce extracellular material faster than chlorine can diffuse through it so they are shielded in slime. By increasing the concentration, chlorine will diffuse farther into the biofilm. When it comes to disinfection of biofilms, high chlorine concentration for short durations is more effective than low concentration for long durations.

2. *Increase the Fluid Shear Stress at the Water-Biofilm Interface:* Simultaneous chlorine sanitization and flushing results in a higher uptake of chlorine by the biofilm and in greater biofilm detachment due to:

   – Increased mass transfer of chlorine from the bulk water to the biofilm.

   – Disruption of the biofilm during chlorination exposes new biofilm surfaces for chlorine attack. Decreased thickness of viscous or laminar sublayer.

3. *Use pH Control.* High pH favors hypochlorite-ion-promoted detachment of mature biofilms, and low pH enhances hypochlorous acid disinfection of thin films. Characklis proposed an interesting procedure would be to alternate between continuous chlorination at pH 6.5 and shock chlorination at pH 8. He doesn't imply that this has been tested.

Chlorine dioxide: Chlorine dioxide has biocidal activities similar to those of chlorine. Because it is unstable, it must be mixed and prepared on-site. Like chlorine, chlorine dioxide is corrosive to metals and must be handled with care.

Ozone: As an oxidizer, ozone is approximately twice as powerful as chlorine at the same concentrations. Like chlorine dioxide, ozone must be generated on-site because of its high reactivity and relative instability. Systems must be designed with appropriate ozone resistant materials. Ozone is usually dosed on a continuous basis at 1-2 mg/l. Success in employing higher dosages on a noncontinuous basis has been limited, possibly because of the limited solubility of ozone in purified water; it is difficult to produce high concentrations of ozone in solution. Although chlorine isn't as powerful as ozone when you compare 1-2 mg/l of each, chlorine can be used in higher sanitizing concentrations with equal disinfecting strength.

Hydrogen peroxide: Hydrogen peroxide is frequently used as a biocide in microelectronic-grade purified-water systems because it produces no by-products; it rapidly degrades to water and oxygen. A 10% by volume solution in purified water appears effective in killing planktonic bacteria, but more studies are needed on the effectiveness against attached biofilm".

Non-oxidizing biocides:

— *Quaternary Ammonium Compounds:* In addition to their biocidal activity, quats are effective surfactants/detergents, which may be an important factor in their use for biofilm inactivation and removal from surfaces. Rinseability can be a problem as removal from a purified-water system often requires exhaustive rinsing.

— *Formaldehyde:* Formaldehyde has been applied to pharmaceutical-grade systems. It is relatively noncorrosive to stainless steel. Its effectiveness against biofilm is questionable and it is a toxic carcinogen.

— *Anionic and Nonionic Surface-Active Agents:* These surfactant or detergent compounds have limited biocidal activity against the bacteria in purified water systems. Applications may be found for these detergents in conjunction with other biocides to improve biofilm and other particulate removal.

Physical Treatments

Heat: Pharmaceutical Water-for-Injection systems use recirculating hot water loops (greater than 80°C) to kill bacteria. According to Mittelman, when these systems are used on a continuous basis, planktonic bacteria are killed and biofilm development is reduced. Biofilms are even found in hot water (80°C). Periodic hot water sanitization can also be used to destroy bacteria in biofilm, but according to Collentro this requires a temperature of 95°C for a period in excess of 100 minutes. This would not be practical in an animal drinking water system!

Mechanical removal: Heavy biofilms cannot be removed from storage tank walls by the use of chemicals alone; mechanical scrubbing or scraping, high-pressure spraying, or a combination is also required. Mechanical removal of biofilm from distribution systems is impractical. For RO system maintenance, we don't routinely scrub storage tanks, but there is usually a continuous low chlorine level in the stored water, so heavy biofilms aren't allowed to develop.

Biocide resistance

Unlike antibiotics used to fight bacteria associated with human, animal, and plant diseases, bacteria do not develop the same type of resistance to industrial biocides. The difference between antibiotics and industrial biocides is that while an antibiotic may have a small number of target sites on or in the bacterial cell, all oxidizing biocides have a multitude of potential target sites. Chlorine, for example, is thought to have more than a hundred potential target sites on or in microorganisms. It is virtually impossible for microorganisms to develop a general resistance to such compounds. However, bacteria in a biofilm can resist biocides because they are shielded in slime.

Biofilm recovery (Regrowth)

Bacteria associated with biofilms are much more difficult to kill and remove from surfaces than planktonic organisms. According to Characklis, numerous investigators and plant operators have observed "a rapid resumption of biofouling immediately following chlorine treatment." Incomplete removal of the biofilm will allow it to quickly return to its equilibrium state, causing a rebound in total plate counts following sanitization.

According to Characklis, biofilm recovery may be due to one or all of the following.

- The remaining biofilm contains enough viable organisms that there is no lag phase in regrowth. Thus, biofilm recovery after shock chlorination is faster than initial accumulation on a clean pipe.
- The residual biofilm on the surface makes it rougher than clean pipe. The roughness of the deposit may provide a stickier surface which adsorbs more microbial cells and other compounds from the water.
- The chlorine preferentially removes extracellular polymers and not biofilm cells, thus leaving biofilm cells more exposed to the nutrients when chlorination ceases.
- Surviving organisms rapidly create more slime (extracellular polymers) as a protective response to irritation by chlorine.
- There is selection for organisms less susceptible to the sanitizing chemical. This is usually the organisms that produce excessive amounts of slime like Pseudomonas.

DETECTION AND ENUMERATION OF BACTERIA

Routine monitoring of bacterial levels is an essential part of monitoring the quality of laboratory animal drinking water. The classic way to enumerate bacteria in water is to do a plate count which is to spread a known volume of sample on the surface of a laboratory medium and count the number of visible colonies that develop after a period of time. However, it should be recognized that plate counts may underestimate the total number of bacteria present in a watering system.

Water samples only collect planktonic or free-floating bacteria. Free-floating bacteria in animal drinking water are either sloughed off of the biofilm or pass through from the incoming water supply. If a plate count test is low, one shouldn't assume that bacteria are not present in the watering system. More than 99% of the bacteria in water systems are in biofilms attached to pipe surfaces. If the integrity of a mature biofilm hasn't been disrupted by recent flushing or sanitization, it may not slough off many cells into the drinking water, but it is still there.

Plate counts are based on the ability of bacteria in a sample to grow on a defined nutrient medium. When bacteria grow on a nutrient, they form distinct colonies. Theoretically, a colony is derived from a single bacteria cell. Some underestimation of bacteria is caused by clumps of bacteria that form only one colony.

Another reason viable counts can be too low in nutrient-poor purified water is that the bacteria are in a starved state and cannot grow on rich nutrient media. Rich laboratory media are toxic to bacteria adapted to living in high-purity water systems. To get higher bacterial recoveries from purified waters, special media (R2A agar), decreased incubation temperatures, and increased incubation times are sometimes used.

Sometimes the results of bacterial plate count testing seem very erratic. Samples taken from one point in the system may vary from less than 10 cfu/ml to TNTC (too-numerous-to-count). Or maybe the counts are usually low, but occasionally a high count appears. Some of this variability can be explained by understanding that biofilms periodically "shed", causing bacterial counts to skyrocket.

Bacteria constitute a very successful life form. In their evolution, they have developed successful strategies for survival which include attachment to surfaces and development of protective biofilms where they behave very differently than free-floating bacteria. Their successful strategies make it difficult for us to control biofilm growth in automated watering systems.

REFERENCES

Fiksel, Joseph, and Vincent T. Covello. (eds.) *Biotechnology Risk Assessment: Issues and Methods for Environmental Introductions.* N.Y.: Pergamon Press. 1986.

Ginzburg, Lev. R. *Assessing Ecological Risks of Biotechnology.* Boston: Butterworth-Heinemann. 1991.

Mannion, Antoinette M. "Sustainable development and biotechnology." *Environmental Conservation,* 19(4):297-306. Winter 1992.

9

Biotechnological Applications of Microbial Metabolism

Of all the life forms occurring in nature within the biodiversity there should be no doubt to realise that microorganisms are the most powerful creatures in existence. They determine life and death on this planet. They can kill merciless humans, animals and plants, but at the same time they can be harnessed to sustain life. Nature has provided us with a perfect balance in Carbon, Nitrogen and Phosphorous cycles to sustain microbial, plant, animal and human life. Any interference in these cycles can swing the pendulum very quickly into the direction of killing or sustaining mankind. It is the microorganism, which determines the growth and existence of plants, animals and humans on this planet. It is therefore of utmost importance that we give microbiology a first priority, as we have to isolate and investigate the biochemistry and behaviour of the microorganism in order to understand how nature works. Only when we obtain this information will we be able to sustain and improve life in our community. The microorganism is much more flexible and adaptable to environmental changes than plants, animals and humans.

Biological resources constitute an asset with a great deal of immediate as well as potential benefit for the quality of life. These resources evolved through evolutionary processes of life on earth involving genetic differentiation and genomic adaptation to a microenvironmental variation of habitats. As a result these biological resources consisting of a tremendous diversity of species of microorganisms, fungi, plants and animals occur in a variety of ecological settings throughout the world, from the richest areas in tropical to the poorest in the temperate and cold regions of the planet earth. Such a variety of life forms is collectively known as biological diversity or biodiversity. Biodiversity therefore encompasses all levels of organisation of life from genes to populations, species, ecological communities and ecosystems. They are essential resources for human survival The human species is therefore an integrated part of this biodiversity.

METABOLIC CLASSIFICATION

During the study of any phenomenon in science, it has been deemed necessary to draw a distinction between the thing being studied and its' environment. Indeed, this obsession with the omnipresence of a 'thin red line' is one of the hallmarks of science if not humanity as a whole. We therefore open this essay with the necessary (and unfortunate) assumption of the existence of things as entities distinct from their environment.

Presumably, these things interact with their environment, and do so in different ways. There exists one particular class of things which has received a disproportionate amount of attention, presumably because the entities paying the attention themselves belong to that particular class. The things in question, which we will call life, are those that possess the unique ability to metabolize. That is, living cells interact with their environment by taking up chemical substances (often subsequently transforming them), while releasing waste products. The thin red line separating life from death is thus demarcated by a cells' outer envelope.

In some cases the process of transformation is dissimilatory and releases chemical energy, part of which is conserved in a form the cell can use. Alternatively, cells can perform transformations that are assimilatory, and thereby synthesize new cell material from chemical substances derived from their environment. These twin processes, the acquisition/storage of energy and the use of stored energy for biosynthesis, are termed catabolism and anabolism respectively. These form the keystones of metabolism.

This essay aims first and foremost to give a comprehensive overview of the remarkably rich metabolic diversity found in the living world. Basic biochemical pathways and processes are discussed, with particular reference to environmental constraints on metabolism. However, it is the similarities - rather than the differences – between metabolic processes found on Earth which this essay hopes to incorporate into a comprehensive definition of what it means to be alive or dead.

An all-encompassing classification scheme is needed as a framework for our discussion on microbial diversity. In addition to requiring energy, all known forms of life are carbon-based and require carbon as their primary macronutrient. It is for this reason that life is commonly categorized on the basis of both energy and carbon sources. Phototrophs obtain their energy from light and convert this to chemical energy as part of a process called photosynthesis. Chemotrophs obtain their energy from chemical compounds. The compounds used by chemotrophs may be either organic (such as the carbohydrates you and the bacteria in your mouth had for breakfast) or inorganic (such as the H_2 and CO_2 metabolized into CH_4 by methanogenic microorganisms in your gut). In these two distinct cases the organisms are said to be chemoorganotrophs or chemolithotrophs, respectively.

Carbon is derived from CO_2 in autotrophs and from pre-formed reduced organic compounds such as sugars in heterotrophs. Lastly, some chemoorganotroph somewhere (perhaps a biochemist relying on large quantities of R-OH compounds for energy) decided to call 'chemoheterotrophs' 'mixotrophs' instead.

BASIC BIOCHEMICAL ENERGETICS

The release and conservation of metabolic energy in living cells occurs as the result of reduction-oxidation reactions. Biological systems are thus governed by couples of electron acceptors and donors. The amount of energy released during such a redox reaction can be quantified by the 'reduction potential' (E0') of a couple. Couples with a high (positive) E0' have a greater tendency to accept electrons, and vice versa.

It will come as no surprise that the role of electron and proton (H^+) flow is paramount in microbial energetics. In fact, oxidation-reduction reactions may be considered as chains of events resulting in just such a flow: (i) the removal of electrons from an electron donor; (ii) the transfer of electrons through electron carrier(s); and (iii) the addition of electrons to an electron acceptor. Examples of common biological electron carriers are the coenzymes nicotinamide-adenine dinucleotides NAD^+ and $NADP^+$, and the flavin-adenine di- and mono- nucleotides FAD^+ and FMD^+. The type of electron carrier used is not really relevant here – of interest is the nature of the electron donor and acceptor. The major types of catabolism known are summarized on the basis of their electron – donor and –acceptor in Table 1.

Table 1. Major types of catabolism, with their corresponding electron –donor and –acceptor. Organic e^- donors (carbohydrates) denoted by $(CH_2O)n$.

Process	Organism	Electron Donor(s)	Electron Acceptor(s)
Respiration	Methanogen	H_2, $(CH_2O)n$	CO_2
	Acetogen	H_2, $(CH_2O)n$	CO_2
	Sulfate reducer	H_2S, CH_3SH	Sulfate, SO_4^{2-}
	Sulfur reducer	H_2S, CH_3SH	Sulfur, SO
	Iron reducer	H_2, $(CH_2O)n$	Ferric Iron, Fe^{3+}
	Denitrifier	N_2, N_2O, NO, NO_2	Nitrate, NO_3^-
	Aerobe	$(CH_2O)n$	Oxygen, O_2
Photosynthesis	Photosynthesizer	Light-driven complexes	Cytochrome, Ubiquinone
Fermentation	Fermenter	$(CH_2O)n$	Internal: Lactate

Much of the transfer of electrons from donors to acceptors serves the purpose of relaying energy into a compact, transportable form. Energy can then later be releqased to carry out biological processes. More often than not the function of energy storage itself is performed by high energy phosphate bonds, in molecules such as adenosine triphosphate (ATP). In some cases, sulfoanhydride (thioester) bonds (found, for example, in derivatives of coenzyme A) serve a similar purpose. Cellular energy may also be stored in an electrochemical form called the Proton Motive Force (PMF), which results from a potential difference between the intra- and extra-cellular environments. The PMF is perhaps best visualized as analogous to the potential energy present in a charged battery. Conversion between PMF-derived and ATP-derived energy occurs through the action of a remarkable membrane-associated enzyme known as ATPase. ATPase functions as a reversible proton channel between the cell exterior and cytoplasm; as protons enter, the dissipation of the PMF drives ATP synthesis from ADP + Pi – and vice versa.

METABOLIC DIVERSITY

Chemotrophs derive their energy from chemical compounds, be they of organic or inorganic origin. The oxidation of organic electron donors for energy generation is known as 'respiration'. In aerobic respiration, oxygen acts as the terminal electron donor. In anaerobic respiration, some molecule other than oxygen has to function as the terminal electron acceptor. Fermentation is a special type of anaerobic respiration in which an internal substrate acts as the electron acceptor.

The use of pre-formed organic fuels requires complex metabolic machinery whose operation serves the purposes of both catabolism and anabolism. That is, the oxidation of organic electron donors by chemoorganotrophs not only adds to the ATP and PMF pool but also manufactures the building blocks of life. A total of twelve essential precursor building blocks can be identified from which life can make further macromolecules.

Figure 1 shows how the central pathways of metabolism responsible for chemoorganotrophy concurrently lead to the manufacture of these twelve precursor molecules. The three pathways shown are (i) The Embden-Meyerhof (Parnas, Glycolysis) pathway; (ii) The Pentose phosphate pathway; and (iii) The tricarboxylic acid (TCA) cycle. All three pathways are operational during aerobic respiration. The TCA cycle in particular has major biosynthetic as well as energetic functions, and for this reason the complete cycle or major portions of it are nearly universal to life. Accordingly, it is not surprising that many organisms are able to use some of the acids produced as electron donors and carbon sources.

In the case of anaerobically respiring organisms the lack of oxygen prevents formation of certain TCA-cycle enzymes, thereby compromising the amount of reducing power generated. Manufacture of the twelve precursor molecules is still made possible through the use of anapleurotic pathways: enzymatic reactions or set of chemical reactions that link metabolic pathways, thereby allowing bypass of certain parts of that pathway or allowing the reversal of carbon flow. A noteworthy example of such an anapleurotic pathway is the glyoxylate cycle, which replenishes acids essential to the function of the TCA cycle using the twin enzymes isocitrate lyase and malate synthase.

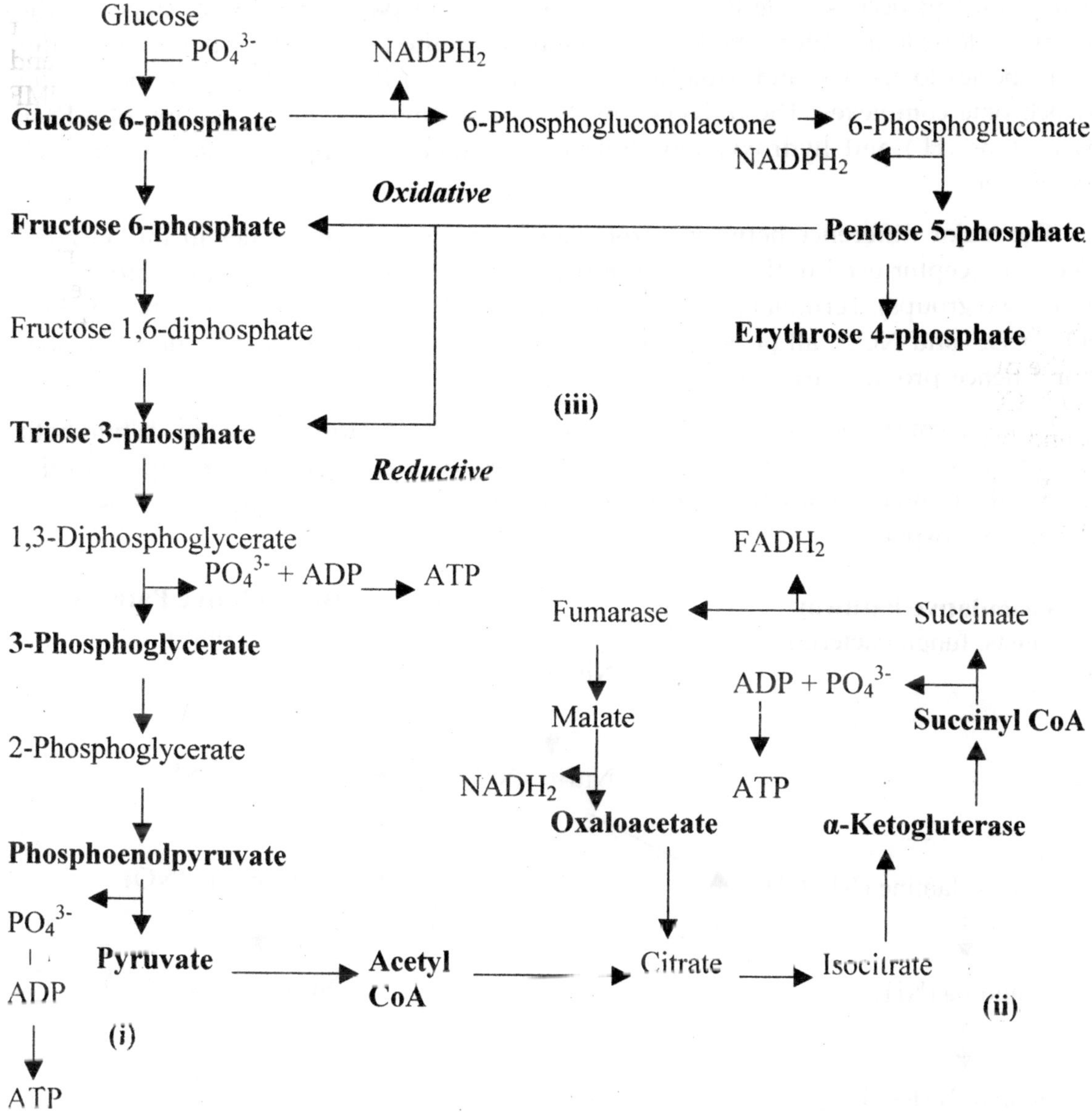

Figure 1. The central fueling and biosynthetic pathways during aerobic respiration in heterotrophs: (i) The Embden-Meyerhof Pathway; (ii) The tricarboxylic acid cycle; and (iii) The pentose phosphate cycle. The oxidative and reductive branches of the pentose phosphate cycle are shown in boldface italics. Anapleurotic reactions and peripheral pathways (including fermentative and respiratory pathways) are not shown. The 12 precursor metabolites are shown in boldface.

Aerobic respiration allows for the continued metabolism of glucose through re-oxidation of the reduced forms of NADH and FADH produced by the central metabolic pathways. This occurs along the so-called 'electron transport chain', with oxygen as the terminal electron acceptor. The key intermediates in the electron transport chain differ from species to species, and usually include a variety of flavin enzymes, quinones and cytochrome complexes. The universal net result, however, is the generation of a PMF as protons (H^+) and hydroxyl ions (OH-) accumulate on opposite sides of the cell membrane.

The major difference between aerobic and anaerobic respirers lies in the terminal electron acceptor used in the electron transport chain (The electron carriers are similar in the two groups). Terminal electron acceptors used by anaerobic respirers include Fe^{3+}, SO_4^{2-}, CO_3^{2-} and NO_3^{2-}, all of which have a less favorable oxidation-reduction potential – and hence produce less ATP - than does O_2.

Figure 2 contrasts assimilative and dissimilative respiratory pathways. While both use nitrogen as a terminal electron acceptor in respiration, the former incorporates the end-product into biosynthetic matter while the latter releases the end-products as waste into the atmosphere.

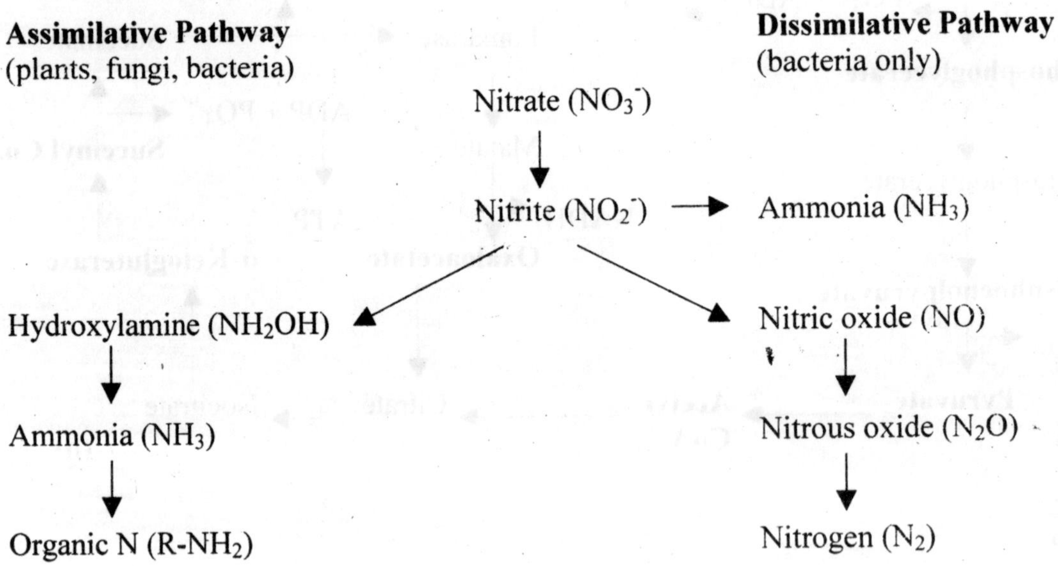

Figure 2. Assimilatory and dissimilatory pathways involving nitrogenous terminal electron acceptors

Two distinct mechanisms in the final step of inorganic nitrogen assimilation have been recognized:

(i) glutamic dehydrogenase: a-ketoglutarate + $NH_3 \rightarrow$ glutamic acid

(ii) L-glutamine synthetase: glutamic acid + NH_3 + ATP $\rightarrow$ glutamine$\rightarrow$glutamate synthetase ('GOGAT'): glutamine + a-ketoglutarate $\rightarrow$ 2 glutamic acid

Fermentation does not utilize an electron transport chain at all, and relies instead on substrate-level phosphorylation. Re-oxidization of NADH is accomplished by an internally balanced oxidation-reduction reaction that uses an organic molecule (commonly lactate) as the terminal electron acceptor.

Chemolithotrophy

As an alternative to aerobic and anaerobic respiration, many microorganisms use inorganic electron donors of geological, biological or anthropogenic origin. Examples of such electron donors include hydrogen sulfide (H_2S), hydrogen gas (H_2), ferrous iron (Fe_2^+) and ammonia (NH_3). Hydrogen bacteria, for example, phosphorylate three molecules of ATP for every molecule of H_2 'respired' using the hydrogenase-mediated reaction:

$$H_2 \rightarrow 2H^+ + 2e- + NAD^- (+ 3ADP + 3Pi) \rightarrow NADH^+ (+ 3ATP)$$

Other examples are the nitrifying bacteria, which generate ATP through oxidization of ammonia (NH_3) to nitrite (NO_2^-) or nitrite to nitrate (NO_3^-). These chemolithotrophs rely on aerobic respiratory processes similar to those found in the electron transport chain of most chemoorganotrophs. However, while chemoorganotrophs generally rely on organic compounds such as glucose for both carbon and energy, chemolithotrophs often have to acquire their carbon elsewhere – usually from atmospheric CO_2 or (more rarely) from CH_4.

Phototrophy

The term 'photosynthesis' refers first and foremost to the conversion of energy in electromagnetic radiation (light) into chemical energy. Additionally, 'photosynthesis' implies an anabolic function; namely, the use of aforementioned chemical energy to 'fix' carbon into structural and functional cell components. The vast majority of photosynthetic organisms use CO_2 as their sole carbon source. Photoheterotrophs, insignificant on a global scale, use organic compounds as a carbon source. Today, CO_2 is readily available in most environments and was probably never a limiting metabolic agent during the Earth's history. In fact, CO_2 is a dominant atmospheric constituent of lifeless terrestrial planets in our Solar System – in the context of astrobiology we can, therefore, safely limit our discussion to photoautotrophs.

In addition to carbon and energy, photoautotrophs require input of electrons from a donor (or 'reducing power') in order to fix CO_2. Care should be taken to distinguish this anabolic energy-tapping electron flow for the purpose of carbon fixation from that involved in catabolic energy generation discussed previously. Commonly, reducing power is generated by the oxidation of water to oxygen in the presence of light. This type of photosynthesis is called oxygenic photosynthesis. The production of reducing power in anoxygenic photosynthesis, on the other hand, rarely requires light and involves an oxidative reaction (but not oxygen production) such as $H_2S \rightarrow SO$ and $SO \rightarrow SO_4^{2-}$. The differences between these types of photosynthesis are summarized diagrammatically in Figure 3.

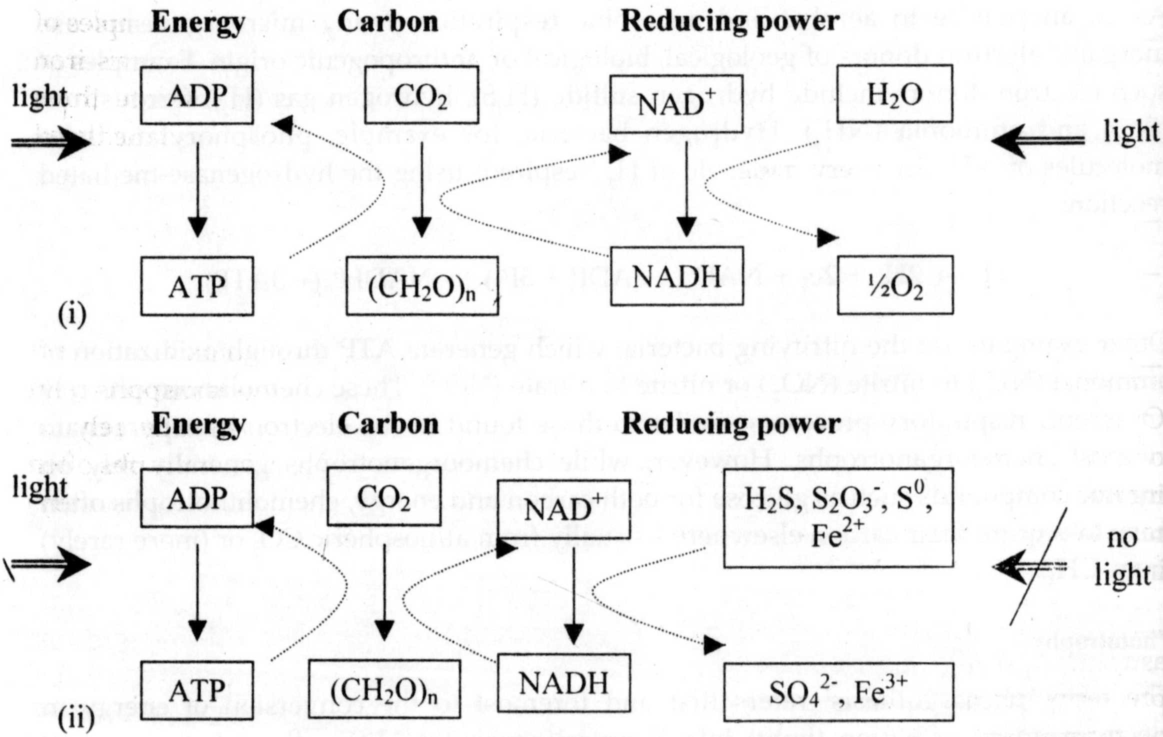

Figure 3. The synthesis of ATP-energy and reducing power in (i) oxygenic and (ii) anoxygenic photosynthesis. Shown are the electron- and energy- carriers, carbon source, and electron -donor and –acceptor.

Photosynthesis can be categorized into distinct catabolic and anabolic phases. During the 'light reaction' phase ATP and NADPH are synthesized, while fixation of CO_2 into cellular carbon takes place during the 'dark reaction' phase.

Two structural types of light-harvesting pigments are commonly used:

— Isoprenoids, such as the carotenoid ß-carotene;

- Tetrapyrroles, which are classed as:
  - Bile pigments such as phycobiliproteins used by cyanobacteria and red algal chloroplasts;
  - Porphyrins such as chlorophylls and bacteriochlorophylls.

It is important to distinguish between oxygenic and anoxygenic photosynthesis on the level of biochemical processes within the cell. All known photosynthesizers make use of a cyclic, anoxygenic pathway called Photosystem I. Although the carriers associated with light-induced electron and energy flow in Photosystem I are often species-specific, a general flow summary is instructive:

- Electromagnetic radiation provides the excitation energy needed to activate the photoreaction center, turning it into a strong electron donor;
- Subsequent electron flow drives the formation of ATP through phosphorylation of ADP;
- Electron flow through a quinone pool leads to charging of NAD(P)H from NAD(P)+;
- Specific cytochrome complexes, with the aid of an external electron donor, help energize another photoreaction center;
- Cyclic phosphorylation through electron flow proceeds.

Oxygenic photosynthesizers make use of an additional pathway called Photosystem II, or the 'Z scheme'. It is this pathway that entails the light-driven breakdown of water into oxygen and hydrogen. This systems entails non-cyclic phosphorylation: the electrons released by this reaction are used to excite a Photosystem II reaction center before passing into and thus linking the Photosystem I cycle described above.

The large array of metabolic types known may, at first, seem overwhelming to the astrobiologist aspiring to draw a thin red line between life and death. However, our preceding discussion reveals several dominant conceptual trends underlying the process of metabolism.

The diversity of metabolic processes encountered on Earth we now turn to the identification of a common ground: what separates metabolism in living things from other chemical processes in the universe? And can we use metabolic characteristics as a stringent, all-inclusive framework in defining life?

'Systems Thinking'

The term 'system' is used in a variety of ways. Whenever we speak of a solar system or a car stereo system, we are referring to a group of related entities. It is the way that these entities, or components, are organized relative to one-another in addition to the individual characteristics of the components themselves which gives rise to the properties and defining features of the system. It is for this reason that such system properties are termed 'emergent' properties. Consider the water molecule: the chemical and physical properties of water do not simply emerge because of the characteristics of the hydrogen and oxygen atoms. Rather, the emergent properties of water are in large part due to the way in which the atoms are 'organized' within the 'system'.

In the context of systems theory, we can extend the above definition somewhat by stating that a true system must incorporate feedback, causal loops that force non-linear behaviors.

Stable non-linear behavior at disequilibrium

It is no coincidence that the physical sciences, for as long as they have been around, have focused on the study of equilibrium systems. In the days before the Silicon Age, any phenomenon representing a perturbation from its' equilibrium state necessitated the use of linear approximation methods such as Taylor Series Expansions. Unfortunately, linear approximation methods are only appropriate for small perturbations. In general, small shifts in the state of the system, for example due to the application of some small flux, will cause the system to respond in such a way as to maximize the generation of entropy. This spontaneous increase in disorder drives the system back to its equilibrium state.

Observations and, more recently, non-linear dynamics and complex systems analysis have taught us that systems experiencing large continuous fluxes – mathematically analogous to a situation 'far' from equilibrium – may form dissipative structures and develop into forms of ever-increasing complexity. It is here that self-organizing open system behavior emerges. Because such phenomena occur far from equilibrium, the classical treatment of employing perturbation methods proves to be wholly inadequate in their study. The farther a dissipative structure is from equilibrium, the greater is its complexity and the higher is the degree of nonlinearity in the mathematical equations expressing it.

In living systems, the state of disequilibrium arises because material and energy do not flow bidirectionally across cell membranes under the classic constraints of statistical mechanics. On the contrary, we have seen in our discussion on metabolism that living cells are highly selective about what they absorb and release. Life employs several different mechanisms to transport solute across their membranes:

Simple ('passive') Diffusion is an energy-independent process that operates without the aid of membrane proteins. Solutes move down a concentration gradient from the environment into the cell cytoplasm, or in the opposite direction. Characteristics of the outer cellular wall and membranes limit simple diffusion to small, chargeless, non-polar or fat-soluble molecules such as O_2, CO_2, NH_3, H_2O, ethanol and certain hydrophobic molecules.

Facilitated Diffusion is an energy-independent process that, unlike simple diffusion, requires the action of stereospecific transmembrane proteins called 'permeases'. Movement still occurs down a concentration gradient, but the dependence on solute-specific permeases allows a certain degree of cellular control on diffusion rates.

Active Transport requires energy expenditure and specific membrane proteins, which have the ability to move solutes from a low concentration extracellular environment to a high concentration intracellular environment. Movement may occur at the expense of the PMF, using a previously established ionic gradient such as H^+ or Na^+. Alternatively, ATP hydrolysis may provide the energy for transportation through a membrane-spanning transport protein.

Group Translocation involves the chemical modification of transported substances during passage across a membrane. An example is the phosphotransferase system (PTS) by which certain sugars (glucose, mannitol, mannose, fructose, N-acetylglucosamine and ß-glucosides) are transported into cells. These sugars are phosphorylated at the cost of a high-energy phosphate bond in the metabolic intermediate phosphoenolpyruvate (PEP).

While mechanism (i) is a normal equilibrium reaction and would still occur in dead cells, mechanisms (ii)-(iv) are examples of energy-driven stable non-equilibrium processes. Despite the constant flux of material and energy through the living system, a constant structure is maintained. This remarkable ability of life to combine the fluidity of change with the stability of structure was aptly termed fliessgleichgewicht, or 'flowing balance', by the German mathematician and naturalist Ludwig von Bertalanffy.

In addition, living units – though complex - display a high degree of order, and manage to build up and maintain this order despite the constant tug of the acclaimed Third Law of Thermodynamics. A living system represents a steep gorge of low entropy in an ever rising landscape of increasing entropy. A subtle example of such self-organization is the universal bias in favor of the lighter 12C isotope during the first enzymatic carboxylation reaction in primary metabolism by CO_2-fixing autotrophs. It turns out that the carbon fixed in this manner is invariably isotopically lighter – that is, more enriched in 12C and relatively less so in 13C – than the inorganic carbon absorbed from the atmosphere. It is, in fact, exactly this time-enduring 'ordered state' property of biological substances that scientists use as a proxy for life in fossil organics billions of years old.

So can we define life purely in terms of entropy and disequilibrium? No. Both whirlpools and flames depend on a constant flux of energy and matter through them, but neither truly develops, evolves or reproduces. Indeed, it is a small matter to simulate non-equilibrium systems in the lab. The cardinal feature of such an open system is, of course, the presence of a constant flux of some sort through the system. A particularly impressive example of such a system occurs during the phenomenon of Rayleigh-Benard Convection. In a classic experiment first described by the French scientist Henri Bénard, a thin layer of liquid is exposed to a constant flux of heat from below. When the temperature difference between the top and bottom liquid surfaces falls within a certain critical range, an ordered pattern of hexagonal cells appears. Hot liquid rises through the cell centers, while cool liquid descends along the 'honeycomb' cell walls. The non-equilibrium that is maintained by the continual flow of heat through the system generates a complex spatial pattern in which millions of molecules move coherently to form the hexagonal convection cells.

The dead world both in- and outside the laboratory abounds with such examples of spontaneous self-organization. The difference with living systems is that they:

(a) Represent networks of production processes in which the function of each component is to participate in the production or transformation of other components within the network; and

(b) Give rise to the creation of a boundary that specifies the domain of the networks' operations and defines the system as a unit.

The anabolic cycles discussed above, in particular the glucose-fixing tricarboxylic acid (TCA) cycle and the CO_2-fixing Calvin-Benson cycle, account for the production and transformation processes referred to in (a). As for (b), the very operation of the electron transport chain fundamental to catabolism relies on the presence of a cell boundary enclosing the living system and allowing for charge and substrate build-up relative to the environment.

DEVELOPMENT OF RURAL AND URBAN SOCIETIES

Since his appearance, man has always lived in an uncertain, sometimes precarious, symbiosis with nature, obtaining his nourishment needed from plants and animals personally accessible to him. In accordance with the climatic and environmental conditions, the search for food (life as Nomades) soon developed into actively growing, storing and preparing the food (life as Settlers and Farmers). Water, sun availability and soil conditions determined the type of original food for the particular society.

It was the farmer, who was responsible for the growth and harvesting of the crops, whereas the families used their own recipes for the conversion of the agricultural products into edible and palatable products for themselves, their animals and later the market place. This practice is still followed today in many societies in SEAsia, Africa and Latin America. The culture of biotechnology originated therefore in the rural areas, where people experimented with the regeneration of soil fertility, breeding of new crop varieties and fermentation for a palatable and well digestible food. Whereas cold and temperate zones used mainly grain for their food and fermentation, tropical zones of SEAsia and Africa produced numerous foods from rice, soybean, cassava [manihot] and other plants.

A second type of fermentation technology was accidentally introduced very soon in form of beverages such as wine, met [honey beer], and beer and in different types of food such as bread, milk products such as cheese, butter and yoghurt. Whereas the societies of cooler climates preferred beer and wine from barley and grapes, respectively, it was pulque from the sweet juice of the Mexican Agave in Aztec countries of Latin America, the saki from rice in SEAsia and the palm wine from palms in some societies of Africa. Other societies in Africa did not encourage this second type of fermentation out of personal and/or religious beliefs.

Since food preparation and fermentation was carried out according to 'local society tradition', handed down from generation to generation, these complex preparations were much more an art than a science. Nevertheless, the traditionally fermented protein-rich foods are highly acceptable to millions of people until today, because they are easily made and are generally more attractive to the consumers than the cooked original substrates.

The organoleptic characteristics of the substrate are improved by the fermentation process. These fermentations also increase the nutritional value of the substrate, since the amount of vitamins are significantly increased as well as the digestibility. If properly fermented, these foods are not hazardous to health since the microorganisms responsible for these processes are not toxin producers It was not before A. van Leeuwenhoek in the 18th century was able to construct the first microscope and Louis Pasteur in the 19th century found the reasons for the fermentations to occur, was the complexity of the traditional food fermentation realised.

The impetus of the industrial revolution during the 18th and 19th centuries transformed the very nature of society in many parts of the world, which are now referred to as the 'developed countries '. These societies were now not only using renewable resources, but also consuming vast amounts of non-renewable resources. The industrial society developed by the accumulation of scientific knowledge, the spread of technological innovations, and the exploitation of enormous natural resources. Traditional vegetable and animal fibres were increasingly replaced by synthetics manufactured by an ever increasing chemical industry from coal and petrochemicals.

This development in the 20th century fundamentally altered the pattern of consumption, land use and international trade, and the distribution of wealth. Longer life expectancy, higher survival rates, and dramatic population increases followed through better housing and sanitation, the production of antibiotics and vaccines. The quality of life was improved by the introduction of petrol and the motor industry among others. The impact on society was dramatic and on culture devastating as a large proportion of the traditional way of life was lost through this development owing to an ever increasing urbanisation.

The majority of societies on this planet, however, were bypassed by the grey revolution and chemical industry development. These countries are in general referred to as 'developing ' or 'less developed ' countries and are situated in two climatic zones, the tropical wet and the tropical arid zones. Ironically, the largest reservoirs of non-renewable resources required for the development of the industrial revolution in the 'developed' countries were found in countries of the tropical arid zone. The exploitation of these non-renewable energy sources situated in predominantly dry dessert areas, where agricultural food production has always been minimal, led to large increases in population which became totally dependent on food imports. The resulting starvation, malnutrition together with an unavailability of antibiotics for the fight against diseases kept living standards and survival rates at a relatively low level.

The green revolution was introduced to secure food for the ever increasing population in all parts of the world. The question therefore arises, can scientific knowledge and technology improve quality of life, life expectancy and thus increase the survival rate for all societies without affecting culture and society in such disastrous ways as occurred in the developed countries?

BIOTECHNOLOGY AND THE CORPORATE WORLD

The reasons for such an impact on culture and society are manifested in the principles of the 'industrial systems' organisation',which dominates today's society in the developed countries. This organisation is based on short-term profit, with a production to sell attitude, with preference given to the production of luxury consumer goods over goods required for basic needs, particularly at the level of the large energy systems, such as coal, hydrocarbons, and nuclear energy. Such a production is an obstacle to the total realisation of individuals and society.

The principal energy sources of antiquity were all derived directly from the sun: human and animal muscle power, wood, flowing water and wind. About 300 years ago, the industrial revolution began with stationary wind-powered and water-powered technologies, which were essentially replaced by fossil hydrocarbons: coal in the nineteenth century, oil since the twentieth century, and now, increasingly, natural gas.

The global use of hydrocarbons for fuel by humans has increased nearly 800-fold since 1750 and about 1^{2}-fold in the twentieth century.

Hydrocarbon-based energy is important for the three main areas of human development: economic, social and environmental. Energy prices have an important effect on almost every major aspect of macroeconomic performance, because energy is used directly and indirectly in the production of all goods and services. Both theoretical models and empirical analyses of economic growth suggest that a decrease in the rate of increase in energy availability will have serious impacts.

The results of these chemical and manufacturing industries are accompanied by ever increasing amounts of effluents of both heat and toxic substances, many of which are non-biodegradable. Modern agriculture is now strongly based on the application of chemical fertilisers and ever-increasing amounts of organic pesticides, mainly as a consequence of an enormous and rapid expansion in world population demanding an ever greater quantity with increasing quality of food and goods of all kinds. This, in turn, encourages the use of still further quantities of non-renewable resources and energy. This development led in the 1970s to a turning point in the perception of man's relation to his natural environment, the biosphere, as well as a shift in man's relationship to the man-made environment, the technosphere. The question was raised whether the earth and its atmosphere can provide an infinite sink and absorb the waste products of industry, agriculture and urban living as they become more and more prevalent. The processes of physical planning are now challenged and well established procedures are under severe scrutiny.

The green revolution was introduced to secure the food for the ever increasing population in all parts of the world as a consequence of the industrial development. Thus, the 'industrial system organisation' was introduced into agriculture in the hope that an equivalent development could be achieved. The introduction of farm mechanisation and monoculture systems initially brought the promised results of higher yields, but drove the small farmer into an already overflowing urbanising area and produced a wonderful breeding ground for insect and other crop pests.

Farm mechanisation also lead to a destruction of the natural soil structure and severe nutrient deficiencies. As a consequence, chemical fertilisers and pesticides were developed and introduced to combat these deficiencies. After the initial boost in productivity it was soon realised that the indiscriminate use of fertilisers killed and changed the microflora in the soil and around the plant causing ever increasing soil infertility and salination. The pesticides in form of chemicals foreign to nature's soil population and thus undegradable, destroyed not only mixed populations of microorganisms necessary for the ripening of the crops and subsequent fermentations, but also killed many insects which served as food for many birds. Furthermore, both types of chemicals often applied in excess appeared in waterways endangering human and animal life. It is becoming more and more clear that many changes in the human life [eg immune system] could be due to the application of these man-made chemicals.

The industrial and agricultural development over the past century has, unfortunately, changed our society from a nature caring and observing society into one full of obsession, greed and disrespect, a so-called 'throw-away society, which does not care anymore about its own environment and the beauty of nature. These phenomena pose a serious threat to sustainable development and species diversity may well be our planet's most important and irreplaceable resource. Industries and urbanised societies started to use Nature as a bottomless sink, into which one can throw anything not wanted. Our planet earth is therefore in an existential crisis, which is of an ecological as well as an economic nature.

HEALTH AND SURVIVAL

Throughout the past century, humankind has made a tremendous effort to understand the biological intricacies of nature. It started with the traditional fermentation of food to the commercial exploitation of all types of biological cells. The most incredible advances occurred since the mid 1940s with the discovery of the life saving antibiotics to the present rapid progress in understanding the genetic basis of the living cells. The latter progress has led to the development of new products and processes useful in human and animal health, food and agriculture, and the environment. It appears, however, that these enormous discoveries could not be integrated into the natural cycles of matter. As a consequence, prevention is being replaced by curing continuously occurring medical and agricultural ailments. This can easily be visualised by the enormous over- and misuse of antibiotics causing a lowering of the immune systems and an ever increasing resistance against these drugs among microorganisms, which in turn requires the never ending search for new antibiotics and vaccine. This is becoming even more difficult as it was found recently that the non-stop battle against parasites appears to be altering the human genome more radically than could ever be imagined, which could explain the reasons why some cultural societies show higher infectious disease rates than others.

The intensification of agriculture during the green revolution with its reliance on antibiotics and hormones in feeding animals in so-called 'animal factories' [eg chicken, pigs, cows] or breeding animals for fast growth as well as on irrigation and chemical inputs in crop fields has led also to serious health and environmental problems. The reason for this unfortunate development must be sought in the fact that research and development, personnel and finance are concentrated in rich countries, led by global corporations and following the global market demand dominated by high-income consumers. As a result, research has neglected opportunities to develop technologies for poor people in developing countries representing approximately 80% of the world population.

According to the UNDP report on Human Development, in 1998 global spending on health research was US$ 70 billion, but just US$ 300 million was dedicated for vaccines for HIV/AIDS and about US$ 100 million to malaria research. Of special concern is the fact that new drugs and vaccines are being developed to export for profit rather than to sell cheaply to local people. Patent, license and royalty have become a tool to create wealth for the developed countries. Although people are the real wealth of nations, it has so far not been possible to create an environment in which people can develop their full potential and lead productive lives in accordance with their needs and interests.

DNA TECHNOLOGY

The third millennium has therefore been marked by a paradox in the life sciences. Modern genetic engineering techniques and sophisticated culture skills honed into a fine art over the last two decades has given rise to new approaches in the medical and agricultural sciences.

In the medical sciences, these new approaches led to significant improvements in the fight against human diseases increasing life expectancies. There is no doubt, that the DNA technology approach in the medical sciences could improve significantly health and living standards of mankind providing the products will be readily available and affordable to all societies. Stem cell research will hopefully correct body malfunctions and cancerous diseases in human life. However, the use of stem cells and cloning techniques requiring human embryos are at present disputed areas of research a both techniques involve the manipulation of human cells. How far should the scientist be allowed to go. Many societies and religious groups are opposed to the manipulation of human cells and this field of biotechnology is becoming a big ethical issue.

In the agricultural sciences, plant cells have been genetically altered for resistance to diseases and pests as well as to increase yields and nutrition for food and energy. This development saves undoubtedly the use of naturally foreign chemicals, but has led to a steady loss of wild strains and o. biological diversity. How different is genetical manipulation to natural selection or old fashioned breeding by farmers, which also alters biological diversity? How will this rapid selection affect our natural cycles of matter and the environment? Is genetic manipulation more favourable to the environment than changing our farm management system, which fosters pest breeding? How will genetic engineering affect traditional food fermentation and will pests overcome the resistance as the bacteria developed resistant strains against antibiotics? How will the culture and society react to the change of dominance from the people of rural communities to corporate bodies? Is it really a wonder that society is asking questions and is divided on the issue of GM crops?

Public debates about the safety of new products introduced in the market go back centuries and were often based less on science than on politics of the time. Similarly today, much of the debate about agricultural biotechnology is steered by myths and misinformation and not by science. The scientific community, with stronger support from governments, must do more to openly address science and technology issues with their publics. Thomas Sinclair asked "What are realistic options for increasing resource availability?" He believes that the solutions are in researching basic questions leading to understanding, improving and managing soils and crops, even though these do not seem to be currently politically attractive to research funders. Thus, genetic engineering need not be yield-increasing technology to have an impact, but should simply need to be yield 'permitting' technology. Such an immediate application to stop current yield losses to pests and abiotic stresses should be enough to stop food shortages in many developing countries, in particular in dessert areas. Parrott indicated that existing transgenic technology to protect crops against many diseases is well established and highly effective, whereas more should be done in the field of abiotic stresses. One should distinguish between wholesale degradation of soils, such as the reduction of a productive forest slope to bedrock or fertile soils to salinity and reduction in biodiversity per se through loss of wild organisms.

Latest GM crop trials in the Uk clearly indicate that one should not generalise and declare all GM is good or all GM is bad and that some GM crops such as oilseed rape and sugar beet fare badly whereas maize appeared to have beneficial results for the environment. Since herbicide-tolerant crops can also be generated without genetic manipulation, they may have the same effect on biodiversity than the GM crops. It is very encouraging to see the successes of soil improvement techniques in some developing countries where improved nutrient management has started to show success in higher crop yields. Poor soils, on the other hand, lead to further yield reduction despite the use of GM crops. The development of knowledge and the application with respect to genetic engineering are socio-technical processes, embedded in a social, economic, cultural and political context. Modern technologies should be blended in with traditional technologies. Furthermore, the future of biotechnology depends very much on quality science education.

A further critical point in the scientific and biotechnological development is, of course the question about the relationship between science, religion and ethics, which certainly affects the impact of science and technology on society. Whereas their appeared to be conflicts in some cases, Watts outlined that the relationship between science and religion needs to be based on mutual respect. It should be realised that science is not as culturally neutral and value free as is sometimes supposed as it is dominated at present by America and other Western countries, in other words Christian countries.

However, the most recent First Arab Human Development Reports lead the way to a better mutual understanding. These conflicts become even greater, spreading through all religions and societies if cloning is expanded to humans in the medical and health areas. In regard to ethics, it has been claimed that ethics is on the corporate payroll and subsequently developed countries have been warned on imposing ethics onto developing countries. On the other hand, stem cell research shows great promises in the battle against human diseases. There is no doubt that the scope of research ethics has to be broaden in order to achieve a mutual agreement and or respect between the scientist and the community. One area is undoubtedly science education together with the realisation by the corporate bodies that it can and should not dictate the way of life of societies. Despite these controversies one should never forget that the seeds of biotechnology are rooted deep in the past, but the fruits of its growth should benefit all societies in one way or the other.

SOCIAL ASPECTS OF BIOTECHNOLOGY

In order to establish the real need for what type of biotechnology is required for developing countries, one has first to realise that there exist three major climatic zones, namely

a) the temperate zones of the developed world;

b) the tropical zones of developing countries; and

c) the arid zones of developing countries

Moreover, there is no escaping the fact that over 90% of biotechnological research and development is occurring in the temperate zones of our world. Secondly, the most serious problems in the developing countries concern:

- *Health.* It has been estimated by UNDP that 2.4 billion people are without access to basic sanitation and 11 million children under five are dying annually from preventable causes.
- *Poverty.* Approximately 1.2 billion people live on less than US$ 1 a day and 2.8 billion on less that US$ 2 a day.
- *Starvation.* The developing world has still 826 million undernourished people, living predominantly in the arid zone areas of Africa and Asia. For example, Sub-Saharan Africa has an infant mortality rate of more than 100 and an under-five mortality rate of more than 170 per 1,000 live births.

There is absolutely no doubt in my mind, that our present biotechnological knowledge is able to abolish the health and poverty problems, with the reduction or even eradication of starvation in arid zones well on the way using modern genetically modified agriculture. It should therefore be possible to create an environment in which people can develop their full potential and lead productive and creative lives in accordance with their needs and interests. Since most of the biotechnology research and development is concentrated in the temperate climatic zones, a closer cooperation has to occur to facilitate an adaptation of the old and newly developed technologies to the appropriate climatic zone, the particular society and the local environment.

What are the most urgent biotechnological issues in developing countries for the improvement of human development? In considering all the available technologies together with those under development, the first and basic priority thinking has to be the fact that 'technologies in general do not transfer from developed to developing countries. Rather they need to be built up in situ using local knowledge and innovative ability after which, if successful, they are being adopted'. Social aspects of psychology, religion and gender are of paramount importance.

Health

It is very hard to understand why our International Agencies have failed to eliminate the health problems in developing countries. Biotechnologists and in particular microbial technologists must fail to comprehend why the numerous existing technologies have neither been supported nor implemented. Basic sanitation should be made available as a first priority in human development. It is well known that the handling of human and animal excreta or manure depends on and varies with the social and religious background of a particular society, but the technologies available today caters for all aspects of human society.

The basis for socio-economical integrated biosystems, recently referred to also as 'ecological sanitation' has been with us for the last century in form of

a) composting toilets wet or dry

b) composting together with household waste and other biomass

c) anaerobic digestion or biogas digesters.

Neither of these systems requires any handling of excreta or manure, as these can be channelled through pipelines to a cesspit, compost or anaerobic digester. Whereas composting and anaerobic digestion takes care of all pathogens, the cesspit requires the addition of lime or ash to help desiccate the manure and raise the pH, which effectively kills off all pathogens within several months. This pathogen-free manure can directly be returned to the soil or, in order to be safe, be added to compost heaps before improving the fertility of soils.

Composting reaches a thermophilic range of 50-800 C and anaerobic digestion reduces the redoxpotential to such an extent that pathogens are killed. In the case of cesspits, compost and anaerobic digestion, the residue can be safely used for soil improvement, replacing chemical fertilisation. Whether this organic fertilisation is used on fields of flower or food crops is dependent on the local society and their religious belief. Technologies are also available to treat and recycle the water [greywater] effectively for reuse.

The best ecological as well as economic sanitation system is undoubtedly the use of anaerobic or biogas digesters. In this case, the families will be able to use the biogas formed for cooking and electricity. These systems are becoming very popular in Bangladesh, Vietnam, Cambodia and China. Biodigesters are also the best socio-economic systems with establishment costs becoming very cheap as has been demonstrated in Vietnam, Bangladesh and Cambodia using polyethylene tubing as construction material.

Sizes of these digesters in use at present range from 6 - 20 m^3, whereas digesters up to 2000 m_3 require concrete or steel construction. It is very distressing to realise that many international organisations, eg FAO, do not condemn the use of raw manure for soil improvement, and do not emphasize the absolute need for treatment before use as was shown in one of the latest electronic conferences on 'Area wide integration of crops and livestock production'. If these old and improved biotechnological techniques are fully supported and properly funded in a way the introduction of GMO crops are funded by UNDP, UNEP, UNIDO, FAO, WHO etc, health in developing countries could be quickly raised to the level of developed countries.

In addition to these disease prevention technologies, International Agencies must also be forced or force corporations to make biotechnological products available to the 2 billion people [one third of the world population], who still have no access to low-cost essential medicines such as penicillin, a technological process developed in the late 1940s. Such access would eliminate outbreaks of measles, cholera, meningitis and haemorrhagic fever

Poverty

The term poverty is very often misinterpreted with starvation. Whereas poverty is flourishing in most developing countries, this is certainly not the case with lack of food causing starvation. However, poverty may cause starvation as the people are not able to buy the available food. Poverty is mainly caused through strong increases in urbanisation, as low income rural farmers stream into the cities to find work and a better income. This depletes very efficient and productive rural agriculture, and reduces the possible maximal agricultural food or crop production.

Whereas the green revolution technologies resulted in increased food production in favourable and irrigated environments, they had little impact on the millions of smallholders living in rainfed and marginal areas, where poverty is concentrated in Asia. The reasons for this trend are manyfold, but can mainly be traced to changes in farm management [single crop production] as well as farm mechanisation [big farms] resulting in a severe reduction of small farm holders, and a severe reduction of funding and investment into agriculture. In order to alleviate poverty and make certain that we are able to continue with feeding an ever increasing population, we need a socio-economic biotechnology revolution realising that we have to learn from the mistakes of the green revolution and secure a proper income to the farmer.

One major problem of the green revolution was that the farmers were made to believe in the production for markets, forgetting their own consumption. It is evident that farmers in some developing countries grow crops solely under contract for supplying processing factories, while they have to buy the food for their own consumption. Traditional food was demoted, whereas canned and bottled food was promoted. Local (traditional) wisdom and knowledge in food preservation and medicine were treated as an 'uncivilized way of life' and is disappearing, so the younger generation is not practising it anymore. Such a new biotechnology revolution has to take into consideration

1. that farming in developing countries is profoundly different from developed countries with crops like cassava, rice, soybean, sago etc;
2. that farms are small and should stay small with minimal mechanisation but more intensive and integrated farming;
3. that single crop production must make way to a multi-product farming, including livestocks on the farm;
4. that we use existing biotechnological techniques to develop biotechnological industries using locally grown biomass such as sago palm and cassava;
5. combine the biomass waste, excessive amounts of agro-industrial wastes with human and animal waste treatments for novel product and renewable energy production.

Such a sustainable socio-economic biotechnology revolution can best be established in form of so-called 'biorefineries', whereby all the biomass is used to improve the living standard of the people. Such biorefineries require a host of different biotechnological and physical techniques ranging from anaerobic digestion of wastes to surplus biomass conversion to renewable energy, food, feed and commodity product formation. All biotechnological and physical techniques are readily available and can immediately be implemented.

The additional incorporation of aspects of modern biotechnological techniques will come as soon as society has learned the advantages of the existing technologies. Some biotechnological issues such as pest-resistant plants and organic fertilisation will involve some GMO plants. However, one should always be aware and not forget, that local breeding experience may be a better way to go initially than GMO introduction.

It is very surprising, for example, that agricultural biotechnology development sofar has totally ignored plants such as cassava and the sago palm as a starch resource, as they are hardly known in developed countries. Cassava can yield up to 65 t/ha with a 65% starch content in marginal soils and the sagopalm can easily produce 25 t of starch/ha in swampy areas unsuitable for any other crop. Since the average intake of food for human is, in general, about 250 kg of grain per year, one hectare of sago plantation can feed 100 people and a 1000 ha sago plantation can subsequently save 100,000 humans from hunger, a clear example of the potential of sago as a major starch crop of the world. Both crops are ideal for obtaining a variety of bioproducts ranging from biofuel, bioplastics, biodetergents, biolubricants to bio-pharmaceuticals.

The establishment of biorefineries will diversify rural farming, keep people employed in the rural areas and will also increase the income of the individual farmer, helping in the alleviation of poverty.

Starvation

The elimination of starvation in the arid zones of the world is the biggest challenge to agricultural biotechnology. Much more effort should be put into the breeding or genetic modification of crops for drought resistance. Improvement of soil condition using treated human and animal manure should go hand-in-glove with the introduction of drought resistant or at least drought tolerant crop varieties. The most beneficial aspect of GMO crops in these areas would immediately improve livestock and food production. However, one has to be aware that different regions have different types of crop demands. Genetic modification for drought resistance should occur with local plants and crops used by the local societies. We should stop introducing GMO plants from temperate zones in developed countries and respect the local food demand and varieties. Such a project would cause much less opposition than genetically modified foreign crops.

The elimination of starvation in arid zones could become an ideal place for combining 'old' biotechnological concepts of soil fertility improvement with 'new or modern' biotechnological concepts of increasing crop and livestock production. Soil fertility improvement should also go hand in hand with a reduction or elimination of rain and other forest clearings, which in turn would stop the expansion of the arid zone areas. Such a combined concept would create more small holding farms with a proper income, helping to stop the development of further poverty.

What are realistic options for increasing resource availability? The solutions are in researching basic questions leading to understand, improve and manage soils and crops. Modern gene technology can certainly take care of crop destruction and spoilage, but what about the soil? Here we have to remember that Nature has given us the perfect answer in the cycles of matter which have worked well for centuries until we used the soil and the environment as bottomless sinks. How do these cycles work and what can we do to regenerate and adapt these cycles to modern need?

CLEAN AND GREEN TECHNOLOGIES

During the past decades, production systems have been based on the assumption that wastes are an unavoidable part of our daily lives but that ecological and environmental destruction could be avoided because of the planet's vast natural resources. However, ecosystems and renewable resources are being destroyed at an increasingly rapid rate and the problems of pollution and waste disposal are growing. It is thus becoming increasingly evident that new production methods must be devised to fulfil society's basic needs. In order to maintain the vitality of the rural sector and preserve the environment, a system must be devised in which energy, food, animal feed, fuel and fertiliser requirements can all be met from renewable resources used at a sustainable level.

In regard to waste management, environmentally friendly waste incineration plants are becoming very popular in developed as well as developing countries. It varies from the largest waste to energy plant in Denmark to the municipal solid waste to energy conversion plant in Kuala Lumpur [Malaysia] operating on 1,500 t/day [= 540,000 t/year] producing 640 kW/day. The Danish Plant has been calculated to save the emission of 80,500 t of CO_2/year. Ecological engineering was used for wastewater treatment using aquaculture principles, which not only reduced carbon, nitrogen and phosphorus, but the additional anaerobic treatment reduced also metals by 48-72%. These examples demonstrate that municipal solid wastes can be incinerated or in connection with industrial wastewater anaerobically digested for recycling energy, fertiliser and water in an environmentally friendly way.

The new socio-economic concept is based on the requirement for full exploitation of a harvested renewable resource and the replacement of monoculture/monoproduct farming with a multiproduct system. Because it produces a variety of products, this system will hopefully enjoy a constant and reliable market demand and will be able to secure income for the rural sector as well as for joint venture industries. It also will help to reduce the vast multiplication rates of pests in mono-culture system operations.

The vast majority of the populations in developing countries live in villages and the importance of rural technology development in those countries cannot be overstated. In the context of technological development it is generally considered imperative to see that such developments are of benefit to their lives as well. Corporate bodies have to realise that they should not and cannot dominate rural technologies. It certainly is recognised that rural life is affected by many factors, such as political and social institutions, rural economic structures, communication, education and technology. An initial step is therefore that building up the rural technology capacity is one of the tools for development. Another step is the recognition that the technology employed or developed should suit locally available resources and skills and be in harmony with local culture.

In order to develop an appropriate biotechnology, resources available together with the social structure of the population are of vital importance. It is often the need and not the economics of the process, which is of importance. Here seems to lie one of the cardinal differences between the thoughts of appropriate technology in developed, from those in developing countries. In this context, a new concept has been developed specifically for rural communities: the so-called integrated rural biotechnology systems. The development of these systems depends primarily on the climatic conditions of the regions. Two of the premises of sustainable development are that economic growth has to be in harmony with the environment and that a rational and sustainable use of natural resources has to be implemented. In congruence with such premises, industrial development has to change from the degradative to the sustainable style, which requires the adoption of cleaner production systems.

The United Nations Environment Program (UNEP) defines the cleaner production concept as 'the continuous application of an integrated preventive environmental strategy to processes, products and services to increase eco-efficiency and reduce risks to humans and the environment'. One of its distinctive features is that reduction of the quantity and toxicity of all emissions and wastes is made before they leave the process stream. In the case of services, environmental concerns should be incorporated into design and delivery.

Eco-efficiency, on the other hand, involves 'the delivery of competitively priced goods and services that satisfy human needs and bring quality of life while progressively reducing ecological impacts and resource intensity, throughout the life cycle, to a level at least in line with the Earth's estimated capacity'. It is becoming very clear that adoption of clean production systems by industries calls for fundamental changes, not only at the technological level but also at the legislative level. Cleaner bioprocesses are under intensive research and development following the general guidelines for cleaner production.

ROLE OF MICROBIOLOGY

Since microorganisms are responsible for diseases and/or sustaining life, we have to start educating people and the whole society on the role of microorganisms in their life. This is a very difficult task as we can not see them in nature [the good ones], but only feel them [the bad ones, if sickness occurs]. We have to educate them that it is the Society itself which has and still is causing not only the destruction of the environment, but also the life and existence of the Society. It is not only the fault of Governments if infectious diseases occur, as we have no one else to blame than ourselves.

With the help of the Society, microbiology and its application, microbial technology, can lead very quickly to meeting and solving many, if not all, of above problems. This cooperation can bring back a clean and sustainable environment without stopping or holding back further development. If one interprets 'sustainability' as the recognition of long-term management and use of natural resources together with the protection of the environment it becomes clear that we have to continue our 'sustainable development' for higher living standards in tune with Nature. This can be achieved using microbial and fermentation technologies involving man, biomass and industry emphasizing the utilisation of renewable resources with a low environmental impact and a high regeneration capacity. Microbial technology must and can provide for:

- Environmental Management through the bioconversion of domestic and animal wastes into two non-polluting fuels such as biogas, ethanol and value-added products such as algae, fresh water fish etc.
- Bioconversion of agro-industrial residues and products into value-added products such as mushroom, biofertiliser through composting, silaging, protein-enriched feed etc.
- Enhancement of soil fertility and stability through the direct application of anaerobically digested sludge material or microbial fertilisers [e.g. sound and responsible farm management, rhizobia and algae etc.]
- Public health programme by eliminating enteric parasites and microorganisms through anaerobic digestion processes or cleaner ecological environment management [e.g. composting, garbage collection etc.]
- Waste water treatment and waste utilisation through microbial-based systems
- Concentration and leaching of valuable minerals and removal of heavy metals from rivers and estuaries from mining waters, low grade ores as well as contaminated rivers
- Substitution of toxic chemicals in using biopesticides [e.g. Bacillus thuringiensis, antagonistic microorganisms]
- Food through improvement of indigenous fermentations, e.g. puto, nata de coco, suka, tempeh, bagoong etc.

Very soon we have to rely totally on microorganisms for our supply on food, feed, fertiliser, energy and disease control. This we can do provided we harnass and protect the biodiversity of microorganisms, plants and animals. The role of microorganisms in health, food, energy and agriculture is unlimited, but we need urgently microbiologists, not for biomedical areas, but for agro-business and food fermentation areas. Here we do not need genetical engineering, but a sound knowledge in microbiology, biochemistry and some aspects of chemical engineering.

The importance of microbial technology for the future of mankind and thus each society can be seen in the establishment of an international network by Unesco, UNEP and ICRO. At present there exist 29 Microbiological Resources Centres (MIRCENs) worldwide with a World Data Centre in Japan. These institutions together with IOBB [International Organisation for Biotechnology and Bioengineering] try to help the exchange, training and research of manpower as well as in an advising capacity.

In the field of microbial biotechnology, the world activities at present center very much around agricultural biotechnologies and waste utilisation. Agriculturally based biotechnology is mainly concerned with renewable biomass resource production, which led not only to plant disease resistance and higher yields, but also to an increasing self-efficiency in some developing countries. This development gives the microbial biotechnologist the task to:

- secure that the farmer is able to stay on his farm by changing mono-products to value-added multiproducts;
- produce products which can be used domestically taking the pressure from a depressed or fluctuating export market;
- introduce flexibility and convert a one-product farming system into a multi-product farming system.

Stabilising the farming community is the only safeguard for the continuous availability of farm products and thus food for the future.

FUTURE PERSPECTIVES FOR LIFE AND HUMAN DEVELOPMENT

The biotechnology issues for developing countries in future requires a change from the presently commercially driven to a more human development, combining 'old' and 'modern' biotechnological techniques for the improvements in the health and living conditions of 80% of our world population. It is very encouraging to observe the establishment of new organisations such as the Program for Research and Documentation for a Sustainable Society (ProSus), ecological sanitation (Ecosan) and many others together with the information distributors Livestock Research and Rural Development (LRRD), Centre for the Analysis and Dissemination of Demonstrated

Energy Technologies (CADDET), Ecosan Newsletters as well as discussion groups such as Integrated Biosystems (IBS) emphasizing the need for such a development.

It is a great opportunity for the International Organisations [UNDP, UNEP, UNIDO, FAO, WHO etc.] to take up the challenge and realising that the so-called 'modern biotechnology' alone cannot solve the problems. As long as the present biotechnology development is driven only by commercial enterprise, human development in developing countries will lag increasingly behind and cannot progress as it would with the application of a total biotechnology concept, such as a socio-economic sustainable bio-integrated system. Sustainable development and human development should not and cannot go separate directions.

If one combines all these efforts reported into an socio-economic strategy, farms and/or farm cooperatives as well as agro-industries are able to combine natural renewable resource production with bioenergy, food, feed, biofuel and fertiliser production. Such a system must be and can be flexible and should be adaptable to local conditions. The new term Bio-Refinery has been given to these systems. In all our consideration one should never forget the natural cycles of matter and the microbial soil population, both so important for the production of our renewable resources, maintenance, environment, and improvement of living standards. Sustainability can be obtained together with higher health and living standards, if appropriate technologies are applied according to the climatic region and local society.

In order to determine what part of biotechnology can be used in rural farming as well as in the biorefineries, we have to analyse what type of renewable resources are available, the climatic zone of the region, the amount of waste produced from humans and animals, and the cultural and societal structure of the local village and regional communities. The demands of communities in developed countries is different from those in developing countries. All over the world, however, rural communities are characterised by certain common factors. In using disease resistant seeds obtained through conventional breeding or DNA-recombinant technology together with proper soil enrichment fertilisation should secure a safe income to the rural farmer as well as secure all the food required for the urban societies.

REFERENCES

Gessner, M.O. and Newell, S.Y., "Biomass, growth rate, and production of filamentous fungi in plant litter", *Pages 390-408 in: Manual of Environmental Microbiology*, Second Edition, ASM Press, Washington, D.C., 2002.

Holmes, B. and Costas, M., "Identification of Enterobacteriaceae by computerized methods", In: Board, R.G., Jones, D. and Skinner, F.A. (Eds.) *Identification methods in applied and environmental microbiology*, Oxford: Blackwell Scientific Publications, 1992.

10

Microbial Leaching Mechanisms

Future sustainable development requires measures to reduce the dependence on nonrenewable raw materials and the demand for primary resources. New resources for metals must be developed with the aid of novel technologies. In addition, improvement of already existing mining techniques can result in metal recovery from sources that have not been of economical interest until today. Metal-winning processes based on the activity of microorganisms offer a possibility to obtain metals from mineral resources not accessible by conventional mining. Microbes such as bacteria and fungi convert metal compounds into their water-soluble forms and are biocatalysts of these leaching processes. Additionally, applying microbiological solubilization processes, it is possible to recover metal values from industrial wastes which can serve as secondary raw materials.

In general, bioleaching is a process described as being "the dissolution of metals from their mineral source by certain naturally occurring microorganisms" or "the use of microorganisms to transform elements so that the elements can be extracted from a material when water is filtered trough it". Additionally, the term "biooxidation" is also used. There are, however, some small differences by definition: Usually, "bioleaching" is referring to the conversion of solid metal values into their water soluble forms using microorganisms. In the case of copper, Copper sulfide is microbially oxidized to copper sulfate and metal values are present in the aqueous phase.

Remaining solids are discarded. "Biooxidation" describes the microbiological oxidation of host minerals which contain metal compounds of interest. As a result, metal values remain in the solid residues in a more concentrated form. In gold mining operations, biooxidation is used as a pretreatment process to (partly) remove pyrite or arsenopyrite. This process is also called "biobeneficiation" where solid materials are refined and unwanted impurities are removed. The terms "biomining", "bioextraction", or "biorecovery" are also applied to describe the mobilization of elements from solid materials mediated by bacteria and fungi. "Biomining" concerns mostly applications of microbial metal mobilization processes in large-scale operations of mining industries for an economical metal recovery.

The area of "biohydrometallurgy" covers bioleaching or biomining processes. Biohydrometallurgy represents an interdisciplinary field where aspects of microbiology (especially geomicrobiology), geochemistry, biotechnology, hydrometallurgy, mineralogy, geology, chemical engineering, and mining engineering are combined. Hydrometallurgy is defined as the treatment of metals and metal- containing materials by wet processes and describes "the extraction and recovery of metals from their ores by processes in which aqueous solutions play a predominant role". Rarely, the term "biogeotechnology" is also used instead of biohydrometallurgy.

One of the first reports where leaching might have been involved in the mobilization of metals is given by the Roman writer Gaius Plinius Secundus (23-79 A.D.). In his work on natural sciences, Plinius describes how copper minerals are obtained using a leaching process. The translation reads approximately as follows: "Chrysocolla is a liquid in the before mentioned gold mines running from the gold vein. In cold weather during the winter the sludge freezes to the hardness of pumice. It is known from experience that the most wanted is formed in copper mines, the following in silver mines. The liquidise also found in lead mines although it is of minor value. In all these mines chrysocolla is also artificially produced by slowly passing water through the mine during the winter until the month of June; subsequently, the water is evaporated in June and July. It is clearly demonstrated that chrysocolla is nothing but a decomposed vein".

The German physician and mineralogist Georgius Agricola (1494-1555) describes in his work de re metallica also techniques for the recovery of copper that are based on the leaching of copper-containing ores. A woodcut from his book illustrates the (manual) transport of metal-containing leachates from mines and their evaporation in the sunlight.

The Rio Tinto mines in south-western Spain are usually considered the cradle of biohydrometallurgy. These mines have been exploited since pre-Roman times for their copper, gold, and silver values. However, with respect to commercial bioleaching operations on an industrial scale, biohydrometallurgical techniques had been introduced to the Tharsismine in Spain 10 years earlier. As a consequence to the ban of open air ore roasting and its resulting atmospheric sulfur emissions in 1878 in Portugal, hydrometallurgical metal extraction has been taken into consideration in other countries more intensely. In addition to the ban, cost savings were another incentive for the development: Heap leaching techniques were assumed to reduce transportation costs and to allow the employment of locomotives and wagons for other services. From 1900 on, no open air roasting of low-grade ore was conducted at the Rio Tinto mines.

Efforts to establish bioleaching at the Rio Tinto mines had been undertaken in the beginning of the 1890s. Heaps (10 m in height) of low-grade ore (containing 0.75% Cu) were built and left for one to three years for "natural" decomposition. 20 to 25% of the copper left in the heaps were recovered annually. It was calculated that approximately 200,000 t of rough ore could be treated in 1896. Although industrial leaching operations were conducted at the Rio Tinto mines for several decades, the contribution of bacteria to metal solubilization was confirmed only in 1961, when Thiobacillus ferrooxidans was identified in the leachates.

Early reports state that factors affecting bioleaching operations were the height of the heap, particle size, initial ore washing with acid, and temperature control to about 50°C. Another critical factor was the supply of water for the leaching heaps. Although usually acidic mine waters were used for ore processing, 4 billion liters of freshwater were required annually.

Although metal leaching from mineral resources has a very long historical record and although the oxidation of reduced sulfur compounds and elemental sulfur resulting in the formation of sulfuric acid was demonstrated already in the 1880s, the oxidation of metal sulfides was not described until 1922 when mobilization of zinc from zinc sulfide was investigated. It was found that the transformation of zinc sulfide to zinc sulfate was microbially mediated. Based on these results, the economic recovery of zinc from zinc-containing ores by biological methods was proposed.

In 1947, Thiobacillus ferrooxidans was identified as part of the microbial community found in acid mine drainage. A first patent was granted in 1958. The patent describes a cyclic process where a ferric sulfate/ sulfuric acid lixiviant solution is used for metal extraction, regenerated by aeration (ferrous iron oxidation by iron-oxidizing organisms), and reused in a next leaching stage.

PRINCIPLES OF MICROBIAL METAL LEACHING

Leaching Mechanisms

Mineralytic effects of bacteria and fungi on minerals are based mainly on three principles, namely acidolysis, complexolysis, and redoxolysis. Microorganisms are able to mobilize metals by:

(1) the formation of organic or inorganic acids (protons);

(2) oxidation and reduction reactions; and

(3) the excretion of complexing agents.

Sulfuric acid is the main inorganic acid found in leaching environments. It is formed by sulfur-oxidizing microorganisms such as thiobacilli. A series of organic acids are formed by bacterial (as well as fungal) metabolism resulting in organic acidolysis, complex and chelate formation. A kinetic model of the coordination chemistry of mineral solubilization has been developed which describes the dissolution of oxides by the protonation of the mineral surface as well as the surface concentration of suitable complexforming ligands such as oxalate, malonate, citrate, and succinate. Proton-induced and ligand-induced mineral solubilization occurs simultaneously in the presence of ligands under acidic conditions

Models of Leaching Mechanisms

Originally, a model with two types of mechanisms which are involved in the microbial mobilization of metals has been proposed:

(1) Microorganisms can oxidize metal sulfides by a "direct" mechanism obtaining electrons directly from the reduced minerals. Cells have to be attached to the mineral surface and a close contact is needed.The adsorption of cells to suspended mineral particles takes place within minutes or hours. This has been demonstrated using either radioactively labeled Thiobacillus ferrooxidans cells grown on $NaH^{14}CO_3$ or the oxidative capacity of bacteria attached to the mineral surface. Cells adhere selectively to mineral surfaces occupying preferentially irregularities of the surface structure. In addition, a chemotactic behavior to copper, iron, or nickel ions has been demonstrated for Leptospirillum ferrooxidans. Genes involved in the chemotaxis were also detected in Thiobaçillus ferrooxidans and Thiobacillus thiooxidans.

(2) The oxidation of reduced metals through the "indirect" mechanism is mediated by ferric iron (Fe^{3+}) originating from the microbial oxidation of ferrous iron (Fe^{2+}) compounds present in the minerals. Ferric iron is an oxidizing agent and can oxidize, e.g., metal sulfides and is (chemically) reduced to ferrous iron which, in turn, can be microbially oxidized again. In this case, iron has a role as electron carrier. It was proposed that no direct physical contact is needed for the oxidation of iron.

In many cases it was concluded that the "direct" mechanism dominates over the "indirect" mostly due to the fact that "direct" was equated with "direct physical contact". This domination has been observed for the oxidation of covellite or pyrite in studies employing mesophilic T. ferrooxidans and thermophilic Acidianus brierleyi in bioreactors which consisted of chambers separated with dialysis membranes to avoid physical contact. However, the attachment of microorganisms on surfaces is not an indication per se for the existence of a direct mechanism. The term "contact leaching" has been introduced to indicate the importance of bacterial attachment to mineral surfaces.

The following equations describe the "direct" and "indirect" mechanism for the oxidation of pyrite:

direct:

$$2FeS_2 + 7O_2 + 2H_2O \xrightarrow{thiobacill} 2FeSO_4 + 2H_2SO_4 \quad (1)$$

indirect:

$$4FeSO_4 + O_2 + 2H_2SO_4 \xrightarrow{T. ferroxidans\ L. ferroxidans} 2Fe_2(SO_4)_3 + 2H_2O \quad (2)$$

$$FeS_2 + Fe_2(SO_4)_3 \xrightarrow{chemical\ oxidation} 3FeSO_4 + 2S \quad (3)$$

$$2S + 3O_2 + H_2O \xrightarrow{T. Thioxidans} 2H_2SO_4 \quad (4)$$

However, the model of "direct" and "indirect" metal leaching is still under discussion. Recently, this model has been revised and replaced by another one which is not dependent on the differentiation between a "direct" and an "indirect" leaching mechanisms. All facts have been combined and a mechanism has been developed which is characterized by the following features:

(1) cells have to be attached to the minerals and in physical contact with the surface;

(2) cells form and excrete exopolymers;

(3) these exopolymeric cell envelopes contain ferric iron compounds which are complexed to glucuronic acid residues.These are part of the primary attack mechanism;

(4) thiosulfate is formed as intermediate during the oxidation of sulfur compounds;

(5) sulfur or polythionate granules are formed in the periplasmatic space or in the cell envelope.

Thiosulfate and traces of sulfite have been found as intermediates during the oxidation of sulfur. Sulfur granules (colloidal sulfur) have been identified as energy reserves in the exopolymeric capsulearound cells of T. ferrooxidans during growth on synthetic pyrite films. "Footprints" of organic films containing colloidal sulfur granules are left on the mineral surface upon detachment of the bacteria.

From the existing data two "indirect" leaching mechanisms have been proposed whereas no evidence for a "direct" enzymatically mediated process has been found.The mineral structure is the determining factor for the prevailing type of leaching mechanism. In the "thiosulfate mechanism" thiosulfate is the main intermediate resulting from the oxidation of pyrite, molybdenite, or tungstenite. Polysulfide and elemental sulfur are the main intermediates in the "polysulfide mechanism" during the oxidation of galena, sphalerite, chalcopyrite, hauerite, orpiment, or realgar. The presence of iron(III) at the beginning of mineral degradation is an important prerequisite.

The following equations summarize the oxidation mechanisms:

Thiosulfate mechanism (found for FeS_2,MoS_2, WS_2):

$$FeS_2 + 6Fe^{3+} + 3H_2O \rightarrow S_2O_3^{2-} + 7Fe^{2+} + 6H^+ \quad (5)$$

Polysulfide mechanism (found for PbS, $CuFeS_2$, ZnS, MnS_2,As_2S_3,As_3S_4):

$$S_2O_3^{2-} + 8Fe^{3+} + 5H_2O \rightarrow 2SO_4^{2-} + 8Fe^{2+} + 10H^+ \quad (6)$$

$$2MS + 2Fe^{3+} + 2H^+ \rightarrow 2M^{2+} + H_2S_n + 2Fe^{2+} \quad (7)$$

$$H_2S_n + 2Fe^{3+} \rightarrow 0.25S_8 + 2Fe^{2+} + 2H^+ \quad (8)$$

$$0.25S_8 + 3O_2 + 2H_2O \rightarrow 2SO_4^{2-} + 4H^+ \quad (9)$$

Several biomolecules are involved in the aerobic respiration on reduced sulfur and iron compounds. It has been found that up to 5% of soluble proteins of T. ferrooxidans is made of an acid stable blue copper protein, called rusticyanin. Additionally, the iron(II) respiratory system contains a (putative) green copper protein, two types of cytochrome c, one or more types of cytochrome a, a porin, and an iron(II)-sulfate chelate. The acid stability of rusticyanin suggests that it is located in the periplasmic space. Figure 1 shows a scheme of the model which combines the electron transport sequence proposed earlier with concepts stemmingfrom the debate on "direct"/"indirect" leaching mechanisms.

Some details of the metal mobilization mechanism, the importance of the presence and attachment of microorganisms and their active contribution have been demonstrated for the leaching of fly ash from municipal waste incineration (MWI). Generally, several mechanisms of metal mobilization can be distinguished: (1) Contact leaching effect on the release of metals. Stock cultures of Thiobacillus ferrooxidans and Thiobacillus thiooxidans were added to ash suspensions and cells were in direct contact with the fly ash. Growth of thiobacilli might be stimulated by increased energy availability from oxidation of reduced solid particles.

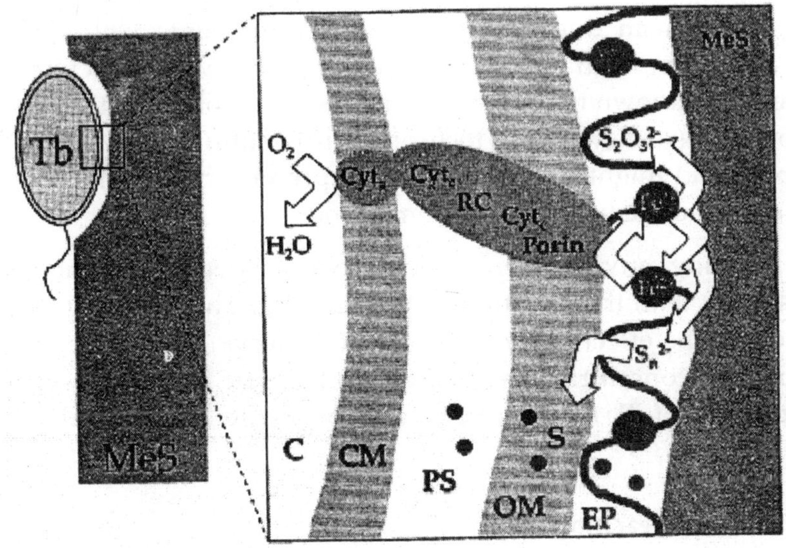

Figure 1. Schematic mechanistic bioleaching model. C: cytoplasm; CM: cell membrane; PS: periplasmatic space; OM: outer membrane; EP: exopolymers; Cyt: cytochrome; RC: rusticyanin; MeS: metal sulfide

(2) Metal solubilization by metabolically active (enzymatic) compounds in the absence of bacterial cells. Stock cultures were filtered to obtain the cell free spent medium.This medium was used for leaching. (3) Metal solubilization by non-enzymatic extracellular metabolic products. Cell free spent medium was autoclaved to obtain a sterile leaching solution without enzymatic activities and to evaluate the leaching ability of acid formed (4) Leaching by fresh medium. Fresh non-inoculated and sterile medium was added to the fly ash suspension and used as control. (5) Chemical leaching due to the preparation of the ash suspension (acidification to pH 5.4). Certain elements such as, e.g., Cd or Zn might be chemically mobilized already during acidification.

MWI fly ash contains reduced copper species (chalcocite {Cu_2S} or cuprite {Cu_2O}) whereas zinc and others are present in their fully oxidized forms.Therefore, copper release from fly ash is directly affected and enhanced by T. ferrooxidans, whereas Zn, as well as Al, Cd, Cr, and Ni, are released primarily due to the acidic environment. Acidification of the fly ash pulp (chemical mobilization) led already to considerable extraction yields for Cd, Ni, and Zn and could slightly be increased using non-inoculated sterile medium as lixiviant. By comparing leached amounts of copper by filtered cell free spent medium with autoclaved sterile spent medium, it was concluded that significant amounts of copper were mobilized - in contrast to other elements - by metabolic products of T. ferrooxidans. Leaching with cell free spent medium indicating a solubilizing mechanism due to extracellular components was significantly more effective than a leaching with autoclaved spent medium where excreted enzymes had been inactivated. It is known that several components involved in the electron transport chain of Thiobacillus (rusticyanin, cytochromes, iron-sulfur proteins) are located in the periplasmic space and might, therefore, also be present in the cell free spent medium catalyzing oxidation of reduced metal compounds.

In many leaching environments conditions (especially iron(II) and iron(III) concentrations) vary with the duration of the leaching. This makes it difficult to assess the importance and the effect of the presence of bacteria. Using an experimental setup to maintain constant concentrations of ferrous and ferric iron, it was possible to show that in the presence of T. ferrooxidans rates of pyrite or zinc sulfide leaching are increased.

Factors Influencing Bioleaching

Standard test methods have been developed to determine leaching rates of iron from pyrite mediated by Thiobacillus ferrooxidans. An active culture of T. ferrooxidans is grown in a defined medium containing (in g LP1):$(NH_4)_2SO_4$ (3.0);K_2HPO_4 (0.5); $MgSO_4{\cdot}7H_2O$ (0.5); KCl (1.0); $Ca(NO_3)_2 \cdot 4H_2O$ (0.01); $FeSO_4 \cdot 7H_2O$ (44.22); and 1 mL 10 N sulfuric acid. Cells are harvested, diluted, and added to pyrite suspensions with a pulp density of 20 g LP1. Total soluble iron as well as sulfate formed during oxidation is periodically determined.

Metal bioleaching in acidic environments is influenced by a series of different factors. Physicochemical as well as microbiological factors of the leaching environment are affecting rates and efficiencies. In addition, properties of the solids to be leached are of major importance. As examples, pulp density, pH, and particle size were identified as major factors for pyrite bioleaching by Sulfolobus acidocaldarius. Optimal conditions were 60 g LP1, 1.5, and ~20 ?m, respectively. The influence of different parameters such as activities of the bacteria itself, source energy, mineralogical composition, pulp density, temperature, and particle size was studied for the oxidation of sphalerite by T. ferrooxidans. Best zinc dissolution was obtained at low pulp densities (50 g LP1), small particle sizes, and temperatures of approximately 35°C.

Metal oxidation mediated by acidophilic microorganisms can be inhibited by a variety of factors such as, e.g., organic compounds, surface- active agents, solvents, or specific metals: The presence of organic compounds (yeast extract) inhibited pyrite oxidation of T. ferrooxidans. Certain metals present in bioleaching environments can inhibit microbial growth, therefore reducing leaching efficiencies. For instance, arsenic added to cultures inhibited Sulfolobus acidocaldarius grown on pyrite and T. ferrooxidans grown on arsenopyrite.Additions of copper, nickel, uranium, or thorium adversely influenced iron(II) oxidation by T. ferrooxidans with uranium and thorium showing higher toxicities than copper and nickel. Silver, mercury, ruthenium, and molybdenum reduced the growth of Sulfolobus grown on a copper concentrate. Industrial biocides such as tetra-n-butyltin, isothiazolinones, N-dimethyl-Nb-phenyl-Nb-(fluorodichloro-methylthio)-sulfamide, or 2,2b-dihydroxy-5,5b-dichlorophenyl-methane (dichlorophen) reduced the leaching of manganese oxides by heterotrophic microorganisms. Biocides were externally added as selective inhibitors to suppress unwanted organisms and to improvemanganese leaching efficiencies. At low concentrations of ~5 mg LP1, however, manganese mobilization was increased by 20%.

Also gaseous compounds can show inhibitory effects on metal leaching: Aqueous-phase carbon dioxide at concentration '10 mg LP1 was inhibiting growth of T. ferrooxidans on pyrite-arsenopyrite-pyrrothite ore. Optimal concentrations of carbon dioxide were found to be in the range of 3 to 7 mg LP1.There are reports on the stimulation of bacterial leaching and the increase of leaching rates by supplementing leaching fluids with carbon dioxide. Concentrations of 4% (v/v) carbon dioxide in the inlet gas of a fermenter showed maximum growth rates of T. ferrooxidans, maximum iron(II), copper, and arsenic oxidation.

Pulp densities of 20 g LP1 delayed the onset of bioleaching of pyrite derived from coal. Increasing pulp densities from 30 to 100 g LP1 decreased rates of pyrite oxidation in Sulfolobus cultures. For fungi such as Aspergillus niger, optimal pulp densities for maximum metal leaching efficiencies were found to be in the range of 30 to 40 g LP1. Quartz particles at pulp densities of 80 g LP1 almost completely inhibited the oxidation of covellite by T. ferrooxidans especially in the absence of iron(II). During bioleaching processes, coprecipitation of metals with mineral phases such as jarosites can reduce leaching efficiencies. In addition, the precipitation of compounds present in the leachates on the minerals to be leached can make the solid material inaccessible for bacterial leaching.

Organic solvents such as flotation or solvent extraction agents, which are added for the downstream processing of leachates from bioleaching, might also lead to inhibition problems (ACEVEDO and. Isopropylxanthate and LIX 984 (used as flotation agent and solvent extraction agent, respectively) prevented the oxidation of pyrite and chalcopyrite by T. ferrooxidans. This fact is of special importance when spent leaching liquors are recycled for a reuse

It has been demonstrated recently that the addition of small amounts of amino acids (cysteine in this case) resulted in an increased pyrite corrosion by T. ferrooxidans as compared to controls without additions. It is suggested that the microorganisms may profit from weakening and break up of chemical bonds mediated by the formation of the cysteine-pyrite complex. This might also be the case under natural conditions by the excretion of cysteine- containing metabolites. An inexpensive alternative to increase metal recovery from ore heaps by the addition of sulfur-containing amino acids such as cysteine has been suggested.

Other metabolites excreted by Thiobacillus might also enhance metal leaching efficiencies: Wetting agents such as mixtures of phospholipids and neutral lipids are formed by Thiobacillus thiooxidans. As a consequence, growth of T. thiooxidans on sulfur particles is supported by the excretion of metabolites acting as biosurfactants which facilitate the oxidation of elemental sulfur. It was also hypothesized that Thiobacillus caldus is stimulating the growth of heterotrophic organisms in leaching environments by the excretion of organic compounds and is supporting the solubilization of solid sulfur by the formation of surface-active agents.

Metal solubilization might also be facilitated by microbial metabolites excreted by organisms other than Thiobacillus which are part of microbial consortia found in bioleaching operations. Microbial surfactants, which show large differences in their chemical nature, are formed by a wide variety of microorganisms. In the presence of biosurfactants which lead to changes in the surface tension, metal desorption from solids might be enhanced resulting in an increased metal mobility in porous media. It has been suggested that this metabolic potential can be practically used in the bioremediation of metal-contaminated soils.

However, there is some evidence that surface-active compounds as well as organic solvents are inhibitory to bioleaching reactions and prevent bacterial attachment. The external addition of Tween reduced the oxidation of chalcopyrite by T. ferrooxidans. Itwas concluded that the need of the microorganisms for surfactants is met by their own formation. In contrast, it was reported that the addition of Tween 80 increased the attachment of T. ferrooxidans on molybdenite and the oxidation of molybdenum in the absence of iron(II).

Bacterial Attachment on Mineral Surfaces

It is known that the formation of extracellular polymeric substances plays an important role in the attachment of thiobacilli to mineral surfaces such as, e.g., sulfur, pyrite, or covellite. Extraction or loss of these exopolymers prevent cell attachment resulting in decreased metal leaching efficiencies. It was concluded that a direct contact between bacterial cells and solid surfaces is needed and represents an important prerequisite for an effective metal mobilization. Interactions between microorganisms and the mineral surface occur on two levels. The first level is a physical sorption because of electrostatic forces. Due to the low pH usually occurring in leaching environments, microbial cell envelopes are positively charged leading to electrostatic interactions with the mineral phase. The second level is characterized by chemical sorption where chemical bonds between cells and minerals might be established (e.g., disulfide bridges). In addition, extracellular metabolites are formed and excreted during this phase in the near vicinity of the attachment site. Low-molecular weight metabolites excreted by sulfur oxidizers include acids originating from the TCA cycle, amino acids, or ethanolamine, whereas compounds with relatively high molecular weights include lipids and phospholipids. In the presence of elemental sulfur, sulfur-oxidizing microorganisms from sewage sludge form a filamentous matrix similar to a bacterial glycocalyx suggesting the relative importance of these extracellular substances in the colonization of solid particles.

BIOLEACHING ENVIRONMENT

A variety of microorganisms is found in leaching environments and has been isolated from leachates and acidic mine drainage. Although environmental conditions are usually described (from an anthropocentric view!) as being extreme and harsh due to pH values and high metal concentrations, these systems can show high levels of microbial biodiversity including bacteria, fungi, and algae. It has long been known that bacteria (Thiobacillus sp.), yeasts (Rhodotorula sp.,Trichosporon sp.), flagellates (Eutrepia sp.), amoebes and protozoa are part of the microbial biocenosis found in acidic waters of a copper mine. Recent detailed investigations based on molecular methods such as DNA-DNA hybridization, 16S rRNA sequencing, RCR-based methods with primers derived from rRNA sequencing, fluorescence in situ hybridization (FISH), or immunological techniques revealed that microbial bioleaching communities are composed of a vast variety of microorganisms resulting in complex microbial interactions and nutrient flows (such as synergism, mutualism, competition, predation).

The organism studied most is Thiobacillus ferrooxidans. Although this is the best known organism from acidic habitats, one may not conclude that this organism is dominant in these ecosystems. It has been found that under specific environmental conditions Leptospirillum sp. is even more abundant than T. ferrooxidans suggesting an important ecological role in the microbial community structure of bioleaching habitats. Thiobacilli are members of the division of Proteobacteria close to the junction between the β and γ subdivision whereas leptospirilli are placed in the Nitrospira division. Genetic studies revealed that the role of T. ferrooxidans in leaching operations has probably been overestimated. Excellent reviews on the genetics of Thiobacilli and leptospirilli have been published recently.

Thiobacillus ferrooxidans belongs to the group of chemolithotrophic organisms. The organism is rod-shaped (usually single or in pairs), non-spore forming, gram-negative, motile, and single-pole flagellated. As carbon source, carbon dioxide is utilized. Ferrou iron is oxidized. Ammonium is used as nitrogen source. Although T. ferrooxidans ha been characterized as being a strictly aerobic organism, it can also grow on elementa sulfur or metal sulfides under anoxic conditions using ferric iron as electron accepto

The genus Thiobacillus represents a versatile group of chemolithoautotrophi organisms. Optimum pH values for growth vary between 2 and 8. It has bee demonstrated that sulfur-oxidizing bacteria are capable of reducing the pH of highl alkaline fly ash suspensions amended with elemental sulfur from approximately 9 t 0.5. It is likely that thiobacilli contribute to increasing acidification of leachin ecosystems in a successive mode: In the initial stages the growth of less acidophilic strain (e.g., *Thiobacillus thioparus*) is stimulated whereas during prolonged leaching the pl decreases gradually supporting growth of more acidophilic strains. This has already bee observed in metal leaching from wastewater sewage sludge.

A variety of thermophilic microorganisms (especially Sulfolobus species) has bee enriched and isolated from bioleaching environments.Temperature optima for growt and metal leaching were in the range between 65 and 85°C. Although copperextractio from mine tailings is more efficient using thermophilic instead of mesophilic organism extremely thermophilic microorganisms show a higher sensitivity to copper and to hig pulp densities in agitated systems limiting, therefore, some practical applications.

Although environmental conditions in leaching operations favour the growth an development of mesophilic, moderately thermophilic, and extremely thermophili microbial communities, metal leaching at low temperatures has also been observe Copper and nickel were leached from pyritic ore samples in significant amounts at °C. However, leaching rates were lower by a factor of 30 to 50 as compared t experiments conducted at 37°C. T. ferrooxidans recovered from mine waters was ab to grow at 2 °C with a generation time of approximately 250 h suggesting psychrotrophic nature of the organism. Bacterial iron mobilization has also bee observed at 0 °C in ore samples obtained from Greenland. Solubilization rates at thes low temperatures were still approximately 25 to 30% of the maximum values observe at 21°C. All these findings may have a potential for practical applications in geographic areas where field operations are subjected to low temperature regimes.

REFERENCES

Peeler, J. T., G. A. Houghtby, and A. P. Rainosek, "The most probable number technique *Compendium of Methods for the Microbiological Examination of Foods*, 3rd Ed., 105-120, 1992.

Peters, Pamela. *Biotechnology: A Guide to Genetic Engineering*. Wm. C. Brown Publishers, 1993.

Seeliger, H.P.R. and D. Jones, "The Genus Listeria", *Bergey's Manual of Systematic Bacteriolog* Volume 2. Williams and Wilkins, Baltimore, 1986.

Turner A. P. F. (ed.). *Biosensors: Fundamentals and Applications*. Oxford University Press. Oxfor 1987.

11

Microbial Technology for Water Treatment

'raditionally, indicator micro-organisms have been used to suggest the presence of athogens. Today, however, we understand a myriad of possible reasons for indicator resence and pathogen absence, or vice versa. In short, there is no direct correlation etween numbers of any indicator and enteric pathogens. To eliminate the ambiguity n the term 'microbial indicator', the following three groups are now recognised:

- General (process) microbial indicators,
- Faecal indicators (such as E. coli)
- Index organisms and model organisms.

direct epidemiological approach could be used as an alternative or adjunct to the use f index micro-organisms. Yet epidemiologic methods are generally too insensitive, miss he majority of waterborne disease transmissions and are clearly not preventative. Nonetheless, the ideal is to validate appropriate index organisms by way of pidemiological studies. A good example is the emerging use of an enterococci guideline or recreational water quality. Often epidemiologic studies fail to show any relationship o microbial indicators, due to poor design and/or due to the widely fluctuating ratio f pathogen(s) to faecal indicators and the varying virulence of the pathogens.

The validity of any indicator system is also affected by the relative rates of removal nd destruction of the indicator versus the target hazard. So differences due to nvironmental resistance or even ability to multiply in the environment all influence heir usefulness. Hence, viral, bacterial, parasitic protozoan and helminth pathogens are nlikely to all behave in the same way as a single indicator group, and certainly not n all situations. Furthermore, viruses and other pathogens are not part of the normal aecal microbiota, but are only excreted by infected individuals. Therefore, the higher he number of people contributing to sewage or faecal contamination, the more likely he presence of a range of pathogens. The occurrence of specific pathogens varies further ccording to their seasonal occurrence.

DEVELOPMENT OF INDICATORS

The Coliforms

The use of bacteria as indicators of the sanitary quality of water probably dates back to 1880 when Von Fritsch described Klebsiella pneumoniae and K. rhinoscleromatis as micro-organisms characteristically found in human faeces. In 1885, Percy and Grace Frankland started the first routine bacteriological examination of water in London, using Robert Koch's solid gelatine media to count bacteria. Also in 1885, Escherich described Bacillus coli (renamed Escherichia coli by Castellani and Chalmers (1919)) from the faeces of breast-fed infants.

In 1891, the Franklands came up with the concept that organisms characteristic of sewage must be identified to provide evidence of potentially dangerous pollution. By 1893, the 'Wurtz method' of enumerating B. coli by direct plating of water samples on litmus lactose agar was being used by sanitary bacteriologists, using the concept of acid from lactose as a diagnostic feature. This was followed by gas production, with the introduction of the Durham tube. The concept of 'coliform' bacteria, those bacteria resembling B. coli, was in use in Britain in 1901.

Therefore, the sanitary significance of finding various coliforms along with streptococci and C. perfringens was recognised by bacteriologists by the start of the twentieth century. It was not until 1905, however, that MacConkey described his now famous MacConkey's broth, which was diagnostic for lactose-fermenting bacteria tolerant of bile salts. Nonetheless, coli-forms were still considered to be a heterogeneous group of organisms, many of which were not of faecal origin. The origins of the critical observation that B. coli was largely faecal in origin while other coliforms were not, could be claimed by Winslow and Walker.

Coliform Identification Schemes

Various classification schemes for coliforms have emerged. The earliest were those of MacConkey which recognised 128 different coliform types, while Bergey and Deehan identified 256. By the early 1920s, differentiation of coliforms had come to a series of correlations that suggested indole production, gelatin liquefaction, sucrose fermentation and the Voges- Proskauer reaction were among the more important tests for determining faecal contamination. These developments culminated in the IMViC (Indole, Methyl red, Voges-Proskauer and Citrate) tests for the differentiation of so-called faecal coliforms, soil coliforms and intermediates ; these tests are still in use today.

Water sanitary engineers, however, require simple and rapid methods for the detection of faecal indicator bacteria. Hence, the simpler to identify coliform group, despite being less faecal-specific and broader (for which Escherichia, Klebsiella, Enterobacter and Citrobacter were considered the most common genera) was targeted.

One of the first generally accepted methods for coliforms was called the Multiple-Tube Fermentation Test.

Most probable number method

In 1914, the first US Public Health Service Drinking Water Standard adopted a bacteriological standard that was applicable to any water supply provided by an interstate common carrier. It specified that not more than one out of five 10 ml portions of any sample examined should show the presence of the B. coli group by the specified Multiple-Tube Fermentation procedure.

Although this test is simple to perform, it is time-consuming, requiring 48 hours for the presumptive results. There are a number of isolation media each with its bias and the bacteria enriched are not a strict taxonomic group. Hence, the total coliforms can best be described as a range of bacteria in the family Enterobacteriaceae varying with the changing composition of the media.

Following presumptive isolation of coliforms, further testing is required for confirmation of the coliform type. During the late 1940s there was a divergence between the UK and US approaches to identifying the thermotolerant or socalled 'faecal' coliforms. In the UK, Mackenzie et al. had shown that atypical fermentors of lactose at 44°C were indole-negative, whereas E. coli was indole-positive. Confirmation of E. coli with the indole test was undertaken in the UK, but lactose fermentation at 44°C alone was used in the US. Thus over a period of some 50 years, water bacteriologists developed the concept of B. coli (later E. coli) as the indicator of faecal pollution, but continued to attach significance to the total lactose fermenters, known variously as 'coli-aerogenes' group, Escherichia-Aerobacter group, colon group or generally referred to as the 'total coliforms' group.

The range of non-faecal bacteria represented in the coliform group and the environmental growth of thermophilic (faecal) coliforms Klebsiella spp. and E. coli have concerned bacteriologists and sanitary engineers since the 1930s. At the other extreme, recent outbreaks of cryptosporidiosis in the absence of coliforms (per 100 ml) are well known, and many earlier classic failures of coliforms to identify waterborne pathogens have also been reported. Despite the obvious failings of the total coliform group to indicate health risk from bacterial pathogens, they provide valuable information on process efficiency which is clearly important in relation to health protection.

Membrane filtration method

Until the 1950s practical water bacteriology relied almost exclusively, for indicator purposes, on the enumeration of coliforms and E. coli based on the production of gas from lactose in liquid media and estimation of most probable numbers using the statistical approach initially suggested by McCrady. In Russia and Germany, however, workers attempted to culture bacteria on membrane filters, and by 1943 Mueller in Germany was using membrane filters in conjunction with Endo-broth for the analysis of potable waters for coliforms. By the 1950s membrane filtration was a practical alternative to the MPN approach, although the inability to demonstrate gas production with membranes was considered a major drawback.

The arbitrary definitions adopted for E. coli and the related coliforms were all based upon cultural characteristics, including the ability to produce gas from lactose fermentation. Hence, the thermotolerant coliforms include strains of the genera Klebsiella and Escherichia, as well as certain Enterobacter and Citrobacter strains able to grow under the conditions defined for thermotolerant coliforms. This phenotypic approach has also resulted in *E. coli* or a related coliform being ignored simply because they failed to ferment lactose, failed to produce gas from lactose or were indole-negative at 44.5°C. The approach had been repeatedly questioned, and was only resolved in the UK in the 1990s.

It has long been recognised that artificial culture media lead to only a very small fraction (0.01-1%) of the viable bacteria present being detected. Since MacConkey's development of selective media for E. coli and coliforms at the beginning of the twentieth century, various workers have shown these selective agents inhibit environmentally or oxidatively stressed coliforms.

Defined substrate methods

Media without harsh selective agents but specific enzyme substrates allow significant improvements in recoveries and identification of target bacteria. In the case of coliforms and E. coli, such so-called defined substrate methods were introduced by Edberg et al. What has evolved into the Colilert® technique has been shown to correlate very well with the traditional membrane filter and MPN methods when used to test both fresh and marine water. Furthermore, these enzyme-based methods appear to pick up traditionally nonculturable coliforms.

These developments have led to further changes in definitions of total coliforms and E. coli. In the UK, for example, total coliforms are members of genera or species within the family Enterobacteriaceae, capable of growth at 37°C, which possess β-galactosidase. In an international calibration of methods, E. coli was enzymatically distinguished by the lack of urease and presence of β-glucuronidase. Furthermore, the International Standards Organisation has recently published miniaturised MPNbased methods for coliforms/E. coli and enterococci based on the defined substrate approach.

Faecal Streptococci and Enterococci

Parallel to the work on coliforms, a group of Gram-positive coccoid bacteria known as faecal streptococci (FS) were being investigated as important pollution indicator bacteria.

Problems in differentiating faecal from non-faecal streptococci, however, initially impeded their use. Four key points in favour of the faecal streptococci were:

(1) Relatively high numbers in the excreta of humans and other warm-blooded animals.

(2) Presence in wastewaters and known polluted waters.

(3) Absence from pure waters, virgin soils and environments having no contact with human and animal life.

(4) Persistence without multiplication in the environment.

It was not until 1957, however, with the availability of the selective medium of Slanetz and Bartley that enumeration of FS became popular. Since then, several media have been proposed for FS and/or enterococci to improve on the specificity.

Taxonomically FS are represented by various Enterococcus spp. and Streptococcus bovis and S. equinus. Of the faecal streptococci, the preferred indicators of faecal pollution are the enterococci. The predominant intestinal enterococci being E. faecalis, E. faecium, E. durans and E. hirae. In addition, other Enterococcus species and some species of Streptococcus (namely S. bovis and S. equinus) may occasionally be detected. These streptococci however, do not survive for long in water and are probably not enumerated quantitatively. Thus, for water examination purposes enterococci can be regarded as indicators of faecal pollution, although some could occasionally originate from other habitats.

Significance of the thermotolerant coliform

Geldreich and Kenner proposed that a faecal coliform:faecal streptococci ratio of four or greater may indicate human pollution, whereas ratios of two or less may indicate animal pollution. There are many factors, however, that can jeopardise the usefulness of this ratio. Foremost are the quicker die-off of coliforms in the environment and different counts from various media used for bacterial isolation. Hence, the use of this ratio is no longer recommended unless very recent faecal pollution is being monitored.

Sulphite-reducing Clostridia and other Anaerobes

Until bifidobacteria were suggested as faecal indicators, C. perfringens was the only obligately anaerobic, enteric micro-organism seriously considered as a possible indicator of the sanitary quality of water. Despite the first isolation of bifidobacteria in the late 1800s and very high numbers in human faeces (11% of culturable bacteria), their oxygen sensitivity has limited their role as useful faecal indicators in waters.

The anaerobic sulphite-reducing clostridia (SRC) are much less prevalent than bifidobacteria in human faeces, but their spore-forming habit gives them high environmental resistance. C. perfringens is the species of clostridia most often associated with the faeces of warm-blooded animals, but is only present in 13-35% of human faeces.Although C. perfringens has been considered a useful indicator species for over one hundred years, its use has been largely limited to Europe, and even then as a secondary indicator mixed in with other SRC. The main criticism of the use of C. perfringens as a faecal indicator is its long persistence in the environment, which is considered to be significantly longer than enteric pathogens. Bonde suggested that all SRC in receiving waters are not indicators of faecal pollution, hence C. perfringens is the appropriate indicator

Bacteriophages

Viruses which infect bacteria, known as bacteriophages or simply as phages, were first described from the intestinal tract of man in the early 1900s. The use of phages as models for indicating the likely presence of pathogenic enteric bacteria first appeared in the 1930s, and direct correlations between the presence of certain bacteriophages and the intensity of faecal contamination were reported.

The evolving role for phages to coliforms, known as coliphages however, has been to model human enteric viruses. The DNAcontaining tailed coliphages (T type) and RNA-containing phages that infect via the F-pili (sex factor) (F-RNA coliphages) have been the most used.

Phages in water environments

Studies on the incidence of phages in water environments have been reported from most parts of the world for some decades now. Unfortunately the data are not particularly consistent and comparisons are generally not meaningful. One reason for this is that there are many variables that affect the incidence, survival and behaviour of phages in different water environments, including the densities of both host bacteria and phages, temperature, pH and so on.

Another important reason is the inconsistency in techniques used for the recovery of phages from water environments, and eventual detection and enumeration of the phages. This is not altogether surprising because virology, including phages, is a young and rapidly developing science. Phages can be recovered and detected by many techniques and approaches, and much of this work is still in a research or developmental stage. A major reason for discrepancies in results is the host bacteria used for the detection of various groups of phages. Nonetheless, international collaboration is now leading to meaningful, universally accepted guidelines for the recovery and detection of phages in water environments (such as those produced by the International Organisation for Standardisation).

Faecal Sterol Biomarkers

The presence of faecal indicator bacteria gives no indication of the source, yet it is widely accepted that human faecal matter is more likely to contain human pathogens than animal faeces. The detection of human enteric viruses is specific, however; the methods are difficult and expensive, and not readily quantifiable. Vivian in his review of sewage tracers, suggested that using more than one method of determining the degree of sewage pollution would be prudent and advantageous. The use of alternative indicators, in this case faecal sterols as biomarkers, in conjunction with existing microbiological indicators, offers a new way to distinguish sources of faecal contamination and monitor river 'health'.

Coprostanol has been proposed as a measure of human faecal pollution by a large number of researchers since the late 1960s, however, coprostanol has never really been embraced as a sanitary indicator for sewage pollution because its presence was not considered as indicative of a health risk. However, Leeming et al. showed that herbivores have a different dominant form (24-ethyl coprostanol) and it was later shown that these differences could be exploited to determine the contribution of faecal matter from herbivore and human sources relative to each other.

PATHOGEN MODELS AND INDEX MICROORGANISMS

The similar morphology, structure and behaviour of F-RNA coliphages, as well as other phages to that of human enteric viruses, suggests that they should be better models for faecal pollution than the faecal indicator bacteria when human viruses are the likely aetiological agents. The same applies to properties such as removal by water treatment processes and resistance to disinfection processes.

While one would expect a poor correlation of phage numbers to the level of human enteric virus titre (phages are always in sewage, but pathogen numbers vary widely based on human infection), what is important for a model organism is that many phages are as resistant as (human) enteric viruses. Laboratory experiments with individual coliphages confirmed that their survival in natural water environments resembles that of enteric viruses and that some phages are at least as resistant as certain enteric viruses to water environments and to commonly used disinfectant such as chlorine.

The value of phages as models/surrogates for viruses has been applied in the routine monitoring of raw and treated drinking water supplies, and in the assessment of the efficiency of domestic point-of-use water treatment units. While they are useful and meet many of the basic requirements as surrogates for enteric viruses.

Phages cannot be regarded as absolute indicators, models or surrogates for enteric viruses in water environments. This is underlined by the detection of enteric viruses in treated drinking water supplies which yielded negative results in tests for phages, even in presence-absence tests on 500 ml samples of water. Phages are probably best applied as models/surrogates in laboratory experiments where the survival or behaviour of selected phages and viruses are directly compared under controlled conditions.

In addition to enteric viruses, parasitic protozoa are important disinfectionresistant pathogens. When sewage is the source of these pathogens, the anaerobic spore-forming bacterium Clostridium perfringens appears to be a suitable index for enteric viruses and parasitic protozoa. Spores of C. perfringens are largely of faecal origin, and are always present in sewage (about 104-105 cfu 100ml-1). Their spores are highly resistant in the environment, and vegetative cells appear not to reproduce in aquatic sediments, which can be a problem with traditional indicator bacteria.

Like spores to C. perfringens, Bacillus spp. spores can also be used as models for parasitic protozoan cysts or oocysts removal by water treatment. Furthermore, since vegetative bacterial cells are inadequate models for disinfection, phages or clostridial spores may provide useful models.

EMERGING MICROBIOLOGICAL METHODS

Fast Detections using Chromogenic Substances

The time required to perform tests for indicator organisms has stimulated research into more reliable and faster methods. One result is the use of chromogenic compounds, which may be added to the conventional or newly devised media used for the isolation of the indicator bacteria. These chromogenic substances are modified either by enzymes (which are typical for the respective bacteria) or by specific bacterial metabolites. After modification the chromogenic substance changes its colour or its fluorescence, thus enabling easy detection of those colonies displaying the metabolic capacity. In this way these substances can be used to avoid the need for isolation of pure cultures and confirmatory tests. The time required for the determination of different indicator bacteria can be cut down to between 14 to 18 hours.

Application of Monoclonal and Polyclonal Antibodies

Antibodies (glycoproteins produced by mammals as part of their defence system against foreign matter) possess highly specific binding and recognition domains that can be targeted to specific surface structures of a pathogen (antigen). Immunological methods using antibodies are widely used to detect pathogens in clinical, agricultural and environmental samples. Antisera or polyclonal antibodies, the original source of immune reagents, are obtained from the serum of immunised animals (typically rabbits or sheep). Monoclonal antibodies which are produced in vitro by fusing plasma cells of an immunised animal (usually a mouse or rat) with a cell line that grows continuously in culture (so that the fused cells will grow continuously and secrete only one kind of antibody molecule, can be much better standardised.

Such monoclonal antibodies have been successfully used for the detection of indicator bacteria in water samples. In these studies the water sample was subjected to a precultivation in a selective medium. In this way the complication of detecting dead cells was avoided. Another option for the detection of 'viable' indicators is the combination of immunofluorescence with a respiratory activity compound. This approach has been described for the detection of E. coli O157:H7, S. typhimurium and K. pneumoniae in water. Detection of Legionella from water samples has also been achieved with antibodies. In general, immunological methods can easily be automated in order to handle high sample numbers.

Antibody technology is often used in medicine with enzyme amplification (ELISA - enzyme linked immunosorbent assay), to allow the development of an antigen signal readable by the naked eye. Such an approach is under development for the rapid identification of coliform microcolonies. As always with immunological techniques, the specificity of the reagents and optimisation of their use is paramount. Although total coliforms are a broad group and likely to be unsuitable immunological targets in environmental waters, E. coli could be identified from other coliforms.

Until reliable index organisms are identified for the parasitic protozoa Cryptosporidium and Giardia, their detection is also relevant when describing methods for important faecal organisms. Current methods for their detection rely on antibodies to assist in the microscopic identification amongst other environmental particles. In addition, magnetic beads coated with antibodies are used for concentration and separation of the oocysts and cysts as described below for immunomagnetic separation (IMS) methods.

IMS/culture and other Rapid Culture-Based Methods

Immunomagnetic separation offers an alternative approach to rapid identification of culturable and non-culturable micro-organisms. The principles and application of the method are simple, but rely on suitable antibody specificity under the conditions of use. Purified antigens are typically biotinylated and bound to streptoavidin-coated paramagnetic particles (e.g. DynalTM beads). The raw sample is gently mixed with the immunomagnetic beads, then a specific magnet is used to hold the target organisms against the wall of the recovery vial, and non-bound material is poured off. If required, the process can be repeated, and the beads may be removed by simple vortexing. Target organisms can then be cultured or identified by direct means.

The IMS approach may be applied to recovery of indicator bacteria from water, but is possibly more suited to replace labour-intensive methods for specific pathogens. An example is the recovery of E. coli O157 from water. Furthermore, E. coli O157 detection following IMS can be improved by electrochemiluminescence detection.

Gene Sequence-Based Methods

Advances in molecular biology in the past 20 years have resulted in a number of new detection methods, which depend on the recognition of specific gene sequences. Such methods are usually rapid and can be tailored to detect specific strains of organisms on the one hand or groups of organisms on the other. The methods have a substantial potential for future application in the field of drinking water hygiene. An international expert meeting in Interlaken concluded that the application of molecular methods has to be considered in a framework of a quality management for drinking water. The new methods will influence epidemiology and outbreak investigations more than the routine testing of finished drinking water.

PCR (polymerase chain reaction)

With the polymerase chain reaction and two suitable primer sequences (fragments of nucleic acid that specifically bind to the target organism) trace amounts of DNA can be selectively multiplied. In principle, a single copy of the respective sequence in the assay can produce over a million-fold identical copies, which then can be detected and further analysed by different methods.

One problem with PCR is that the assay volume is in the order of some micro-litres, whereas the water sample volume is in the range of 100-1000 ml. Bej et al. have published a filtration method to concentrate the sample, but another problem is that natural water samples often contain inhibitory substances (such as humic acids and iron) that concentrate with the nucleicacids. Hence, it is critical to have positive (and negative) controls with each environmental sample PCR to check for inhibition and specificity.

It may also be critical to find out whether the signal obtained from the PCR is due to naked nucleic acids or living or dead micro-organisms. One solution has been established by using a three-hour pre-incubation period in a selective medium so that only growing organisms are detected. Other options under development include targeting short-lived nucleic acids such as messenger or ribosomal RNA.

A most important advantage of PCR is that the target organism(s) do not need to be culturable. A good example is the specific detection of human Bacteroides spp. to differentiate human faecal pollution from that of other animals.

FISH (fluorescence in situ hybridisation)

This detection method uses gene probes with a fluorescent marker, typically targeting the 16S ribosomal RNA (16S rRNA). Concentrated and fixed cells are permeabilised and mixed with the probe. Incubation temperature and addition of chemicals can influence the stringency of the match between the gene probe and the target sequence. Since the signal of a single fluorescent molecule within a cell does not allow detection, target sequences with multiple copies in a cell have to be selected (e.g. there are 102-104 copies of 16S rRNA in active cells). A number of FISH methods for the detection of coliforms and enterococci have been developed.

Although controversial for many pathogens, low-nutrient environments may result in cells entering a non-replicative viable but non-culturable (VBNC) state. Such a state may not only give a false sense of security when reliant on culture-based methods, but may also give the organisms additional protection. An indication of VBNC Legionella pneumophila cell formation was given by following decreasing numbers of bacteria monitored by colony-forming units, acridine orange direct count, and hybridisation with 16S rRNA-targeted oligonucleotide probes. It was concluded that FISH detection-based methods may better report the presence of infective pathogens and viable indicator bacteria.

Future Developments

The future holds numerous possibilities for the detection of indicator and pathogen index organisms. On the horizon are methods based on microarrays and biosensors. Biosensors in the medical area have largely been based onantibody technology, with the antigen triggering a transducer or linking to an enzyme amplification system. Biosensors based on gene recognition, however, look very promising in the microarray format for detecting micro-organisms.

Microarrays using DNA/RNA probe-based rRNA targets may be coupled to adjacent detectors. Eggers et al. have demonstrated the detection of E. coli and Vibrio proteolyticus using a microarray containing hundreds of probes within a single well of a conventional microtiter plate. The complete assay with quantification took less than a minute.

DNA sensing protocols, based on different modes of nucleic acid interaction, possess an enormous potential for environmental monitoring. Carbon strip or paste electrode transducers, supporting the DNA recognition layer, are used with a highly sensitive chronopotentiometric transduction of the DNA analyte recognition event. Pathogens targeted to date include Mycobacterium tuberculosis, Cryptosporidium parvum and HIV-1.

CURRENT APPLICABILITY OF FAECAL INDICATORS

Many members of the total coliform group and some so-called faecal coliforms (e.g. species of Klebsiella and Enterobacter) are not specific to faeces, and even *E. coli* has been shown to grow in some natural aquatic environments. Hence, the primary targets representing faecal contamination in temperate waters are now considered to be *E. coli* and enterococci. For tropical waters/soils, where *E. coli* and enterococci may grow, alternative indicators such as Clostridium perfringens may be preferable.

Numerous epidemiological studies of waterborne illness in developed countries indicate that the common aetiological agents are more likely to be viruses and parasitic protozoa than bacteria. Given the often lower persistence of vegetative cells of the faecal bacteria compared to the former agents, it is not surprising that poor correlations have been reported between waterborne human viruses or protozoa and thermotolerant coliforms. Such a situation is critical to understand, as evident from recent drinking water outbreaks where coliform standards were met. Nonetheless, water regulatory agencies have yet to come to terms with the inherent problems resulting from reliance on faecal indicator bacteria as currently determined.

Fortunately, new index organisms for some pathogens look promising as performance organisms in the HACCP-type management approaches. Examples of such index organisms are *C. perfringens* and the phages. *C. perfringens* for parasitic protozoa, but only if derived from human faecal contamination. Their resistance to disinfectants may also be an advantage for indexing disinfectantresistant pathogens. In Europe, the European Union (EU) recommends the absence of *C. perfringens* in 100ml as a secondary attribute to drinking waters, while in Hawaii, levels are laid down for marine and fresh waters. Also F-RNA coliphages or Bacteroides fragilis bacteriophages are preferred to assess the removal or persistence of enteric viruses. As these index organisms are relatively untested worldwide, extensive trials are necessary before their general acceptance in microbial risk assessment. It should be noted that useful index organisms in one system are not necessarily of value in a different environment.

A further confusion over the use of indicator organisms arises from the fact that some indicator strains are also pathogens. This is perhaps best illustrated by the toxigenic *E. coli* strains. *E. coli* O157:H7 has been responsible for illness to recreational swimmers and several deaths have been documented through food- and waterborne outbreaks. Such toxigenic *E. coli* are also problematic to detect, as they may form viable but non-culturable cells in water.

REFERENCES

Andria M. Costello, Ann J. Auman, Jennifer L. Macalady, Kate M. Scow, and Mary E. Lidstrom, "Estimation of Methanotroph Abundance in a Freshwater Lake Sediment", *Environmental Microbiology,* 2002.

Jürg Keller, Zhiguo Yuan and Linda L. Blackall, "Integrating process engineerinq and microbiology tools to advance activated sludge wastewater treatment research and development", *RelViews in Environmental Science & BiorTechnology,* 2002.

Newell, S.Y., "Fungi in marine/estuarine waters", Pages 1394-1400 in: Bitton, G., ed. *The Encyclopedia of Environmental Microbiology,* Wiley, New York, 2002.

Patrick Dabert, Jean-Philípe Delgenés, René Moletta & Jean-Jacques Godon, "Contribution of molecular microbiology to the study in water pollution removal of microbial community dynamics", *ReNiews in Environmental Science & Bio/Technology,* 2002.

12

Environmental Applications of Microbial Biotechnology

Microbes are everywhere in the biosphere, and their presence invariably affects the environment that they are growing in. The effects of microorganisms on their environment can be beneficial or harmful or inapparent with regard to human measure or observation. Since a good part of this text concerns harmful activities of microbes (i.e., agents of disease) this chapter counters with a discussion of the beneficial activities and exploitations of microorganisms as they relate to human culture. The beneficial effects of microbes derive from their metabolic activities in the environment, their associations with plants and animals, and from their use in food production and biotechnological processes.

NUTRIENT CYCLING AND THE CYCLES OF ELEMENTS THAT MAKE UP LIVING SYSTEMS

At an elemental level, the substances that make up living material consist of carbon (C), hydrogen (H), oxygen (O), nitrogen (N), sulfur (S), phosphorus (P), potassium (K), iron (Fe), sodium (Na), calcium (Ca) and magnesium (Mg). The primary constituents of organic material are C, H, O, N, S, and P. An organic compound always contains C and H and is symbolized as CH_2O (the empirical formula for glucose). Carbon dioxide (CO_2) is considered an inorganic form of carbon.

The most significant effect of the microorganisms on earth is their ability to recycle the primary elements that make up all living systems, especially carbon (C), oxygen (O) and nitrogen (N). These elements occur in different molecular forms that must be shared among all types of life. Different forms of carbon and nitrogen are needed as nutrients by different types of organisms. The diversity of metabolism that exists in the microbes ensures that these elements will be available in their proper form for every type of life. The most important aspects of microbial metabolism that are involved in the cycles of nutrients are discussed below.

Primary production involves photosynthetic organisms which take up CO_2 in the atmosphere and convert it to organic (cellular) material. The process is also called CO_2 fixation, and it accounts for a very large portion of organic carbon available for synthesis of cell material. Although terrestrial plants are obviously primary producers, planktonic algae and cyanobacteria account for nearly half of the primary production on the planet. These unicellular organisms which float in the ocean are the "grass of the sea", and they are the source of carbon from which marine life is derived.

Decomposition or biodegradation results in the breakdown of complex organic materials to forms of carbon that can be used by other organisms. There is no naturally-occurring organic compound that cannot me degraded by some microbe, although some synthetic compounds such as teflon, styrofoam, plastics, insecticides and pesticides are broken down slowly or not at all. Through the metabolic processes of fermentation and respiration, organic molecules are eventually broken down to CO_2 which is returned to the atmosphere.

Waste management, whether in compost, landfills or sewage treatment facilities, exploits activities of microbes in the carbon cycle. Organic (solid) materials are digested by microbial enzymes into substrates that eventually are converted to a few organic acids and carbon dioxide.

Nitrogen fixation is a process found only in some bacteria which removes N_2 from the atmosphere and converts it to ammonia (NH_3), for use by plants and animals. Nitrogen fixation also results in replenishment of soil nitrogen removed by agricultural processes. Some bacteria fix nitrogen in symbiotic associations in plants. Other Nitrogen-fixing bacteria are free-living in soil and aquatic habitats.

Oxygenic photosynthesis occurs in plants, algae and cyanobacteria. It is the type of photosynthesis that results in the production of O_2 in the atmosphere. At least 50 percent of the O_2 on earth is produced by photosynthetic microorganisms (algae and cyanobacteria), and for at least a billion years before plants evolved, microbes were the only organisms producing O_2 on earth. O_2 is required by many types of organisms, including animals, in their respiratory processes.

ASSOCIATIONS WITH ANIMALS AND PLANTS

Microbes invariably enter into beneficial, sometimes essential, associations with all higher forms of organisms, including insects, invertebrates, fish, animals and plants. For example, bacteria and other microbes in the intestines of animals and insects digest nutrients and produce vitamins and growth factors. In the plant world, leguminous plants (peas, beans, clover, alfalfa, etc.) live in intimate associations with bacteria that extract nitrogen from the atmosphere and supply it to the plant for growth.

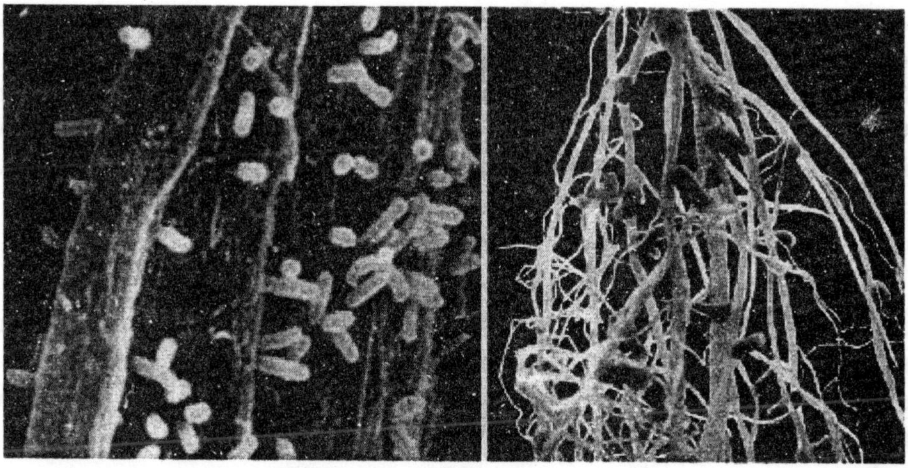

Figure 1. The mutualistic association between nitrogen fixing bacteria and leguminous plants. Left. Nitrogen-fixing Rhizobium bacteria colonized on the root hairs of clover plants. Right. Nodules containing Rhizobium bacteria on the plant roots. In the nodule, the bacteria fix nitrogen which they share with the plant. In exchange, the plant supplies the bacteria with a source of carbon and energy for growth.

The microbes that normally live in associations with humans on the various surfaces of the body (called the normal flora), such as Lactobacillus and Bifidobacterium, are known to protect their hosts from infections, and otherwise promote nutrition and health.

PRODUCTION OF FOODS AND FUELS

In the home and in industry, microbes are used in the production of fermented foods. Yeasts are used in the manufacture of beer and wine and for the leavening of breads, while lactic acid bacteria are used to make yogurt, cheese, sour cream, buttermilk and other fermented milk products. Vinegars are produced by bacterial acetic acid fermentation. Other fermented foods include soy sauce, sauerkraut, dill pickles, olives, salami, cocoa and black teas. Yeast are also involved in fermentations to convert corn and other vegetable carbohydrates into ethanol to make beer, wine or gasohol, but bacteria are the agents of most other food fermentations.

MEDICAL, PHARMACEUTICAL AND BIOTECHNOLOGICAL APPLICATIONS

In human and veterinary medicine, for the treatment and prevention of infectious diseases, microbes are a source of antibiotics and vaccines. Antibiotics are substances produced by microorganisms that kill or inhibit other microbes which are used in the treatment of infectious disease. Antibiotics are produced in nature by molds such as Penicillium and bacteria such as Streptomyces and Bacillus. Vaccines are substances derived from microorganisms used to immunize against disease. The microbes that are the cause of infectious disease are usually the ultimate source of vaccines. Thus, a version of the diphtheria toxin (called toxoid) is used to immunize against diphtheria, and parts of Bordetella pertussis cells are used to vaccinate against pertussis (whooping cough). The use of vaccines such as smallpox, polio, diphtheria, tetanus and whooping cough has led to virtual elimination of these diseases in regions of the world where the vaccines have been deployed.

BIOTECHNOLOGY

Microbiology makes an important contribution to biotechnology, an area of science that applies microbial genetics to biological processes for the production of useful substances. Microorganisms play a central role in recombinant DNA technology and genetic engineering. Important tools of biotechnology are microbial cells, microbial genes and microbial enzymes.

The genetic information for many biological products and biological processes can be introduced into microbes in order to genetically engineer them to produce a substance or conduct a process. The genes can come from any biological source: human, animal, plant or microbial. This opens the possibility for microbial production of foods, fuels, enzymes, hormones, diagnostic agents, medicines, antibiotics, vaccines, antibodies, natural insecticides and fertilizers, and all sorts of substances useful in our civilization and society. Also, the microbial genes that encode for these substances, most of which are unknown, are a tremendous resource of information for application in medicine, pharmacy, agriculture, food science and biotechnology.

BASIC RESEARCH

Microorganisms, in particular the bacterium, Escherichia coli and the yeast, Saccharomyces, have been used as model organisms for basic research and the study of cellular life. Hundreds of thousands of scientific papers have been published on these two organisms. Because of cell theory and the unity of biological processes in all organisms, this information provides us with insight and understanding of life at all levels, including human.

HARMFUL EFFECTS OF MICROBES

The primary harmful effects of microbes upon our existence and civilization is that they are an important cause of disease in animals and crop plants, and they are agents of spoilage and decomposition of our foods, textiles and dwellings.

A microbe which is capable of causing infectious disease in an animal or plant is called a pathogen. Four groups of microbes contain pathogens: bacteria, fungi, protozoa and the viruses. Only the archaea and algae are lacking pathogens. Pathogens are the cause of infectious diseases.

Historically, infectious diseases are the most significant cause of death in humans. Until the beginning of the Twentieth Century, it is estimated that more than half the people who ever lived died from smallpox, caused by a virus, or malaria, caused by a protozoan. Bacteria, too, have been the cause of some of the most deadly diseases and widespread epidemics of human civilization. Bacterial diseases such as tuberculosis, typhus, plague, diphtheria, typhoid fever, cholera, dysentery and pneumonia have taken a huge toll on humanity.

Deaths from infectious diseases declined markedly in the United States during the 20th century. This contributed to the nearly 30-year increase in life expectancy during this period. In 1900, the three leading causes of death were pneumonia, tuberculosis (TB), and diarrhea and enteritis, which (together with diphtheria) caused one third of all deaths. In 1997, heart disease and cancers accounted for 55% of all deaths, with 4.5% attributable to pneumonia, influenza, and human immunodeficiency virus (HIV) infection. However, one of the most devastating epidemics in human history occurred during the 20th century: the 1918 influenza pandemic that resulted in 20 million deaths, including 500,000 in the United States in less than 1 year - more than have died in as short a time during any war or famine in the world.

HIV infection, first recognized in 1981, has caused a pandemic that is still in progress, affecting 33 million people and causing an estimated 13.9 million deaths. This illustrates the volatility of infectious disease and the unpredictability of disease emergence and points us to the challenges ahead.

Progress in the 20th century is based on the 19th century discovery of microorganisms as the cause of many serious diseases (e.g., cholera and TB). Disease control resulted from improvements in sanitation and hygiene, the discovery of antibiotics, the implementation of universal childhood vaccination programs, and technological advances in detecting and monitoring infectious disease. Water purification, immunization (vaccination), and modern antibiotic therapy (all developments in the field of bacteriology) have dramatically reduced the morbidity and the mortality of infectious disease during the Twentieth Century, at least in the developed world where these are acceptable cultural practices. However, many new microbial pathogens have been recognized in the past 30 years and many "old" bacterial pathogens, such as Staphylococcus aureus and Mycobacterium tuberculosis, have emerged with new forms of virulence and new patterns of resistance to antimicrobial agents.

Microbes are also the cause of many diseases in plants, which, if crop plants or forest resources, may have important economic or social consequences.

Microbes are the agents of food spoilage and decomposition of clothing and sheltering materials. The factors that allow microbes to accomplish biodegradation and carbon cycling are at work on everything organic, which includes foods and grains stored in granaries, supermarket or refrigerator, as well as natural structural materials and textiles used for our shelters and clothing. Nothing lasts forever, and the microbial decomposition of everything organic will occur in time. Fungi and bacteria are the major microbial agents of decomposition in aerobic environments. Bacteria take over in environments that lack oxygen.

ENVIRONMENTAL BIOTECHNOLOGY

Over the past few decades enormous quantities of industrial pollutants have been released into the environment. A large number of them, particularly those structurally related to natural compounds, are readily degraded or removed by microorganisms found in soil and water. However, superimposed on the wide variety of pollutants present in the environment is an increasing number of novel industrial compounds rarely found in nature. These xenobiotic compounds are usually removed slowly and tend to accumulate in the environment. Due to the high degree of toxicity, their accumulation can cause severe environmental problems.

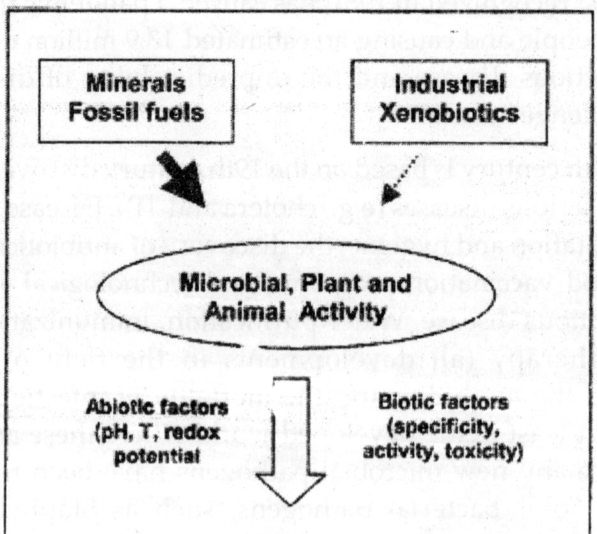

Figure 2. Main sources of environmental pollutants and factors in.uencing their nature removal from the environment.

Because of the problems associated with pollutant treatment by conventional methods, such as incineration or landfills, increasing consideration has been placed on the development of alternative, economical and reliable biological treatments. Although natural microorganisms collectively exhibit remarkable evolutionary capabilities to adapt to a wide range of chemicals, natural evolution occurs at a relatively slow rate, particularly when the acquisition of multiple catalytic activities is necessary. In these cases, the acceleration of these events via genetic engineering/processing engineering is helpful since the desirable traits can be carefully designed and controlled. The drive toward this goal represents the essence of environmental biotechnology.

Environmental biotechnology refers to the utilization of microorganisms to improve environmental quality. Although the field of environmental biotechnology has been around for decades, starting with the use of activated sludge and anaerobic digestion in the early 20th century by civil engineers, the introduction of new technologies from modern microbiology and molecular biology has enabled engineers and scientists to tackle the more contemporary environmental problems such as detoxification of hazardous chemicals. Chemical engineers are uniquely poised to contribute in this emerging area since many of the potential solutions require a combined perspective from modern biology and process engineering, two areas where chemical engineers excel.

For example, the realization of environmental biotechnology into practical solutions requires the implementation of process design, which is the foundation of the chemical engineering discipline. The same is also true at the cellular level, where the functions of cells are determined primarily by networks of specific catalytic reactions. The nature and activities of these networks are dictated by the genetic information, thereby defining the ways in which engineers can influence cellular functions and metabolic capabilities toward the designed of improved biocatalysts for environmental remediation. Although, superficially these strategies seem distant from traditional chemical engineering, a deeper inspection reveals that the design of biological catalysts, based on defined techniques from biochemistry and biology is indeed parallel to our understanding of chemical kinetics, transport, separation, and control. In addition, advances in genomics and proteomics are providing opportunities to predict, in a quantitative manner, the potential manipulations necessary.

Even though chemical engineers are well prepared to contribute new research directions in environmental biotechnology, only a few are working in this area today. Fortunately, the number of chemical engineers showing interest is growing every year and with the recent research emphasis on biotechnology, it is easy to envision that many others will join this exciting research area in the near future. In this article, we will attempt to highlight opportunities available for chemical engineers to make significant contributions and their future challenges.

Engineering Biosorbents for Heavy Metal Removal

Immobilization of heavy metals into biomass or precipitation through reduction to lesser bioactive metal species, such as metal sulphide are the major mechanisms employed by nature (microorganism, animals and plants) to counteract heavy metal toxicity. These natural mechanisms can be easily exploited to optimize biosorbents that are more efficient for heavy metal removal. In one example, a sulphide-dependent metal removal strategy was developed by engineering the sulfate reduction pathway into a robust bacterium E. coli.

The resulting strains produced significantly more sulphide and removed more than 98% of the available cadmium under anaerobiosis. Further improvement in metal precipitation was achieved by engineering effective sulfate reduction under aerobic conditions. E. coli expressing both serine acetyltransferase and cysteine desulfhydrase overproduced cysteine and converted it to sulphide. The resulting strain was effective in aerobically precipitating cadmium. This aerobic approach of metal precipitation is particularly attractive as large-scale processes could be implemented under aerobic conditions. The challenges are to incorporate these genetic modifications into a robust environmental microbe that could survive and thrive under the required operation conditions.

Similar success in engineering enhanced biosorbents has been achieved by displaying metal-binding peptides onto the cell surface. One example was recently reported by creating a repetitive metal-binding motif consisting of (Glu-Cys)nGly.These peptides emulate the structure of phytochelatins, metalchelating molecules that play a major role in metal detoxification in plants and fungi. The phytochelatin analogs were presented on the bacterial surface, enhancing Cd^{2}-and Hg^{2}-bioaccumulation by 12-fold and 20-fold, respectively. Unlike nature metal-binding peptides, these "de novo" designed metal-binding peptides are attractive as they offer the potential of improved affinity and selectivity for heavy metals. To this end, the use of molecular modelling and evolutionary strategies may enable the rapid discovery of novel peptide sequences that are superior metal chelators.

In addition to peptides, metalloregulatory proteins are another group of useful metal-binding moiety with striking affinity and specificity. The highly specific nature of these proteins is the result of a cleverly designed genetic circuit that is tightly under their control. Examples are MerR and ArsR, which are regulatory proteins used for controlling the expression of enzymes responsible for mercury and arsenic detoxification, respectively. The high affinity and selectivity of MerR toward mercury has been exploited for the construction of microbial biosorbents specific for mercury removal. Presence of surface-exposed MerR on an engineered strain enabled 6-fold higher Hg^{2}-biosorption. Hg^{2}-binding via MerR was very specific with no observable decline even in the presence of 100-fold excess Cd^{2-} and Zn^{2-}. Similarly, cells overexpressing ArsR accumulated 5- and 60-fold higher levels of arsenate and arsenite with no observable binding to other heavy metals. As nature provides a wide range of metalloregulatory proteins for other important metal pollutants, this strategy may be used as a universal approach to enable selective binding of target metals of interest.

Designer Strains for Enhanced Biodegradation

Using well-established tools from metabolic engineering and biochemistry, efforts have been made on engineering microbes to function as "designer biocatalysts," in which certain desirable traits are brought together with the aim of optimizing the rate and specificity of biodegradation pathways. One common bottleneck is the transport of pollutants across the cell membrane, which limits the overall rate in many microbial biodegradation. An example is for a class of neurotoxic organophospates, which are used extensively as pesticides and chemical warfare agents. Although an enzyme, organophosphorus hydrolase (OPH), has been shown to degrade these pesticides effectively, the use of whole cell detoxification is limited by the transport barrier of substrates across the cell membrane.

Display of OPH onto the cell surface has been employed to bypass this transport barrier, resulting in 7-fold faster degradation compared to whole cells expressing OPH intracellularly. This simple approach typified the unique combination of chemical engineering principle with modern genetics, and has been similarly employed for other useful environmental applications such as the display of metal-binding proteins described earlier. Although only fairly simple enzymes or peptides are successfully displayed so far, continued development in this area should pave the way for the successful display of more complex enzymes, such as dioxygenases or monooxygenases, enabling a broader class of pollutants to be targeted.

Recruiting different pathways into a designer microbe is another powerful approach to enhance biodegradation. Very often, these pathways are combined with other existing pathways to enable complete biodegradation. For example, construction of a hybrid strain which is capable of mineralizing components of a benzene, toluene, and p-xylene mixture simultaneously was attempted by redesigning the metabolic pathway of Pseudomonas putida. A hybrid strain carrying both the tod and the tol pathways was constructed and was found to mineralize a benzene, toluene, and p-xylene mixture without accumulation of any metabolic intermediate. Since the number of known biodegradation pathways is increasing everyday, this in combination with the increasing number of genome sequence elucidated for environmental microbes, should allow us to rationally combine useful pathways across species into any desirable combinations using tools available from metabolic engineering. In this respect, it will also be interesting to see whether multiple enzymes can also be displayed onto the surface, allowing sequential degradation to occur without any uptake limitation. The challenges here are to devise strategies that will allow not only multiple enzyme display, but also the display of complex enzymes without compromising integrity and viability.

Another promising approach for success remediation is the introduction of biodegradation pathways into microbes that thrive in the contaminated environment. Deinococcus radiodurans is a soil bacterium that can survive acute exposures to ionizing radiation of 15,000 Gy without lethality. A recombinant D. radiodurans strain expressing toluene dioxygenase was shown to effectively oxidize toluene, chlorobenzene, and TCE in a highly irradiating environment. The recombinant strains were also tolerant to the solvent effects of toluene and TCE at levels exceeding those of many radioactive waste sites. The prospect of using this strategy to alleviate the toxicity of radionuclides and heavy metals, and to provide efficient treatment for a variety of organic wastes is very promising.

Similarly selective advantages can be achieved by exploiting the synergistic plant-microbe relationship in a rhizoshere. This strategy was recently reported using a wheat rhizosphere system for the detoxification of soil-borne trichloroethylene (TCE). The toluene o-monooxygenase (Tom) gene was introduced into Pseudomonas fluorescents 2-79, a bacterium that colonizes the wheat root, enabling the establishment of a bacterium-plant-soil microcosm. Treatment of TCE-contaminated surface and near-surface soil was demonstrated, with more than 63% of the initial TCE removed within 4 days.

The most attractive aspect of this technology is the low cost associated since only expenses required for planting is necessary. Our group is actively pursuing this strategy by engineering both metal-binding and TCE capabilities into a single rhizobacterium, allowing it to retain TCE degradation in the presence of high level of metal contamination. Since over 40% of all superfund sites in the U.S. are co-contaminated with organic pollutants and heavy metals, the use of plant-microbe rhizoremediation will provide an ecologically sound and safe method for restoration and remediation. This will also represent an excellent opportunity for chemical engineers working primarily with microorganisms to collaborate with others focusing on plant, combining the unique features of rhizoremediation with phytoremediation.

A notable opportunity that has been so far overlooked by most chemical engineers is the production of valuable products and energy directly from wastewater. Of particular interest is the possibility of biohydrogen and bioelectricity production. In most cases, only natural microorganisms are exploited, resulting in fairly modest yields. However, this poor conversion also represents an excellent opportunity to employ the tools from metabolic engineering, protein engineering, molecular evolution, and system biology for the discovery of novel microorganisms with significantly improved efficiencies.

Enzyme Engineering for Improved Biodegradation

The ever-increasing information regarding the structure and function of enzymes and pathways involved in biodegradation of recalcitrant pollutants offers opportunities for improving enzymes or entire pathways by genetic engineering. Control mechanism and enzyme properties can be tailored by site directed mutagenesis, which is often guided by computer-assisted modelling of the three-dimensional(3-D) protein structures. For example, site directed approaches have been applied to enlarge the binding pocket of haloalkane dehalogenase, resulting in several-fold faster dechlorination of dichlorohexane. However, no mutant tested could utilize the more bulky substrates, such as TCE, suggesting limitations using this structural based approach. Perhaps the use of computational methods to predict subtle and distal changes in the protein backbone without perturbing the overall protein structure could be used to further improve enzyme function and stability.

Site-directed or rational approaches can often fail because it is known that mutations far from the active site can modulate catalytic activity or substrate recognition but are difficult to predict a priori. These methods are also restrictive because they allow the exploration of only a very limited sequence space at a time. This is clearly indicated by the creation of several chimeric enzymes guided by sequence comparison between two similar biphenyl dioxygenases. Although the resulting variants were capable of hydroxylating both double ortho- and para-substituted PCBs, combining the substrate range of the two parental enzymes, no new activity was observed. In this case, irrational approaches such as DNA shuffling, which allow the cross-breeding of genes between diverse classes of species, can be a preferable alternative to direct the evolution of enzymes or pathways with highly specialized traits.

In two independent studies, the substrate range of biphenyl dioxygenases toward PCBs has been successfully extended using directed evolution. Variants were obtained by random shuffling of DNA segments between the large subunit of two wild type biphenyl dixoxygenases. Several variants had extended substrate ranges for PCBs exceeding those of the two parental enzymes. Similar attempts to extend the substrate specificity of toluene ortho-monooxygenase (TOM) and OPH have been successful. In both cases, DNA shuffling was combined with simple plate screening assays, resulting in rapidly degradation of virtually nondegradable substrates. These examples are perhaps the best reminder, suggesting that other important biodegradation enzymes could be similarly improved with this strategy since the number of related dioxygenases, monoxygenases, and hydrolases for different pollutants are virtually unlimited.

Molecular evolution is probably the most useful way for evolving biodegradation enzymes for extended substrate specificities since microbial degradation of xenobiotics is usually by cometabolism and does not exert a natural selective pressure on bacteria. Computational methods that are useful to guide experimental design for directed evolution may be used to predict the optimal number of mutants that must be screened. Moreover, an optimal design of the parental DNA sequence set will allow a more focused probing of sequence space in only those regions that are likely to yield functional hybrids and should lead to a more efficient utilization of experimental resources, saving time and effort by reducing the number of evolutionary cycles.

Evolutionary and Genomic Approaches to Biodegradation

Evolutionary approaches are extremely useful for optimization of an entire biodegradation pathway comparing to step-by-step modifications offered by rational design. This was recently demonstrated by the modification of an arsenic resistance operon using DNA shuffling. Cells expressing the optimized operon grew in up to 0.5 M arsenate, a 40-fold increase in resistance. Moreover, a 12-fold increase in the activity of one of the gene products (arsC) was observed in the absence of any physical modi.cation to the gene itself.

The authors speculate that modifications to other genes in the operon effect the function of the arsC gene product. Such unexpected but exciting results are more likely to be realized using irrational approaches. This strategy is particularly attractive since the ultimate goal of many remediation approaches is for complete mineralization of the pollutants, and the concurrent optimization of an entire pathway will allow the efficient search for the correct coordination between a complex set of biodegradation reactions. Along the same line, recent advances in genome shuffling between species, which allow the exchange and recombination of diverse pathways into a single species, will further accelerate the discovery of novel microbes that are useful for the remediation of even a complex mixture of pollutants.

The availability of bacterial genomes relevant to biodegradation in recent years has allowed the feasibility to study the complex interactions between cellular reactions from a genomic and proteomic level. A quantitative understanding of how cells function requires every gene and protein to be placed in their dynamic context, which entails the integrated consideration of many interacting components. From this perspective, a system biology approach is necessary to predict the functioning of an organism in a complex environment and to describe the outcome of the thousands of individual reactions that are simultaneously taking place in a microbial cell. So far, such prokaryotic models have been limited primarily to E. coli and a few pathogens. However, similar modelling approaches should be able to predict contaminant bioremediation by microorganisms that are known to predominate in polluted environments.

Recently, de Lorenzo and coworkers-presented a pioneering study on the characteristics of the "global biodegradation network", in which they considered the global pool of known chemical reactions implicated in biodegradation regardless of their microbial hosts. The characteristics of this network support an evolutionary scenario in which the reactions evolved from the central metabolism toward more diversified reactions, allowing us to understand the evolution of new pathways for the degradation of xenobiotics and provide the basis for predicting the abilities of chemicals to undergo biological degradation, and for quantifying the evolutionary rate for their elimination in the future. This type of analysis, when coupled with the predictive approach for microbial catabolism using the University of Minnesota Biocatalysis/Biodegradation Database (UM-BBD) as a knowledge base and various sets of heuristic rules, will lead to untapped and improved strategies for bioremediation. This represents an excellent opportunity for chemical engineers who are already involved with system biology, and will undoubtedly evolve into an important research direction within the next 5 years.

Process Engineering for Improved Biodegradation

Scientists have generally considered biological treatment processes too inefficient to challenge chemical treatment processes, particularly in treating large volumes of waste. The typically long contact time of 10 s or even longer between the pollutants and microbes would require an impractically large process for a practical treatment plant. However, it is well known that biological processes are generally safer, greener, and cheaper to run. Recently, advances in process design have brought this dream into reality by converting a chemical scrubber that removes hydrogen sulphide (a gas that smells like rotten-egg) into a biotrickling filter, which reduces hydrogen sulphide into non-smelly sulfate. To achieve a gas contact time that is comparable with that of a chemical scrubber (1.6-2.3 s), the researchers passed contaminated air through the biofilter at a high velocity using a packing material with a high surface area. The unusually high air velocity resulted in a higher gas-film mass-transfer coefficient and outstanding H_2S removal.

After almost 2 years of continuous operation, the bioscrubbers still retained H_2S removal efficiencies of more than 98%. This finding is important as it suggests that substantial improvement in process performance could be achieved using a very simple design improvement, based on chemical engineering principles. The obvious challenges are whether similar conversions of chemical treatment processes into biological processes could be achieved, offering the same treatment capacity that are much cheaper and safer. Considering the cost benefit and environmental impact of switching to biological treatment processes, this is one opportunity that chemical engineers cannot afford to miss.

Undoubtedly, the numbers of problems that define environmental biotechnology are not restricted to those discussed in this article. However, it is our opinion that these perhaps represent the best opportunities for chemical engineers to make significant contributions, because of the unique intertwining of molecular biology, microbiology, reaction engineering, transport phenomena, and process design. Because of the interdisciplinary nature of this research area, a successful chemical engineer in the field must receive training in an interdisciplinary environment where expertise from different areas is available, a trend that is consistent with the evolving nature of chemical engineering education.

Although the ability to predictively design microbes or enzymes for any given remediation remains an overwhelming task, the increasing understanding of fundamental mechanistic principles generated from both genomic research or directed evolution will likely lead to the emergence of novel solutions for improved bioremediation. Chemical engineers must embrace this important opportunity in a fashion similar to the efforts invested on the early evolution of metabolic engineering and tissue engineering, two important areas that are now partially defined by chemical engineers.

ENVIRONMENTAL BIOCATALYSIS

Advances in both chemical catalysis and biocatalysis are determinants in reducing the environmental footprint of chemical processes and petroleum-based technologies. In the field of chemical catalysis (i.e. using catalysts of nonbiological origin) the principle of environmental catalysis is well established and has been accepted for decades: in the 60s and 70s it flourished, producing excellent research on how to decrease, catalytically, the amounts of contaminants in the fuels derived from petroleum, and in the 80s and 90s, increasing interest about catalysis and the environment was documented in many scientific scenarios. Nowadays, leading scientists are expressing serious reservations about the long-term health risks and problems with the weather patterns (including global warming and pollution) that have traditionally been associated with chemical processes and 'dirty-fuels'. Furthermore, biocatalysis, either using whole microorganisms or just enzymes (known as white biotechnology), is implicated in many spheres of human activity in terms of:

- environmentally friendly processes;
- the limited opportunities for the production of renewable and clean energies; and
- remediation of many compounds that are unfriendly or even toxic to the environment by the present ecological standards of our societies.

In this context, biocatalysis fully participates in the 'green chemistry' concept that was introduced in the 90s, and its effect on sustainability is now established beyond question. We would like to draw the attention of the biotechnology community to the current, developing concept of environmental biocatalysis in the context of the three topics defined above. Moreover, we want to stress that environmental biocatalysis provides a different landscape compared with the well-known concept of environmental microbiology (e.g. the use of mixed cultures for environmental aims), raising specific questions on both the enzymatic remediation of contaminants for preserving the environmental health of our ecosystems and the new advances in green chemistry for more benign processes for the production of new high-value compounds.

Although the benefits of biocatalysis for environmentally friendly fine-chemical transformations have been generally accepted, critics cite several important issues regarding the economic use of enzymes for processes involving clean energy production and bioremediation. Recent studies rely on the tools of modern molecular biology, such as protein engineering, (meta) genomics and proteomics, to reduce the cost and the use of chemicals for future developments in biocatalysts, and, at the same time, to decrease overall costs and to increase yields and efficiency. In this review we discuss the growing field of environmental biocatalysis, paying particular attention to the needs and means by which enzymatic processes can be beneficial to the environment.

Biocatalysis platform for green Processes

'Green chemistry' is defined as the design, development and application of chemical processes and products to reduce or eliminate the use and generation of substances hazardous to human health and the environment. Biocatalysts (either enzymes or whole-cells) constitute a greener alternative to traditional organic synthesis that offers appropriate tools for the industrial transformation of natural or synthetic materials under mild reaction conditions, low energy requirements and minimizing the problems of isomerization and rearrangement.

In addition, biocatalysts are biodegradable and can display chemo-, regio- and stereo-selectivity, resulting in decreased by-product formation and avoiding the need for functional-group activation, protection or deprotection. Large-scale industrial applications of biocatalysts include, for example, the thermolysin-catalysed synthesis of the low calorie sweetener aspartame, the production of acrylamide and nicotinamide (assisted by nitrile hydratases), the synthesis of the non-cariogenic sweetener isomaltulose by sucrose mutases and, more recently, the production of biopolymers such as polylactic acid (Figure 3). Good examples of the replacement of traditional organic processes by a greener biocatalytic alternative include the industrial synthesis of semisynthetic penicillins and cephalosporins, the transformation of natural and synthetic fibers, the pulp kraft-bleaching and recycling of paper, and the multi-step synthesis of polyketide and glycopeptide antibiotics.

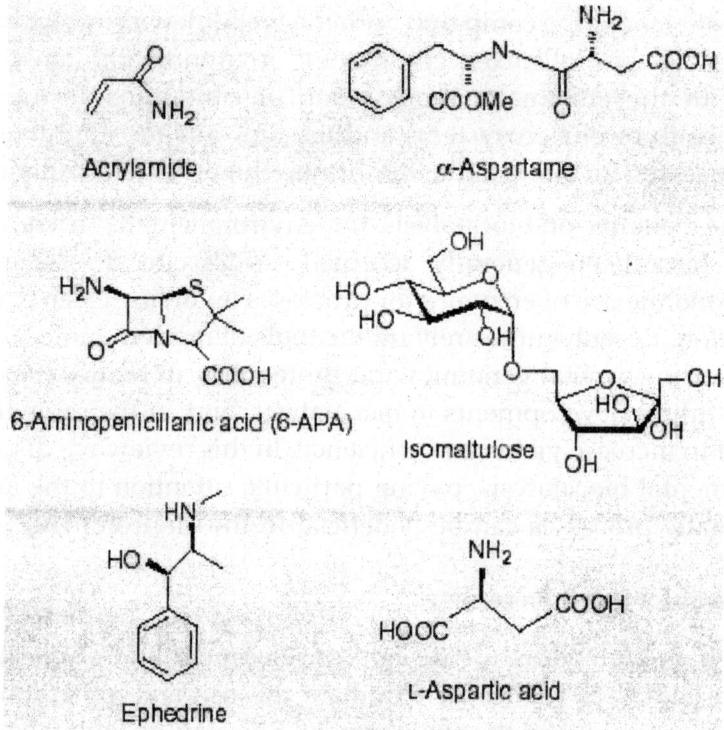

Figure 3. Examples of molecules manufactured using large-scale industrial applications of biocatalysts.

To overcome the problem of competing chemical reactions, the current frontiers for biocatalysis are protein activation and stabilization, and reaction specificity. For example, by using protein engineering and high-throughput screening it is currently possible to create stereoselective enzymes with broadly extended substrate specificities, which are useful in organic synthesis ; stable hydroxynitrile hydrates for the synthesis of, for example, substituted R-mandelic acids ; and hydrolases capable of enantioselective carbon-carbon bond formation or selective oxidation processes. Moreover, in synthetic reactions, involving either oxidoreductases, expensive redox co-factors or where whole cells are needed, biotechnological developments - from upstream (strain, cell and organism development) and midstream (fermentation and other unit operations) to downstream processes will significantly benefit from the application of white biotechnology in chemical transformations.

Additionally, combinatorial immobilization techniques are providing effective methods for optimizing operational performance, in addition to the aiding the recovery and reuse of biocatalysts. Although some enzymatic and/or microbiological processes are well established in industry, we strongly believe that biocatalysis is still in its infancy, and its future will depend on the search for novel and versatile enzymes, which are able to catalyze reactions that are difficult to perform by chemical methods. Indeed, metagenomics technology has revolutionized the possibilities of biocatalysis, and we can now access the genomes and genes, and the enzymatic activities encoded by them, of unculturable microorganisms. The success of metagenomics in finding new enzymatic activities has unequivocally demonstrated the power of this approach, showing that metagenomics is not a future opportunity anymore, rather it is a current reality for the production of new products and processes that were until recently hidden from us.

Bioremediation of Persistent Contaminants

Removal of widely dispersed, anthropogenic, organic pollutants is considered as one of the main concerns for the reasonable and sustainable development of planet Earth in the 21st century. Compared with traditional physico-chemical methods, bioremediation is generally the safest, least disruptive and most cost-effective treatment. This biotechnological tool uses whole microorganisms, naturally occurring or introduced, or isolated enzymes to degrade persistent contaminants into non- or less-toxic compounds and, in many cases, can be combined with complementary physical, chemical or mechanical processes to improve the reliability and effectiveness of detoxification. For example, since November 2002, microbial bioremediation using naturally occurring microbes in combination with mechanical approaches are currently being used as the major mechanism of removing low and high molecular weight, polycyclic aromatic hydrocarbons (PAHs) from the Prestige ship spill on the north coast of Spain. In fact, there are many possibilities in this area of biotechnology, either by in situ bio-stimulation (e.g. bioventing and natural attenuation)or ex situ technologies (e.g. biocells and landfarming).

Microbial Bioremediation

It is generally assumed that one of the major concerns of bioremediation with microbes, particularly in maritime areas, is that the microorganisms must be able to resist a variety of adverse conditions that are far removed from the ideal conditions of the laboratory. Additionally, when using inoculated microorganisms, particularly genetically modified organisms (GMOs), two major problems should be considered: the weakness and low level of fitness and growth of the inoculated microorganisms in competition with the indigenous population, and the possibility of altering a given ecosystem by introducing GMOs - international legislation is strict on this point.

The controlled release of Pseudomonas fluorescens HK44 containing a plasmid that encodes a naphthalene catabolizing enzyme represents the first and only genetically engineered microorganism approved for bioremediation testing. Whatever the case, and regardless of whether the organisms are genetically modified or not, microbial bioremediation is limited by factors such as mass transfer (low contaminant bio-availability), aeration, nutrient level at contaminant sites (might require bio-stimulation techniques) and problems with the thermal conditions. All of these issues should be addressed on a caseby- case basis to obtain the maximum microbial growth to metabolize the toxicants.

Enzymatic Bioremediation

Admittedly, we should recognize that, in the past few years, enzymatic bioremediation has become an attractive alternative to further support the bio-treatment techniques currently available - enzymes provide simpler systems than a whole organism. Most xenobiotics can be submitted to enzymatic bioremediation, for example, polycyclic aromatic hydrocarbons (PAHs), polynitrated aromatic compounds, pesticides such as organochlorine insecticides, bleach-plant effluents, synthetic dyes, polymers and wood preservatives (creosote, pentachlorophenol) (Figure 4).

It is just a matter of searching for the microorganisms capable of feeding on a particular pollutant, and then focusing the effort on identifying the enzyme(s) responsible. Historically, the most studied enzymes in bioremediation are bacterial mono- or di-oxygenases, reductases, dehalogenases, cytochrome P450 monoxygenases, enzymes involved in ligninmetabolism (such as laccases, lignin- and manganese peroxidases) from white-rot fungi, and bacterial phosphotriesterases. Moreover, new developments in the design and application of enzymatic 'cocktails' for the biotreatment of wastewaters have recently emerged owing to the effort of many companies and administrations such as American Industry Enzyme Technologies and Australian Orica. From an environmental point of view, the use of enzymes instead of chemicals or microorganisms undoubtedly presents some advantages :

- the biotransformation does not generate toxic sideproducts as is often the case with chemical and some microbiological processes, and the enzymes are digested, in situ, by the indigenous microorganisms after the treatment;
- the requirement to enhance bio-availability by the introduction of organic co-solvents or surfactants is much more feasible from an enzymatic point of view than using whole cells; and
- the potential to produce enzymes at a higher scale, with enhanced stability and/or activity and at a lower cost by using recombinant-DNA technology.

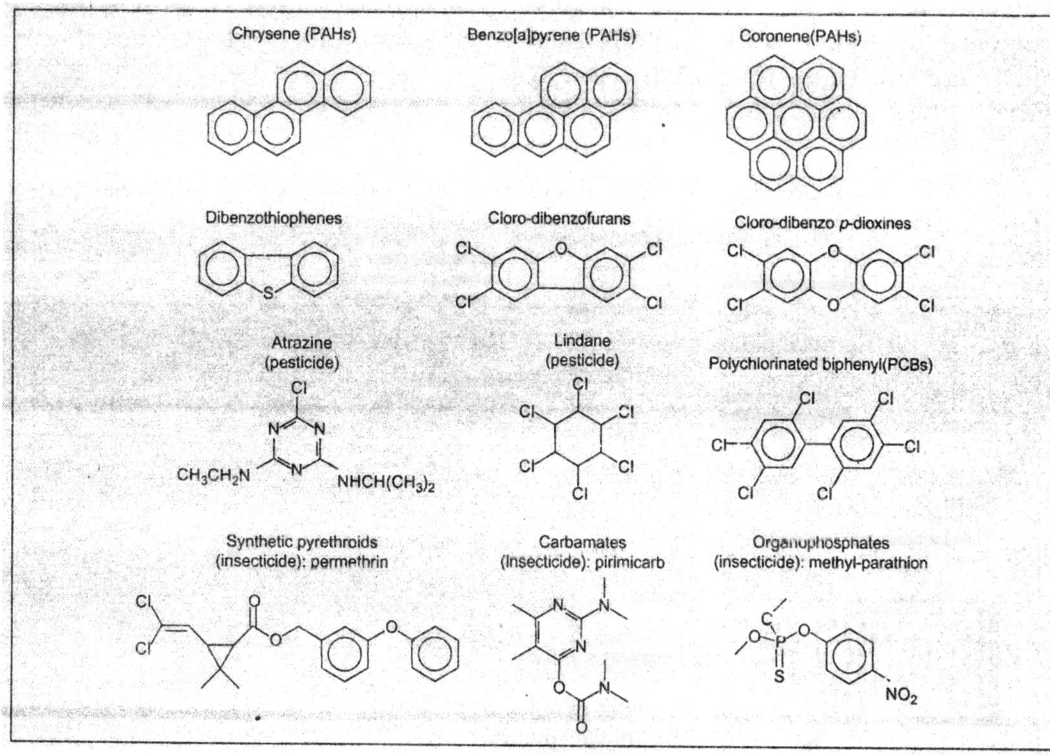

Figure 4. Some xenobiotics amenable to enzymatic bioremediation.

When performing enzymatic bioremediation, it is imperative that the optimal conditions for the enzyme are maintained during operational conditions. This requires cheaply produced enzymes (i.e. heterologous expression), with high substrate affinity (Km in the micromolar range), supporting thousands of product turnovers. At the same time, enzymes should display robustness under an array of external factors and low dependency on expensive redox cofactors (i.e. NAD(P)H), which might prove prohibitive in a commercial setting. Many of these shortcomings have been successfully resolved by using directed enzyme evolution (Figure 5) or by semi-rational approaches (i.e. combinatorial saturation mutagenesis of several hot-spot residues).

Obviously, a convenient method of assessing bioremediation applications should be the improvement of several enzymatic properties simultaneously, such as stability and activity. By directed (or forced) evolution, new features can be conferred to enzymes that, somehow, do not demonstrate affinity against novel or poorly degraded xenobiotics such as insecticides, herbicides, fungicides and mycotoxins. One example of this is atrazine (2- chloro-4-ethylamino-6-isopropylamino-1,3,5-s-triazine), a class of herbicide that first appeared in the 50s. In 2001 Raillard and co-workers applied directed evolution to shuffle two highly homologous triazine hydrolases, and the mutants used triazines that were originally not substrates for parent types.

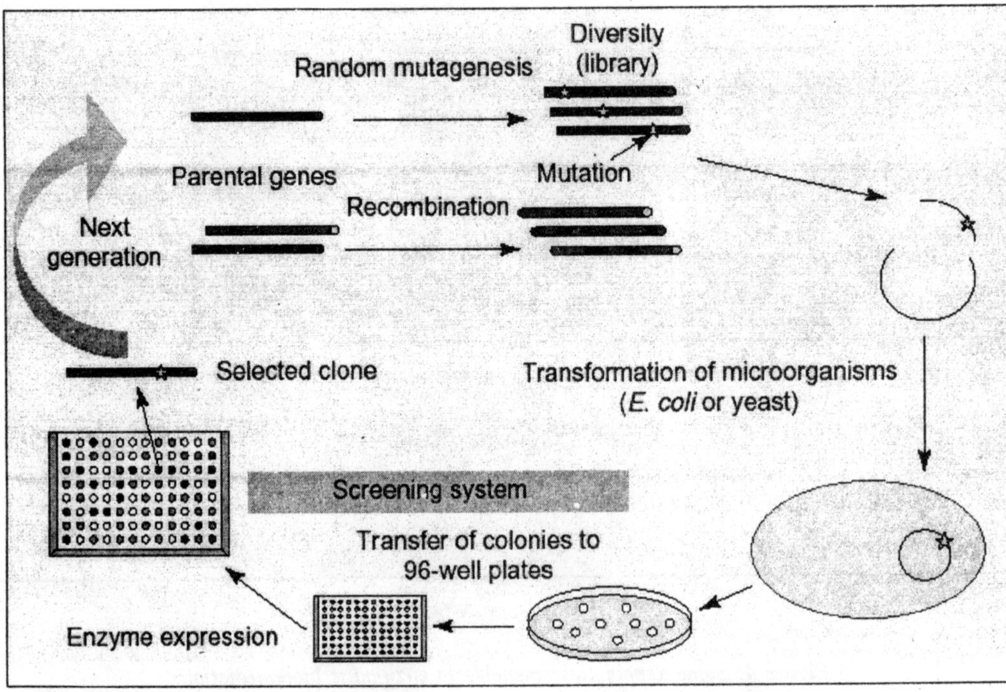

Figure 5. Typical experiment of laboratory evolution. After a few rounds of in vitro evolution, tailor-made enzymes with improved properties are ready for trials or field applications. The development of the high-throughput technology, along with the application of computational methodologies, will further push the engineering of enzymes by directed evolution and semi-rational approaches.

Directed evolution (two rounds of DNA shuffling) was also used to tailor a highly efficient phosphotriesterase to degrade organophosphates (OP) such as methyl parathion, a highly toxic neurotoxin used in insecticides and chemical warfare agents. Sutherland et al., through directed evolution and semirational approaches, designed a phosphotriesterase that is capable of hydrolyzing aliphatic OP compounds not degraded by natural enzymes. A similar approach has been used to engineer a biphenyl dioxygenase that attacks polychlorinated dibenzofuran in the lateral position, thus proposing a new pathway of degradation for this molecule. More recently, we have applied in vitro evolution procedures to obtain an improved version of a hexachlorocyclohexane dechlorinase (LinA), which is the primary biocatalyst, under aerobic conditions, to eliminate chlorine atoms from lindane (ghexachlorocyclohexane), a widely used and recalcitrant pesticide. This might constitute a platform for lindane-degradation schemes.

Laccases (belonging to the multicopper blue family) have also been extensively investigated for new and challenging decontamination programs because they affect the oxidation of many aromatic compounds towards more benign and less toxic products. Indeed, laccases can be involved in the detoxification of phenols, trichlorophenols, organophosphorus pesticides, azo dyes and, interestingly, PAHs such as benzo[a]pyrene, a class of highly mutagenic and carcinogenic xenobiotics, which are widely distributed in terrestrial and aquatic environments. Recently, we were involved in the functional expression of a thermophilic laccase in Saccharomyces cerevisiae. After ten rounds of laboratory evolution, we improved the enzymatic activity up to 170-fold, along with improving performance at high temperatures. This system can now be scaled up for promising applications in decontamination schemes.

Owing to the hydrophobicity and low aqueous solubility of xenobiotics, such as PAHs, enzymatic oxidations (e.g. by laccases) can be performed in the presence of organic solvents to minimize mass transfer limitations; however, laccases in organic solvents are fairly unstable and end up denatured or inhibited. Keeping this in mind, we have recently engineered a thermophilic laccase, by in vitro evolution, to be highly active and stable in the presence of increasing concentrations of acetonitrile and ethanol. Moreover, five rounds of sequential error-prone PCR, in vivo shuffling and saturation mutagenesis have led to the discovery of laccase mutants with several fold improvement in turnover rates at high concentrations of organic solvents (M. Alcalde, M. Zumarraga, H. Garcia- Arellano, F.J. Plou and A. Ballesteros, unpublished).

It should also be noted that to exert their remarkable action on PAHs or on lignin and lignin-related compounds, laccases need the presence of redox mediators (either synthetic or, more recently, from natural sources). Thus, future researchers should pay specific attention to the development of novel laccases with low dependency on redox mediators and/or higher redox potential (so far, in the range 0.4-0.8 V) to convert this biocatalyst into an efficient environmentally useful tool. To accomplish the goals described above, dramatic improvements in certain core technologies, such as (meta)genomic analysis of samples from extreme environments (i.e. contaminated soils), will be needed. Such developments will assist in the location of a greater repertoire of novel genes and/or enzymes that can be used as parent types for directed evolution in such a manner that, eventually, more robust and efficient biocatalysts can be tailored.

Enzymes for Clean Energy Production

Nowadays, exciting new opportunities for biocatalysis in the production of renewable and clean energy sources, such as biodiesel, bioethanol and biohydrogen, are rapidly emerging. Based on the premise that these alternatives can contribute to a cleaner environment, particularly when using renewable agricultural products, the demand for these energy sources is increasing. Today, bioenergy accounts for approximately 15% of the energy consumption of the world, and we believe that in the next 20 to 40 years we will be able to convert biomass into our transportation fuels. According to the enzymatic platform and the biomass used, we can distinguish three main groups of clean energies.

Biodiesel

The conversion of vegetable oils to methyl- or other shortchain esters in a single transesterification reaction using lipases has led to the production of high-grade biodiesel. This technology overcomes the disadvantages of chemical transformations based on acid- or base-catalysts because it reduces the consumption of energy and the need for separation of the catalyst from the reaction mixture, which is costly and chemically wasteful.

However, there are pros and cons to the use of biodiesel. Biodiesel is renewable, has low emissions per volume and is exempt from diesel tax, through special legislation, in several European countries, making processes involving biocatalysis more competitive. Moreover, the efficient, solventfree synthesis of oleic acid short-chain alcohol esters have been achieved with immobilized lipases such as those from Pseudomonas cepacia, Rhizomocur miehei and Candida antarctica. The limitations include the relatively high production cost, moderate reaction yields and the difficulties found during purification of the unreacted substrates, which obviously will require new future advances.

Bioethanol

Before the discovery of petroleum, natural carbohydrates were used for the production of food, clothing and energy. Ethanol fuels can be derived from renewable resources including agricultural crops, such as corn, sugar cane and sugar beet, or from agricultural byproducts, such as whey (from cheese making) and potato processing waste streams. Ethanol can be used either as a 100% replacement for petroleum fuels or as an extender: it can replace the toxic oxygenate methyl tert-butyl ether.

The best currently available technology for conversion is the acid hydrolysis of the biomass into sugars; however, alternative technologies, using enzymes such as a-amylases, glucoamylases, invertases, lactases, cellulases and hemicellulases, to hydrolyze starch, sucrose, lactose, cellulose or hemicellulose into fermentable sugars are in development. These sugars can be further fermented with bacteria, yeasts and fungi to produce ethanol, avoiding the use of strong acids and resulting in a cleaner stream of sugars for fermentation with fewer by-products. Again, the environmental benefits stem from the greater utilization of natural, renewable resources, safer factory working conditions, reduced harmful automobile emissions, and the consumer benefits from a safer alternative to the existing supply of liquid fuels such as gasoline.

Biohydrogen and Biofuel Cells

The possibility of using molecular hydrogen as a renewable, efficient and pollution-free energy source is also gaining attention. Hydrogen is colorless, odorless, tasteless, non-toxic and, on combustion, it produces water as the only by-product, making it different from every other common fuel we use today. Hydrogen obtained from biomass has the potential to compete with hydrogen produced by other methods such as from natural gas, which requires the catalytic conversion of hydrocarbons or electrochemical or photochemical water splitting.

Although certain microbes, such as Rhodobacter spheroids, have been used, successfully, in the production of hydrogen from fruit and vegetable waste, the process is currently still at the laboratory stage, and work needs to be done on increasing cost efficiency and applications. For these reasons, most research has concentrated on the use of hydrogenases for the production of hydrogen, for example, by fermentation of sugars or, more attractively, from waste. However, at present, typical production ranges are only between 0.37 and 3.3 moles hydrogen per mole of glucose, and these low yields, in addition to the cost, might explain why hydrogen is not our primary fuel. This has prompted the search for new hydrogenases using genome database mining and metagenomics, although the latter has not been used for enzymatic screening so far.

Hydrogenases, laccases and other redox enzymes also have broad applications as electrocatalysts, particularly in the development of biofuel cells. In this field, recent investigations have demonstrated that hydrogenases, which convert hydrogen to generate an electric current, possess similar energy conversion efficiency to noblemetal-based commercial methods. In this context, an enormous effort is being made to incorporate laccases into the design of biofuel cells - laccases are one of the few enzymes that can accept electrons from the cathodic compartment of a biofuel cell. In the immediate future, we will see growth in the use of enzymes in biological hydrogen and energy production but this will require the collaboration of biologists, chemists and engineers to integrate their knowledge to improve enzyme efficiency.

MICROBIAL ECOLOGY

Microbial ecology and environmental biotechnology are blossoming fields that are taking advantage of profound advances in biology, materials, computing, and engineering. Although traditionally microbial ecology and environmental biotechnology have been distinct disciplines, their futures are intimately linked. Together, they offer much promise for helping society deal with some of its greatest challenges in environmental quality, sustainability, security, and human health. What are these exciting fields? How do they connect to each other? And what is their common vista? How will microbial ecologists and environmental biotechnologists invest their resources so that the two fields, working as partners, create the greatest scientific advancements and benefits to society? That was the question for the symposium held last April at Arizona State University. The symposium's 10 speakers, who are also the authors of this article, are all considered international leaders working at the interface of microbial ecology and environmental biotechnology. These experts participated in focused discussion sessions aimed at arriving at a consensus view on the most important directions for the partnership of the two fields.

Microbial ecology is a long-standing scientific discipline that is undergoing remarkable, even revolutionary, changes. The core of the field aims to understand microbial communities, which are self organizing and self-sustaining assemblages of different microorganisms, and how these communities interact with their environment.

The science of microbial ecology tries to answer four fundamental questions. First, what microorganisms are present in the community? The phylogenetic makeup (identity and number) of the microorganisms present in a community is called a community structure. Next, what are the capabilities of the microorganisms for carrying out reactions that transform the community's environment? The catalogue of catalytic capabilities is called the community's phenotypic potential. Third, what reactions and transformations are the community members actually performing? The realization of phenotypic potential is called the community's function. Finally, what are the interrelationships among the community's members and between them and their environment? Interrelationships involve spatial organization (i.e., who is near whom) and materials that the microorganisms exchange. Understanding these interrelationships is the ultimate goal of microbial ecology.

Answers to these four questions usually change over time and especially in response to perturbation to the environment. The ability of microbial communities to respond to environmental changes (natural or anthropogenic) is called community resilience and stability, and it influences each of the four questions.

The beginnings of microbial ecology can be traced back at least to the late 1940s and 1950s; great conceptual advancements were made in the 1960s and early 1970s. However, microbial ecology struggled as a scientific discipline for decades because of a simple reality: Tools to address the four fundamental questions either did not exist at all or were unreliable. The small sizes and simple shapes of microorganisms made morphology a tedious and insufficient gauge of identity. Selective culturing on the basis of metabolic function was a giant step forward, conceptually and in practice, but often failed or gave biased results. Thus, brilliant microbiologists laboured to gain the faintest glimpse of the diversity and organization of microbial communities. They knew that so much more could be learned, if they only had the tools.

The use of molecular biology tools, beginning around 1985, started to change the prospects for microbial ecologists. Selective and reliable amplification of defined DNA with the polymerise chain reaction (PCR) and hybridization with DNA oligonucleotides made it possible to interrogate directly the genetic information of individual microorganisms and entire communities. The small-subunit ribosomal RNA (SSU rRNA, also known as 16S rRNA for prokaryotes) was the first target for hybridization and is still the most widely used. This rRNA gives information on the phylogenetic identity of the microorganisms, and this addresses the first fundamental question of microbial ecology.

Other targets within the genetic system of microorganisms answer different questions. For example, amplifying and detecting specific genes in the genome of microorganisms in the community defines the phenotypic potential, and this addresses the second question about the capability of microorganisms. Messenger RNA (mRNA) reveals which genes are being expressed and, therefore, which functional proteins likely are formed. This gives information on the expressed phenotypic potential, or the third question about the community's function. Assessing RNA targets by using microscopic visualization, such as with fluorescence in situ hybridization (FISH), adds information on spatial organization. Recently, microbial ecologists have begun to clone and sequence large genome fragments from various microbial communities ; therefore, it is now possible, at least for low-diversity communities, to reconstruct entire genomes of uncultured community members. In addition, microautoradiography, coupled with FISH (i.e., MAR-FISH) and stable-isotope probing, makes it possible to detect the function in the community, even when the researcher has no idea what genes or mRNA should be detected.

Environmental biotechnology manages microbial communities that provide services to society. Prominent and emerging services include removing contaminants from water, wastewater, sludge, sediment, or soil; capturing valuable products from renewable resources (e.g., biomass), particularly energy carriers but also nutrients, metals, and water; sensing contaminants or pathogens in the environment or, perhaps, in humans; and protecting the public from dangerous exposure to pathogens.

These services are essential if modern human society is to be safe, sustainable, and secure. Microbiological communities, properly managed, can provide these services reliably, continuously, economically, and without creating other hazards.

Environmental biotechnology is almost a century old, although this name is relatively new. Past names include biological treatment, biological processes, bioprocess engineering, bioremediation, and bioenvironmental systems. The new name reflects that environmental biotechnology is adapting and benefiting from the modern tools of molecular biology, as well as from other advances in science and technology.

The scientific foundation of environmental biotechnology is microbial ecology. Because environmental biotechnology ultimately aims to manage microbial communities for the good of society, a deep understanding of microbial communities-that is, microbial ecology-is essential. Environmental biotechnology addresses real-world goals by managing microbial communities, rather than finding or creating the solves-all-problems "superbug".

A field that has been around for a century usually does not show a burst of creativity, but environmental biotechnology is doing just that today. The reason is the convergence of a strong "needs pull" with a strong "science push". Today, environmental biotechnology gets a powerful pull from the needs of human society arising from increasing challenges to achieve sustainability, environmental quality, security, and human health.

The sustainability of modern human society depends on extracting essential materials from renewable resources and reducing reliance on nonrenewable resources. At the top of the list of at-risk resources are two that environmental biotechnology addresses directly: water and energy. Today, water providers are forced to tap sources of lower quality: polluted ground and river water, eutrophic lake and reservoir water, and wastewater. These poor-quality sources need substantial treatment to eliminate public-health risks, unpleasant taste and odor, and discoloration. Experts on the Middle East predict that water-not oil-will be the greatest future cause of strife in this region.After water, energy is the most precious resource, and future sources must shift from fossil fuels to renewable ones. Environmental biotechnology is at the heart of upgrading poor water sources for human use and for converting renewable energy sources- particularly biomass and sunlight-to useful forms, including natural gas, hydrogen, and electricity.

Organics, nutrients, and metals that are not captured but are instead discharged to the environment become pollutants, not resources. Environmental biotechnology has a long-standing role in treating wastewater and other contaminated water, air, and solids. Increasing population, urbanization, and economic activity heighten the need to apply environmental biotechnology to preserve (or improve) environmental quality, along with capturing valuable resources. Innovative environmental biotechnology approaches appear to be well suited for improving environmental quality in developed and developing countries alike. The readiness of developing countries to use these technologies lies in their improved reliability and operability in a decentralized setting.

Security and human health also relate to environmental biotechnology. Infectious diseases from pathogenic microorganisms remain the main cause of death worldwide, and terrorist threats to disseminate pathogens heighten the danger. Microbial systems can monitor for pathogens or chemical toxins in hospitals, water supplies, the air, and, perhaps, even humans. Microbial systems also contribute to cures or therapies by producing drugs or enzymes to fight diseases.

Environmental biotechnology can contribute in all these ways because microbial communities have seemingly infinite ways to live, even in environments that appear to be bizarre or hostile. Their ability to organize and sustain themselves provides human society with a cornucopia of metabolic potential to find services to benefit society. Microbial ecology, the core scientific discipline, allows us to understand microorganisms as part of their communities: "to think like the microorganisms". Armed with this deep insight, the environmental microbiologist can create sustainable systems that "work for the microorganisms so that they work for us".

On the science side, the great advances in microbial ecology, previously summarized, stand at the head of the science push and tend to garner the greatest credit. However, microbial ecology is not the only strong science push behind environmental biotechnology at the beginning of the 21st century.

The products of modern materials science provide another hearty push, perhaps as important as from microbial ecology. The materials push began around 1970, when lightweight, high-strength plastics made possible biological towers for wastewater treatment, which were the first biofilm processes with high surface areas and small footprints. From the 1980s through the 1990s, lightweight biofilm carriers in the form of gravel-sized pellets made even more compact, high-rate processes possible. From the late 1990s to today, microfiltration membranes have been replacing gravity separators for activated sludge, improving effluent quality, reliability, and compactness. More recently, the membrane biofilm reactor makes it possible to use H2 to reduce NO3 -, ClO4 -, and a large range of oxidized contaminants in drinking water, groundwater, and wastewater. Recent advancements in nanoscale (1-100-nm) materials and biomicroelectromechanical systems (bioMEMS) technologies surely will provide similar opportunities.

Another science push comes from mathematical modelling, the ultimate tool for integrating the large number of microbiological, chemical, and physical processes that occur in any microbial community of environmental significance. A model uses mass balance equations to represent the significant components in the community. Creating a model demands that the modeller identify the important system components, for example, the critical types of microorganisms, the substrates they consume, and the products they produce. Likewise, the modeller must represent the important reactions, such as the synthesis of new biomass, consumption of substrates, and generation of products, with mathematical expressions that capture what is known about the microorganisms and how they function. On the one hand, modelling is an exercise in integrating all that we know about microorganisms so that we can test and exploit our understanding of how the microbial communities work as a functioning ecosystem. On the other hand, modelling is the best way to find out what we do not understand about the microorganisms and their interrelationships with each other and their environment. When we cannot model a system, we can identify what we do not understand, and this helps us direct fundamental research to the most important questions.

Whether the goal is gaining scientific understanding or applying that understanding to create high-value services or products, the most fundamental questions are at the interface of microbial ecology and environmental biotechnology. Perhaps the overarching question is, "How do microbial communities self-assemble to achieve and maintain a function?"

The environmental biotechnologist depends on self-assembly once the proper conditions are in place. Therefore, self-assembly needs to be based on identifiable principles that can be put into practice. The overarching question then leads to corollary questions: What are the underlying principles of self-assembly? What community structure is optimal for a biotechnological application? What environmental conditions trigger assembly of the desired community with its desired function? Are the conditions and outcomes predictable, reproducible, and controllable?

Self-assembly also implies that microbial communities behave as a kind of multicellular organism, one in which the whole is more than the sum of its parts. Corollary questions along this line of inquiry include: Does the community have a multicellular "life of its own"? How do the microorganisms communicate and coordinate to gain the advantages of a multicellular state? What are the advantages, if any, to the community and to humans?

The consensus vista for the partnership of microbial ecology and environmental biotechnology shows three "peaks" in the distance, representing different themes: more powerful analytical tools, integrated "-omics" approaches, and research that is more theory- or process-driven. More powerful analytical tools. The strongest science push in microbial ecology comes from the explosion of new molecular tools. These tools have propelled the field forward, which makes it possible to answer questions that we could not think to ask in the not-so-distant past. Nonetheless, existing tools are inadequate for studying the complexity of microbial ecosystems and supporting the design and operation of innovative environmentalenvironmental biotechnology approaches. Shortcomings include methods that produce results too slowly and with too much effort, have biases, are too expensive, offer insufficient quantification, and lack coverage over the ranges of structure and function that are important in relevant microbial communities.

To overcome these shortcomings, developers of new molecular tools need to meet four criteria. First, the methods must be high-throughput, generating relevant data in minutes to a few hours. Second, the outputs must be quantitative enough to provide the information that sorts out the "lead actors" from the "bit players" for a given function and sufficiently sensitive to find the important microorganisms or reactions even when others dominate in numbers. Third, the outputs must provide the type of information that reveals structure and function in parallel, because they are intimately connected. Finally, methods must put more emphasis on the structure and function of eukaryotes and phages in order to yield a more complete picture of community structure and function. Ideally, more than one method should be applied to generate data (e.g., PCR-based and PCR-independent approaches).

Integrate the "-omics" approaches. Most molecular interrogation has been directed toward DNA, focused on selected genes, or (recently) aimed at high-throughput genomics. Although the potential for expanding capabilities in environmental genomics is enormous and must be pursued with vigour, resources also need to be invested in developing and using tools on the basis of the other molecules within the cells, or the other omics disciplines. The primary other -omics are transcriptomics, which is the study of mRNA as an indicator of gene expression; proteomics, which focuses on protein (enzyme) identification, characterization, and quantification as they relate to cellular function; lipomics, which targets the study of membrane lipids; and metabolomics, which studies the metabolic intermediates of cellular functions.

Each of the several -omics approaches provides different but complementary information that reveals distinct information about communities. Information from genomic and proteomic studies forms a comprehensive picture of the cellular response to perturbations or other changes in the environment; however, metabolomics reveals the ultimate functional response and is most likely to predict community phenotype.

A great challenge is to borrow strategies that were developed for the analysis of single organisms and apply them to complex microbial communities containing many and often unknown species. Initial success stories of monitoring gene expression and protein formation are encouraging and provide important guidance for future designs.

The hallmark of high-value -omics technology is the integration of the high-throughput data. An essential element is bioinformatics tools, which offer database support for data management and mining and biostatistics to maximize the benefit of these expensive and labour-intensive methods. Our current format for disseminating scientific results (e.g., refereed journal papers) is not ideally suited to handle the massive data sets of -omics-based experiments. Long-term stewardship and maintenance of these data over time must be addressed.

Make the research more theory- or process-driven. Molecular-based research in microbial ecology has developed an image of "stamp collecting", or the gathering of a lot of data because it is possible with the available tools. This enthusiasm for data collection is natural in an emerging field and was necessary when so little information was available before. Having any information is a giant leap forward, and no one knows a prior what will be the most interesting findings. Therefore, collecting microbiological stamps has been essential and will never disappear.

After 15 or so years of profitable stamp collecting, microbial ecology and environmental biotechnology are poised to advance to a higher plane of molecular-based research. This next step will emphasize research driven by ecological theory, process needs, or both.

The field of macroecology has developed elaborate theories about ecosystem succession, diversity, invasion, and stability, such as the hypothesis that highly diverse ecosystems are more stable. Statistical and deterministic modelling tools have been developed to design experiments and interpret results. For the most part, these theories have not penetrated microbial ecology. Without a doubt, the application of ecological theory from macroecology will demand a major modification to take into account the fundamental differences between systems of macroecology and microbial ecology. For example, microbial ecosystems deal with huge numbers of cells (e.g., more than trillions of bacteria per liter of water) that often cannot be differentiated as individuals; this contrasts with much smaller numbers of identifiable individuals in a macroecosystem. In addition, macroecosystems are based to a large degree on specific prey-predator relationships, whereas microbial ecosystems more often target the parallel metabolism and exchange of materials, despite predation being important (albeit poorly understood). Thus, the simplistic application of macroecology theory to microbial ecosystems would likely backfire. Nonetheless, proper application of ecological concepts-such as diversity, stability, competition, redundancy, and allied tools-provides a powerful intellectual framework for designing research and understanding the meaning of the results.

The next plane of research also needs to be process- driven rather than tools-driven. Now that a wide array of tools is available and continuing to expand, the researcher can ask critical questions about the process and then pick the right tools to find the answers. This approach allows researchers to take a more holistic approach toward understanding community structure and function. The researcher can form a hypothesis and bring in a range of molecular, modelling, and other tools to gain information that tests the hypothesis. A powerful research tool is to perturb the microbial community and then observe whether the organisms respond in the predictpredicted manner. In this hypothesis-driven approach, the questions lead to selection of the tools; this is a welcome reversal from the common situation in which tools dictate what questions can be addressed.

Theory- and process-driven research must recognize the role of scale-spatial and temporal. Microorganisms are generally on the order of a few micrometers in size, and microbial aggregates are only occasionally >1 mm. The technological processes in which microbial communities are used are very much larger. Even a bench-scale bioreactor has dimensions in centimeters, and full-scale systems measure in meters, often in tens of meters. Research that focuses on the micrometer-to-millimetre scale relates to what happens at the scale of meters, but the application is not always simple.

In such a situation, modeling can be of special value, because it takes into account transport processes and heterogeneity. In parallel with spatial scales, the range of temporal scales is extreme. A classic microbiology experiment might last 1 day, and an experiment on microbial ecology might last for a few months and have controlled operating conditions; however, a real-world process operates continuously for years and with unpredictable changes in temperature, substrate loading, and other factors. It seems logical that increasing time horizons and variability affect community structure and, perhaps, function. These effects need to be addressed directly in research design and how results are interpreted.

Skeptics demanding proof that "molecular research is really improving environmental technology" have confronted all of the participants in the vista symposium. Skeptics want specific examples of cause and effect in which research led directly to a new or improved process. Perhaps they seek a single "magic bullet". Up to now, direct, incontrovertible proof has been scarce, but it is beginning to emerge.

The demand to see cause and effect is justifiable, and the partnership of microbial ecology and environmental biotechnology needs to satisfy the demand. Achieving the three-peak vista will go a long way toward satisfying critics and attracting new supporters of the partnership. The immediate and direct pay out from the science research is thinking like the microorganisms. Then, good engineering and modern materials can translate the understanding into systems that manage the microbial communities to provide new and better services: working for the microorganisms so that they work for us. It is a two-part process that grows and harvests the fruit of microbial ecology and environmental biotechnology.

Despite these impressive achievements that are beginning to emerge and the high potential for the partnership, roadblocks are inevitable because of the inherent complexities of environmental microbial ecosystems. First, the numbers of different microbial strains are enormous-de facto infinity. Furthermore, only a tiny fraction of the strains have been cultured and characterized. Second, microorganisms can evolve rapidly-we face a perennial "moving target". Third, environmental microbial ecosystems are physically complex aggregates that change in time and space. Fourth, the environmental matrices of sludge, wastewater, sediment, and soil are "dirty"-a special problem for extracting cellular material and avoiding interferences. Fifth, a microbial ecosystem needs to be understood in terms of its structure, function, and interactions with other ecosystems in order to achieve society's goals. In summary, the study of microbial ecology in the context of environmental biotechnology is substantially more difficult than its study in medical settings, where many of the tools originated. Thus, connecting microbial ecology to environmental biotechnology always will demand a lot of hard work, persistence, and creativity as roadblocks are identified and overcome.

Modern-day microbial ecology and environmental biotechnology are young disciplines, still finding their identities separately and as partners. As with any emerging field, its researchers and the society thasociety that supports them must have a degree of faith that an investment of resources will yield a large benefit, because this is the way breakthrough advances in science and technology often have been made. Scientists first construct a "cathedral of knowledge", which inspires and enables creativity that leads to the great technological breakthroughs. Perhaps microbial ecology and environmental biotechnology are putting up the buttresses of the cathedral now.

MICROBIAL MODEL SYSTEMS IN ECOLOGY

Although many biologists have embraced microbial model systems as tools to address genetic and physiological questions, the explicit use of microbial communities as model systems in ecology has traditionally been more restricted. Here, we highlight recent studies that use laboratory-based microbial model systems to address ecological questions. Such studies have signifi-cantly advanced our understanding of processes that have proven difficult to study in field systems, including the genetic and biochemical underpinnings of traits involved in ecological interactions, and the ecological differences driving evolutionary change. It is the simplicity of microbial model systems that makes them such powerful tools for the study of ecology. Such simplicity enables the high degrees of experimental control and replication that are necessary to address many questions that are inaccessible through field observation or experimentation.

Ecologists are faced with the challenge of understanding the structure and function of systems, the component parts of which interact with each other in complex and diffuse ways at different scales of space and time. The approaches used by ecologists to make sense of such complexity vary in their degree of abstraction from nature and include field observations, experimentation in the field or laboratory and mathematical modeling. Studies of model systems (simplified representations of more complex systems) have played a particularly important role in ecology. In spite of the significance of model systems in ecology, the explicit use of microbial model systems has been relatively rare.

This limited use of microbial model systems stems, at least in part, from the historical division between ecologists and microbiologists. Throughout much of the 20th century, communication between the two disciplines was rare, and the study of general ecology developed separately from the ecological study of microorganisms. Even the field of microbial ecology, in spite of its crossdisciplinary name, developed as a distinct subdiscipline of microbiology, isolated from general ecology.

Recently, both general ecologists and microbial ecologists have become increasingly interested in bridging the gap between these disciplines. Research programs that integrate theory and microbial model systems with observational studies or experiments in the field are becoming more common.

The growing popularity of microbial model systems is due, in part, to the degree of experimental control they offer. In addition, the abundance of genetic and physiological information available for commonly used microorganisms, combined with their small size and short generation times, enables the design of replicated experiments across a wide range of spatial and temporal scales. Microorganisms are also amenable to genetic manipulation and to prolonged storage in a state of suspended animation. These advantages enable the ecologist to deconstruct the complexity of nature into its component parts and to explore the role of each part in creating patterns in nature, first in isolation, then in combination.

In spite of these advantages and the historic importance of microbial model systems in the development of ecology, some ecologists remain skeptical about what microbial laboratory systems can tell us about the natural world. They are concerned, for example, that such laboratory model systems are overly simplified, contrived and too small in spatial and temporal scale to be useful. However, these criticisms reflect confusion about the purpose of microbial model systems. Laboratory model systems are not intended to be miniature versions of field systems, and laboratory ecologists do not intend to reproduce nature in a laboratory model system. Rather, the purpose of laboratory model systems is to simplify nature so that it can be more easily understood. The ultimate test of our ecological understanding is if we can predict the behavior of an ecological system, whether in the laboratory or the field. If we cannot accurately predict the behavior of a simplified laboratory system, it is unlikely we understand enough to make predictions of field systems.

Here, we discuss recent studies that illustrate the utility of microbial model systems for studying ecology, focusing on their successes in addressing ecological issues, such as the effect of spatial interactions on community dynamics, the interplay between ecology and evolution, and the generation and maintenance of biological diversity. We take a broad approach, discussing select studies that illustrate the strengths of microbial model systems, as well as their limitations. Overall, we argue that the simplicity of microbial model systems provides a stark contrast to the complexity of the natural world, enabling researchers to test competing hypotheses about ecological processes and to establish the plausibility of mechanisms presumed to be operating in field systems.

Local interactions lead to patterns at large spatial scales One of the central questions in ecology is how the sum of repeated local interactions gives rise to ecological patterns at larger spatial scales. For example, distributions of organisms in the rocky intertidal form large-scale regional patterns that are the result of ecological processes acting at multiple spatial scales, ranging from localized competition to regional recruitment. The small size of microorganisms provides a unique opportunity to explore questions of scale in controlled, replicated experiments. Although small in absolute terms, microbial microcosms are orders of magnitude larger than the organisms they contain, making it possible to explore ecological patterns occurring at several spatial scales.

The importance of local interactions in governing patterns of diversity at larger spatial scales is emphasized in a recent study by Kerr et al. on the coexistence of toxin-producing, toxin-sensitive and toxin-resistant bacteria. These three populations of Escherichia coli exhibit nontransitive competitive relationships, similar to the game rock-paper-scissors. Kerr et al.showed that coexistence of all three types is favored when competition and dispersal occur locally (e.g. when the community is grown on the surface of an agar plate and propagated in a way that preserves spatial structure). Diversity is rapidly lost when these processes occur globally (e.g. in a well mixed flask where spatial structure does not develop). Many biological communities exhibit nontransitive interactions. Studying such interactions in microbial systems can provide valuable insight regarding their ecological and evolutionary roles in structuring communities.

Another goal of ecologists is to understand the processes governing the spatial and temporal distribution of diversity. The amount of energy available in an ecological system is thought to be a key determinant of diversity. Several studies have demonstrated a hump-shaped relationship between diversity and productivity in macroorganisms and some microorganisms, but the mechanisms underlying such trends are poorly understood.

Recent work on the relationship between diversity and productivity in a microbial model system provides another example of how local interactions, in this case competition among niche specialists in a heterogeneous environment, can explain patterns of diversity occurring at larger spatial scales. The familiar hump-shaped relationship between diversity and productivity was observed when populations of the bacterium Pseudomonas fluorescens were allowed to compete in the spatially heterogeneous environment of an unshaken microcosm. In this microcosm, microenvironments developed as a result of gradients in oxygen and the production of metabolic byproducts. In homogeneous (shaken) environments, such gradients did not develop and the hump-shaped diversity-productivity pattern was not observed. This study enabled the identification of key processes underlying diversity patterns, a difficult task in many field systems where such patterns can occur on regional scales (c. 106 km^2). The role of habitat patchiness in facilitating the persistence of extinction-prone populations across large spatial scales has played a central role in the development of metapopulation theory. Metapopulation models predict a lower probability of extinction for many small patchily distributed populations connected by dispersal than for a single large population of the same size, because dispersal enables locally extinct populations to be recolonized.

However, for many populations, determining interpatch dispersal and identifying the extinction and recolonization patterns predicted by classic metapopulation models remains challenging. Holyoak and Lawler compared predator-prey communities of protists in small patches linked by dispersal (i.e. arrays of connected microcosms) with communities in isolated large patches (i.e. single undivided microcosms of the same total volume as the array of small microcosms) to demonstrate that populations persist longer in spatially subdivided habitats. Furthermore, by quantifying densities of predators and prey in individual patches, the authors identified many of the features predicted by metapopulation theory, including extinction-prone patches and asynchronous population dynamics among patches.

History of the Use of Microbial Model Systems

Microbial microcosms have played a central, if sometimes underappreciated, role in the history of ecology. The earliest published record of a microbial microcosm experiment is that by W.D. Dallinger who, in his address as the president of the Royal Microscopy Society in 1887, described his attempt to discover 'whether it was possible by change of environment, inminute life-forms, whose life-cycle was relatively soon completed, to superinduce changes of an adaptive character, if the observations extended over a sufficiently long period'. Dallinger demonstrated that the evolution of ecological specialization is underlain by a cost of adaptation and that evolution was amenable to study in the laboratory.

Almost 20 years later, L.L. Woodruff conducted experiments with hay infusions and concluded that interactions among organisms were an important driving force in the successional sequence of protozoans that he observed. Later, G.F. Gause conducted several more ecological experiments on competition and predation usingmicrocosms comprising bacteria, yeast and protozoa. Several key principles in ecology are attributed to these studies. For example, through estimating growth parameters for each species grown alone, Gause was able to predict which species would be competitively dominant. The interpretation of this work by G. Hardin ultimately led to the niche exclusion principle. Furthermore, Gause's work on predator-prey dynamics using Didinium nasutum and Paramecium caudatum demonstrated the importance of spatial refugia and immigration for the maintenance of predator and prey. Since Gause's pioneering experiments, microbial microcosms have been used to study various topics in ecology such as succession, the diversity-stability relationship, predator-prey dynamics, the coexistence of competitors and the coexistence of generalists and specialists.

Ecological patterns and processes at multiple temporal scales The results of many field-based studies in ecology are limited by the temporal resolution of the experiments. However, many processes that might be important in structuring plant and animal communities, such as succession, coevolution, invasion and climate change, occur over much longer timescales than those of the average research grant or doctoral dissertation project, and short-term trends often yield different conclusions than long-term analyses. Because of their rapid generation times, microbial model systems can be used to address ecological questions over multiple temporal scales, and to explore community-level responses to environmental change.

The problem of temporal scale hinges on whether the observed responses to perturbation reflect equilibrium properties of the community or transient dynamics. This issue has particularly limited our understanding of the relationship between community diversity and ecosystem function. Since Odum, Elton and Macarthur first reasoned that more complex communities are more stable, researchers have sought empirical tests for these predictions and a more mechanistic understanding of the diversity-stability relationship.

Although large-scale field experiments are undoubtedly key to understanding the relationship between diversity and many important ecosystem properties, their experimental design and execution presents logistical challenges. Recently, researchers usedmicrobial model systems to test these predictions. McGrady-Steed et al.assembled laboratory communities of algae, bacteria, protists and small metazoans that differed in their initial diversity, both within and across different trophic levels. The different communities rapidly stabilized within 40-80 generations of the dominant organisms, at diversities that were lower than initial levels. Measuring the functional attributes of the stable communities after six weeks revealed that the more diverse communities showed both less variation in CO_2 flux and more resistance to invasion by an exotic species than did the less diverse communities. These results provide support for the biological insurance hypothesis, which posits that the redundancy within functional groups is important to overall ecosystem performance. Similarly, Naeem and Li used replicate microcosms comprising algae, bacteria and protists with differing degrees of diversity within functional groups to test the hypothesis that more diverse communities should exhibit more predictable ecosystem properties. Measures of biomass and density were more predictable as the number of species per functional group increased (i.e. the standard deviation of density measures decreased as the number of species pe functional group increased, irrespective of environmental conditions). Both studies demonstrate a positive relationship between diversity and measures of stability and show that patterns observed among macroorganisms are also relevant for microorganisms.

Microbial microcosms can also be used to investigate the effects of simulated climate change on communities. Petchey et al., for example, explored the effects of ecosystem warming on community structure and function using communities of eukaryotic microorganisms. Communities differing in trophic structure and diversity were gradually warmed by 28C per week (or 0.1 to 0.28C per generation, roughly scaling to temperature changes that long-lived organisms might experience based on global warming predictions). In warmed communities, herbivores and predators tended to go extinct more frequently than in unwarmed communities, suggesting that global warming could result in significant extinctions of organisms in higher trophic levels.

Many ecologists remain skeptical that microbial model systems can tell us something useful about ecological processes in 'natural' communities. This is due, in large part, to the strong tradition of field research in ecology. Recent criticisms include assertions that microcosms, in general, and microbial model systems, in particular, are too contrived, too simple, too small in spatial and temporal scale, and fundamentally different from macrobial systems.

Microbial model systems are too simple and lack generality Microbial model systems are simple. Indeed, they must be if these experiments are to be informative tests of ecological theory. This simplicity is a strength of laboratorymodel systems and its purpose is to simplify nature so that aspects of it can be better understood. That laboratory systems lack generality is amisconception. Laboratory experiments with microorganisms usually address fundamental ecological questions using simple systems and, because of this, they potentially have more generality than do studies of more complex and often more idiosyncratic field systems.

Nature of Microbial Model Systems

E. Hutchinson criticized laboratory studies as being highly artificial and essentially 'a rather inaccurate analogue compute using organisms as its moving parts'. However, the experimental organisms and their interactions are not creations of the experimenter, neither are they under the direct control of the experimenter. Although a researcher can control the initial composition of a community, the subsequent dynamics result from ecological interactions and natural selection. Thus, outcomes not predicted by simulation models are often observed. Furthermore, most studies use species that co-occur in a particular habitat and, in that sense, they are no more artificial than exclosure experiments in the field.

Among ecologists, there is some hesitation to accept microbial model systems because microbes are thought to have a unique biology. However, both prokaryotic and eukaryotic microorganisms share the fundamental properties of macroorganisms. Thus,microorganisms are valid model organisms for questions that are concerned with these fundamental properties, such as trade-offs in life-history traits and resource competition. Arelated criticism is that it is inappropriate to use asexual genotypes as analogs for species in a community. However, most theory only assumes that 'species' do not exchange genes; thus, for many questions, asexual genotypes and species are equivalent.

Experiments with microbial model systems are too small in scale Laboratory studies have been criticized for being too small in spatial scale and too short in temporal scale. However, based on a literature search, Ives et al. concluded that microbial microcosm studies might be of longer average duration, in terms of generations of the organisms involved, than most field studies. Similarly, the size of a 30-mL chemostat relative to the size of *E. coli* cells contained therein is orders of magnitude larger than the ratio for growth chambers in greenhouses and enclosure studies in the field. One of the advantages of using microorganisms is that such relatively large temporal and spatial scales are possible. Furthermore, scale is an issue that confounds ecological inference in many experimental systems and is not unique to microbial model systems.

Direct manipulation of ecological complexity One of the more significant benefits of microbial model systems is that the degree of complexity is determined by the experimenter and is not imposed upon the experiment by nature. This benefit is particularly apparent in studies that have used microbial model systems to explore food-web theory.

Consider the large body of theory about the causes and implications of food-chain length. Food-chain models suggest that productivity is an important determinant of food-chain length, and that food-chain length influences the population-level responses of trophic levels to changes in productivity. Experimental tests of this theory are especially challenging, in part, because it is difficult to simplify natural food-web relationships into the clearly defined trophic categories required of food-chain theory. Microbial model systems sidestep this problem by defining, a priori, all components of a food chain. The properties of these food chains can then be studied in detail, and compared with the patterns observed in natural food webs.

For example, Kaunzinger and Morin explored the effect of productivity on the length and stability of microbial food chains of different lengths (one-, two-, and three-level food chains) by manipulating the resource concentration available to the primary producer. Foodchain length increased with productivity, with the longest food chains persisting only at the highest resource concentrations. The abundance of individuals within any given trophic level changed with productivity in a manner consistent with the theory of trophic cascades: the increased production associated with increasing the resource concentration available to primary producers percolates up through the food chain to increase the population density of the top consumer and those trophic levels an even number of levels below it. These results provide some of the most compelling experimental evidence for the interplay among primary productivity, food-chain length and population regulation. An intriguing problem in the study of trophic interactions has been that of omnivory.

Although omnivory is commonly observed in nature, many models predict that it should be rare overall and that coexistence of omnivores and their prey and/or competitors should depend on productivity. Furthermore, whereas many population models predict destabilization of predator-prey dynamics with resource enrichment, some theoretical models predict that the presence of omnivory can actually stabilize these dynamics. The difficulties of testing these predictions in most systems (e.g. accurate descriptions of current food web structure are rarely paired with data on population dynamics) can be avoided by using microbial model systems in the laboratory.

For example, Morin explored the effect of omnivory on population dynamics using communities of protists and bacteria. His model communities comprised bacteria, the bacteriovorous ciliate *Colpidium striatum* and the omnivorous ciliate Blepharisma americanum, which could consume bacteria and *C. striatum*. Both ciliates coexisted at high productivity levels, but at low productivity levels, the omnivore was excluded through resource competition with its prey and/or competitor, as predicted by theory. In addition, the interaction between *B. americanum* and *C. striatum* showed increased stability with resource enrichment, suggesting that the response of food webs to enrichment can depend on the amount of omnivory. Studying these interactions in simplified microbial model systems enables identification of key determinants of food web structure and characterization of the consequences of particular food web configurations.

Evolution of Ecological Characters

In microbial model systems, the traditional distinction between ecology and evolution is blurred, enabling researchers to study the evolution of important ecological characters. For example, Bell and colleagues addressed the evolution of the ecological niche in a series of experiments with the unicellular alga *Chlamydomonas reinhardtii*, which grows as an autotroph in the light, but can also grow as a heterotroph in the dark when an exogenous source of carbon is provided. Theory predicts that the breadth of adaptation will evolve to match the amount of environmental variation. Indeed, this is exactly what happens: light- or dark-specialists evolve when selection occurs solely in the light or the dark, respectively, whereas broadly adapted generalists evolve in environments that vary in time.

Recent laboratory experiments with the alga Chlorella vulgaris as prey and the rotifer Brachionus calyciflorus as predator demonstrated that ecological dynamics are ultimately inseparable from evolution. Classic predator-prey models predict inherent oscillations in predator and prey dynamics and a one-quarter-cycle phase shift between predators and prey. When the rotifer-algal system was cultured in laboratory chemostats, the authors observed cycle periods that were longer than predicted and were almost exactly out of phase. Subsequent experiments manipulating the degree of clonal diversity in the alga population demonstrated that these dynamics were a consequence of rapid prey evolution. The results of these experiments highlight the importance of considering real-time evolution in understanding ecological processes such as population dynamics.

Post-hoc analysis of ecological mechanisms Many microbial populations can be stored indefinitely in ultra-low temperature freezers and revived as required. The results of completed experiments can thus be dissected a posteriori to gain a more detailed understanding of the mechanisms underlying the ecological and evolutionary processes at work. Rainey and Travisano, for example, documented a striking example of adaptive radiation in the soil bacterium Pseudomonas fluorescens occupying a spatially structured (static) microcosm. They showed, through competition experiments using strains that were archived during diversification, that diversity in these cultures is maintained by negative frequency-dependent selection, exactly as predicted by theory.

A second example comes from a study of the stability properties of predator-prey communities. Schrag and Mittler documented the stable coexistence over 50 generations of bacteriophage and bacteria under continuous culture conditions, a situation where theory predicted that coexistence was not possible. They found that a small fraction (,5% or less) of the bacterial population was phage-sensitive and so enabled the phage to persist. Furthermore, the phage-sensitive population was sustained because the culture environment was much less homogeneous than was first thought - populations of bacteria grew on the walls of the culture vessel, as well as in the bulk fluid. Thus, stability was conferred on the community through the spatial structure of the culture vessel itself. This was confirmed by manipulations of the degree of spatial structure. Increasing the surface area of the culture vessel by the addition of glass beads increased the duration of coexistence of phage and bacteria. Removing the effect of the wall-associated populations by transferring cultures into clean vessels daily led to the rapid extinction of phage. Thus, although most basic models of predator-prey interactions fail to predict the stable coexistence of predator and prey that is observed in many environments, these studies identified the importance of spatial refuges in maintaining coexistence.

The experiments discussed here underscore the utility of microbial model systems for answering ecological questions that can be difficult to address using field systems. Moreover, the rapidly increasing pool of genetic information concerning many microorganisms provides the opportunity to understand ecological processes at all scales of biological organization. For example, the work on adaptive radiation in P. fluorescens, discussed above, has been extended to identify the loci responsible for niche specialization, with the ultimate aim of providing a comprehensive explanation for adaptive radiation in this system.

The studies reviewed represent a subset of those that have used microbial model systems to address ecological questions. Microbial microcosm studies have also been successfully used to explore ecosystem-level selection, the effect of resource supply ratios on the outcome of competition, the ecology and evolution of mutualisms, predator-prey coevolution, and the ecology and evolution of sociality.

However, microbial model systems are not appropriate for all ecological questions. For example, the small scale of microorganisms makes it difficult to manipulate explicitly environmental heterogeneity at relative scales similar to those in experiments with plants and animals. Heterogeneity in resources or conditions on such scales does develop in microbial model systems (e.g. gradients in oxygen availability or metabolic byproducts), and such heterogeneity is important for subsequent ecological and evolutionary dynamics. However, researchers are usually forced to describe this heterogeneity after the fact. Another limitation of microbial systems is that evolution of organisms in some microbial microcosms can occur over the order of days, often changing interaction dynamics before characterization has been completed by the researcher. Furthermore, some questions in ecology, such as those related to extinction and genetic drift, may require small populations, which are difficult to work with in microbial systems. Some of the unique aspects of microorganisms, such as clonal reproduction, unicellularity and lack of morphological diversity, also make microbial model systems inappropriate for addressing questions concerning age-based phenomena or behavioral ecology. That said, several questions that were once thought to be outside the realm of microorganisms have since been addressed using particular microbial model systems.

Finally, there are limits to our ability to extrapolate from microbial experimental systems to larger and often more complex systems. Such limitations are not unique to microbial systems, but are shared by ecological studies in general. Identifying the appropriate experimental scale and the limits of extrapolation from this scale are critical aspects of conducting research in all areas of biology, not just microbial systems. Microbial model systems offer a complementary approach to field and laboratory studies of macroorganisms. While there are practical limitations to microbial model systems, their simplicity makes microbial microcosm experiments especially powerful for determining the biological plausibility of theoretical predictions. Topics that are particularly ripe for exploration include studies of the genetic basis of ecological traits, and the evolution of self-supporting ecosystems. Microorganisms also offer a unique opportunity to explore more practical problems in fields such as toxicology and agriculture. Finally, microbial model systems can reveal much about the natural history of microbes themselves, which is essential given that microorganisms are important in nutrient cycling, industry and medicine.

REFERENCES

Alexander, M. *Biodegradation and Bioremediation*. Academic Press. New York, 1994.

Astrid A. Van de Graaf et al, "Anaerobic Oxidation of Ammonium Is a Biologically Mediated Process", *Applied and Environmental Microbiology*, 1995.

Cookson, J. T. *Bioremediation Engineering: Design and Application*. McGraw-Hill. New York. 1994.

Dekker, J., and G. Comstock. "Ethical and Environmental Considerations in the Release of Herbicide Resistant Crops." *Agriculture and Human Values*. Summer 1992.

Edgington, Stephen M. "Environmental Biotechnology." *Bio/Technology*. December 1994. p. 1338-42.

Gessner, M.O. and Newell, S.Y., "Biomass, growth rate, and production of filamentous fungi in plant litter", *Pages 390-408 in: Manual of Environmental Microbiology*, Second Edition, ASM Press, Washington, D.C., 2002.

Mannion, Antoinette M. "Sustainable development and biotechnology." *Environmental Conservation*, 19(4):297-306. Winter 1992.

Margaret G. Mellon. *Biotechnology and the Environment : A Primer on the Environmental Implications of Genetic Engineering*.

Peters, Pamela. *Biotechnology: A Guide to Genetic Engineering*. Dubuque. IA: Wm. C. Brown Publishers, 1993.

13

Industrial Microbiotechnology

Industrial microbiology is a field concerned with the development of technologies to control and manipulate the growth and activities of selected biological agents to create desirable products and economic gain or to prevent economic loss. In addition to bacteria and yeasts, animal and plant cell cultures are now used to produce sophisticated products such as monoclonal antibodies, immunomodulating compounds, and complex plant metabolites. Although fermented products have been consumed for thousands of years, only in the nineteenth century was microbial activity associated with the fermentation process. Soon after that discovery, microorganisms, especially bacteria, were selectively introduced on the commercial level. Techniques were developed gradually for pure-culture fermentation and strain improvement, but the major advance in industrial microbiology occurred during World War II with the large-scale production of penicillin by submerged-culture fermentation. In the 1950s, industrial microbiology shifted its focus to the production of therapeutic agents, especially antibiotics. Advances in molecular biology have greatly increased the potential applications of industrial microbiology in areas such as therapeutics, diagnostics, environmental protection, and agriculture. The techniques of genetic engineering, along with technology developments in bioprocessing, make possible large-scale production of complex natural compounds that would otherwise be very difficult to obtain. With the exception of the food industry, few commercial fermentation processes use wild strains of microorganisms. Of the many thousands of microbial species, few are used commercially, and fewer still are used as hosts for genetically engineering genes. Process development occurs in large part by strain improvement directed at increasing product yield, enhancing growth on cheaper substrates, and simplifying purification.

The pharmaceutical industry is an important user of microbes. For many years, microbes have produced antibiotics and steroid hormones. Genetic engineering has made it possible for bacteria to produce a wide variety of mammalian substances that are medically important. In agriculture, bacteria of the genus Rhizobium are added to legume seeds where, following nodule formation on the host plant, they trap or fix atmospheric nitrogen and reduce the need for ammonia fertilizer. Several specialty chemicals are most economically produced by microbes. These include some amino acids and vitamins. Industrial solvents like alcohols and acetone can be produced in microbial fermentations. However, at present it is cheaper to make them chemically from petroleum. Many of the most important industrial metabolites are secondary metabolites (see diagram in the next page), produced in the stationary phase of the culture after microbial biomass production has peaked. These compounds are not essential for growth of the microbe. Their synthesis is usually regulated by the cell. Therefore, to obtain high yields, environmental conditions that elicit regulatory mechanisms such as repression and feedback inhibition must be avoided.

In addition, mutant strains that overproduce the compound are selected. After a suitable microorganism has been identified from laboratory studies for an industrial process, there are still a number of scale-up problems to solve. These include provision of adequate aeration and mixing throughout the large fermentor. The difficulties involve the enormous volume of the vessel, areas where mixing is less efficient, and the high biomass content of the fermentor. High biomass is desirable to increase the product formed, but it creates an enormous demand for oxygen.

Furthermore, a strain that worked well on a small scale may not be as efficient under the different conditions experienced in the large fermentor. For industrial processes, fermentor of capacity up to 400,000 litres is used. The process may be aerobic or anaerobic. In general, aerobic processes are more difficult to run because it is difficult to adequately aerate a large vessel that contains high biomass concentrations. The fermentors are constructed of stainless steel, and have an external jacket by which it can be sterilized initially and cooled during the fermentation. The acetic acid bacteria from an alcoholic fluid such as wine or cider can produce vinegar, if oxygen is provided. There are several industrial methods for bringing these reactants together. Citric acid is produced by the fungus Aspergillus niger. Industrial fungal fermentations may occur on the surface of a medium or submerged in the liquid.

Yeast cells are the most extensively used microbe in industry. They are grown for use as baker's yeast in bread dough production. Under anaerobic conditions, they ferment sugars to alcohol, and form the basis for the wine and beer industries. Most industrial processes use various strains of one yeast species, Saccharomyces cerevisiae. Industrial Strain Development Strain development is achieved by either a traditional mutation and selection program or direct genetic manipulation. The recombinant DNA approach has succeeded in introducing new genetic material into a convenient host microorganism and amplifying genetic material. About 20% of the synthesizing capacity of a bacterium can be devoted to a single polypeptide or protein.

Industrial microorganisms are initially selected from natural samples, or taken from a culture collection because they have been shown to produce a desired product.

However, the strain is then modified to improve the product yield. This entails rounds of mutation, with careful selection for the rare clones that produce more or improved products. The selected strain is unlikely to survive well in nature, because the selection process has altered the regulatory controls in the cell to create metabolic imbalances. Other desirable characteristics are (1) rapid growth, (2) genetic stability, (3) non-toxicity to humans, and (4) large cell size, for easy removal from the culture fluid.

Industrial microbiologists may aim to produce (1) microbial biomass itself, (2) specific enzymes, or (3) metabolites. Enzymes that degrade polymers are especially important. The metabolites may be major metabolic products of catabolism, or compounds normally produced in trace amounts by natural isolates. To develop the procedures for obtaining new microbial metabolites, some of the steps include: (1) screening (selection for the production of new metabolites with new isolates and/or new test methods), (2) chemical modification (modification of known microbial substances), (3) biotransformation (change in a chemical molecule by means of a microbial or enzyme reaction), (4) interspecific protoplast fusion (recombination of genetic information from rather closely related producer strains. New or hybrid substances are expected), and (5) gene cloning (genes may be transferred between unrelated strains which are producers of known substances).

The major motivation for industrial strain development is economic, since the metabolite concentrations produced by wild strains are too low for economical processes. Depending on the system, it may be desirable to isolate strains that require shorter fermentation times, which do not produce undesirable pigments, which have reduced oxygen needs, which exhibit decreased foaming during fermentation, or which are able to metabolize inexpensive substrates. Wild strains frequently produce a mixture of chemically closely related substances. Mutants who synthesize one component as the main product are preferable, since they make possible a simplified process for product recovery. Changes in the genotype of microorganisms can lead to the biosynthesis of new metabolites. Thus, mutants, which synthesize modified antibiotics, may be selected

INDUSTRIAL MICROBIOLOGY PRODUCTS

The microorganisms utilised may be natural isolates, laboratory selected mutants or microbes that have been genetically engineered using recombinant DNA methods. For example, most antibiotics come from microbial fermentations involving a group of organisms called actinomycetes. Yeasts are used in baking, in the production of alcohol for beverages and in fuel production (gasohol). Other groups of microorganisms form products that range from organic acids to enzymes used for the production of various carbohydrates, amino acids and detergents. For example, aspartame, a sweetener, is derived from amino acids produced by microorganisms. Industrial microbiology also deals with products associated with the food and dairy industries, with the prevention of deterioration of processed or manufactured goods and with waste disposal systems.

Many industrial microbiologists/biotechnologists are responsible for the discovery, development or implementation of certain processes and the quality of resultant products. Such products and processes include:

- *Antibiotics/Antimicrobials.* Both natural and chemically enhanced microbial products can be used to control human, animal and plant diseases. Using traditional genetics or recombinant DNA techniques, the microorganism can be modified to improve the yield or action of antibiotics and other antimicrobial agents. New research directions are aimed at discovering microbial metabolites with pharmacological activities useful in the treatment of hypertension, obesity, coronary heart disease, cancer and inflammation.

- *Vaccines.* Vaccines are essential to protect humans and animals from microbial diseases. Recombinant DNA technology has allowed the production of novel vaccines that offer protection without the risk of infection (e.g. hepatitis B vaccine).

- *Health-Care Products.* The development and production of diagnostic assays that utilise monoclonal antibody or DNA probe technology are essential in the manufacture of health-care products such as rapid tests for strep throat, pregnancy or AIDS. Other examples include the use of microbial cells to produce human or animal biologicals such as insulin, growth hormone and antibodies. The industrial microbiologist/biotechnologist may screen new microbial sources (e.g., marine microorganisms) for their ability to produce new pharmaceuticals.

- *Foods/Beverages produced by microbial activity.* Yogurt, cheese, chocolate, butter, pickles, sauerkraut, soy sauce, food supplements (such as vitamins and amino acids), food thickeners (produced from microbial polysaccharides), alcohol (beer, whiskeys, wines) and silage for animals are all products of microbial activity. The industrial microbiologist/biotechnologist may be involved in producing concentrated microbial inocula for fermentations or the maintenance of fermentation systems utilised in production facilities.

- *Foods/Beverages cured or improved by microbial activity.* Production of coffee, tea, cocoa, summer sausage, vanilla, cheese, olives and tobacco all require microbial activity.

- *Food flavoring agents and preservatives.* Organic acids such as citric, malic and ascorbic acids and monosodium glutamate are microbial products commonly used in foods.

- *Foods.* Mushrooms, truffles and some red and green algae are consumed directly. Yeasts are used as food supplements for humans and animals.

- *Agriculture.* Conventional, recombinant DNA and monoclonal antibody techniques are used to improve microbial inoculants which serve as fertilizer supplements by fixing atmospheric nitrogen and as plant pest controls.

- *Enzymes.* Industrial applications of enzymes include the production of cheese, the clarification of apple juice, the development of more efficient laundry detergents, pulp and paper production and the treatment of sewage. These processes have been dramatically enhanced by the use of recombinant DNA techniques to design enzymes of increased activity, stability and specificity. Carbohydrates. Molecular sieves (e.g., dextran) for purification/separation processes; and thickening agents (e.g., xanthan) that are stable at high temperature are examples of microbial carbohydrates. The latter are used for secondary oil recovery in oil fields, lubricants in drilling oil wells, gelling agents in foods and thickeners in both paints and foods.
- *Carbohydrates.* Molecular sieves (e.g., dextran) for purification/separation processes; and thickening agents (e.g., xanthan) that are stable at high temperature are examples of microbial carbohydrates. The latter are used for secondary oil recovery in oil fields, lubricants in drilling oil wells, gelling agents in foods and thickeners in both paints and foods.
- *Organic chemicals.* Compounds such as acetone, methanol, butanol and ethanol all have multiple applications in industrial settings. Microbes will increasingly be used to supplant or replace those processes which rely on petroleum/natural gas for the production of these compounds.
- *Oil recovery/mining.* Oil recovery may be facilitated by the development of unique bacteria which produce a surfactant that forces trapped oil out of rocks. Extraction of minerals from low-grade ores is enhanced by some bacteria (microbial leaching). In addition, selective binding of metals by biohydrometallurgical processes is important in recycling of metals such as silver and uranium.
- *Contamination control.* Development of assays to detect microbial contaminants in food; evaluation of natural or synthetic agents for the prevention of disease, deterioration or spoilage; determination of minute quantities of vitamins or amino acids in food samples; development of preservatives for control of food spoilage; and development of procedures for control of deterioration in cosmetics, steel, rubber, textiles, paint and petroleum products rely on extensive knowledge of industrial microbiology and biotechnology.
- *Waste and wastewater management.* Isolating or developing microbial strains capable of degrading and detoxifying hydrocarbon and halogenated hydrocarbon waste (for example organophosphates, acetylcholinesterase inhibitors) of industrial, agricultural or military origin is essential in waste management.
- *Environmental science.* Understanding the role of microorganisms in the degradation and transformation of chemicals in the environment is a key factor in assessing the environmental safety of new and existing chemicals. Engineering microorganisms may solve contamination and recycling process problems.

Biotechnology holds clear promises of becoming the next general purpose technology. Biotechnology can be considered a platform technology comprising a large array of sub-technologies based on a series of key scientific advances in the second half of the 20th century that established new insights into the biological basis of life. Biotechnology develops and builds on knowledge and capabilities in a diverse set of disciplines, especially molecular biology, but also chemistry, biochemical engineering, microbiology, computing and programming.

Biotechnology is "technology", not just "science", because it represents huge commercial opportunities for innovation. But, contrary to other technologies, the activities underlying the development of biotechnology strongly overlap with those underlying fundamental science. Biotechnology thus both reflects a new scientific and technological paradigm; a context which until now especially has had substantial impacts on the organisation and dynamics of the pharmaceutical industry. But biotechnology can also be conceived as an "industry", a collection of firms that primarily specialise in research and development of biotechnology. Especially in the USA, a virtual bubble of entrepreneurial energy fed into establishing dedicated biotechnology firms (DBFs) during the 1980s and 1990s, founded by university scientists and based on university science. More than elsewhere in the world institutional changes were implemented in the USA both in regard to intellectual property rights (IPR) in life sciences and financing of purely science-based firms, mobilising high-powered incentives among university scientists to "go commercial".

Incumbents in pharmaceuticals have had to acquire and assimilate biotechnology capabilities and to engage in cooperative relations with DBFs, universities and other research institutions in order to survive, and the same seems slowly to become the case, even if at a much smaller scale, in these and other industries. This development is likely to come to resemble the diffusion of information and communication technology (ICT) during the last 20 years, where other industries than the core ICT industries have come to account for an increasing share of overall ICT innovations.

From its embryonic beginnings the biotech industry is now about 30 years old. However, as Maryann Feldman points out in this issue, "[b]iotech is still at an early stage of development and there are many competing hypotheses about its future development". This issue presents five papers representing cutting-edge research on the industrial dynamics of biotechnology which in numerous ways challenge conventional wisdom. By the year 2025, it's estimated in a recent study that the market for biotechnology-related products in the European Union will be worth ECU250 billion (US$278 billion) – equivalent to around three million jobs. 70% of these jobs will be in the agricultural and food sectors. Already in the Netherlands, 20% of consumer spending and 18% of the workforce are in the agro-food industries. Across Europe, biotechnology-based products in the agricultural sector (mostly the use of enzymes in animal feed) are currently worth US$5.6 billion per year, and in the food sector (use of enzymes to make bread, beer and cheese; production of vitamins, amino acids, starch and glucose) are worth a further US$24.5 billion. A growing area of industrial biotechnology is the development of products and techniques for cleaning up pollution caused by agriculture, industry or urbanisation – this is known as environmental biotechnology or 'bioremediation'.

Many industrial biotechnology products, such as beer, antibiotics or enzymes, are produced by growing microorganisms, under clean, highly controlled conditions in containers called fermenters. Most fermenters are stainless steel or glass tanks containing anywhere from 10 to 10000 litres of liquid. The liquid contains all of the nutrients that living cells need to grow – carbon and nitrogen sources, vitamins, minerals, phosphorous and oxygen. The concentration of all of the nutrients and oxygen and the temperature are all controlled by computer. Cells from other organisms – plants or animals – can also be grown in this way, and used to produce large amounts of valuable products. A valuable purple dye called shikonin is produced in this way in Japan from a plant that was almost extinct: the dye was originally extracted from the roots of 7-year-old plants, but can now be harvested from cells in culture after just 20 days.

Growth in biotechnology blurs the lines of differentiation between different fields of science. For instance, biotechnology advancements are integrating with semiconductor technologies such as micro-electromechanical systems resulting in the evolution of biochips. Biochips provide pharma companies with sophisticated tools for understanding gene expressions, biological mechanisms, speedy development of drugs, and accurate diagnosis. DNA and protein biochips streamline drug development and significantly lower average drug screening costs from $2 to $0.0001. In addition, with the transition of the medical lab-on-chip (LOC) from a concept to a reality, the focus has been on developing high density and smaller-sized biochips wherein less than 300 wafers may accommodate the entire human genome. As opposed to conventional laboratory analysis, LOCs eliminate the need for trained personnel and expensive equipment to provide significant time and cost efficiencies.

The biotechnology industry by providing a comprehensive view of emerging technologies and applications such as bioanalytics, biopharmaceuticals, glycobiology, nanobiotechnology, and industrial biotechnology. Analysis of the technological trends, drivers, challenges and recent developments will assist in creating effective marketing and production strategies. Participants can identify potential collaborators, stay ahead of the competition, and stay shoulder-to-shoulder with critical developments in their industries.

The rapid evolution of glycomics as a natural extension of proteomics provides a better understanding of glycoproteins; thereby, helping the development of novel biodrugs. "Improved understanding of the structural and functional data in glycomics coupled with phenomenal developments in genetic engineering techniques drives the usage of recombinant glycoproteins as therapeutic agents and as an alternative to small molecules," says the analyst of the study.

Demand for therapeutic glycoproteins and monoclonal antibodies is expected to increase creating an urgent need for higher production capacity. Additionally, alternate manufacturing mediums such as transgenic animals, plants, and mammalian and fungal cell lines are also on the rise. Fungal cell lines, in fact, provide considerable time and cost benefits over mammalian cell lines. The latter is not only a lengthy process but may also alter the properties of the final therapeutic glycoprotein. Conversely, fungal cell lines such as engineered yeast expression systems facilitate production of humanly glycosylated protein, provide faster fermentation and generate higher product yield.

MICROBES AND FOOD INDUSTRY

Cheeses are simply a form of spoiled milk, many of which are covered with the very same molds that we throw out in disgust when we find them growing on our bread or tomatoes. The French and Germans consider snails such a gastronomical delight that they fight local snail-wars over the right to scour the woods for these slimy gastropods. Horse's milk and meat is eaten by more people than eat beef and cow's milk. Insects are a staple of the diet of most non-Westerners. In various parts of the world one can experience the delights of fried grasshoppers, a variety of fat, juicy fried worms and crunchy crickets or water bugs. Some insects are even eaten alive as their flavor is considered the best in this state. Almost all peoples make and enjoy fermented beverages, some of which are produced in unusual ways such as having old women chew up and spit the raw material into containers to aid in the fermentation process. Certain countries consider dog and cat meat a special treat, and monkeys and rats are a common food for many humans around the world. The expensive steaks you purchase in fine restaurants are juicy and tender because they have been hung in the cold room long enough to allow a thick layer of mold to cover the sides of beef; the mold releases proteases that tenderize the meat.

The discovery of the cheese-making process is very old and certainly was accidental. Early man learned to carry his water, beer and milk in natural containers like animal stomachs, bladders and lengths of intestines tied at the ends. These were tough, water-proof and light, and they could easily be tied around ones neck, shoulder or waist. The stomach of young cattle contains an enzyme, rennin, that cleaves the casein protein of milk making it easier to curdle when microbes convert the lactose sugar in milk to acid; *This is the basis of cheese making*. A likely scenario is that a calf's stomach, full of milk, was left in a cool corner of a cave or hut for several weeks during which time the milked curdled, the liquid evaporated and microbes contaminating the milk grew. The molds and/or bacteria that grew on and in the curd as it continued to dry produced a unique flavor. When the owners finally returned they found a furry chunk of what once had been milk and being hungry they gave it a try and found out that it didn't taste too bad. This experience probably happened numerous times given our ancestors propensity to carry milk & other liquids in the stomachs of dead animals. People quickly realised that the semidried curd (the precipitated milk protein) was lighter to tote around than the milk and that it lasted a relatively long time before it spoiled so badly you couldn't stand to eat it. Finally someone came up with the idea of making it happen on purpose and the cheese industry was born.

Intentional alcohol production by humans is known to have been around for at least 10,000 years. It is not illogical to imagine that it is even older than that; probably coinciding with man's use of containers to carry around liquids described above. Like modern man, our ancestors relished honey and certainly raided wild bee hives for the sweet nectar they contained. Because of the liquid nature of honey early humans undoubtedly placed the honey in whatever containers they could use; i.e., animal stomachs, bladders etc. After a successful raid on a hive our ancestors must have, like Winnie-the-Poo, sat around the fire and enjoyed dipping their fingers in the "honey pot". Certainly they quickly discovered that by adding some water to the container they could make a sweet drink and certainly an occasional bag of honey-drink was left unattended for a period of time sufficient to allow fermentation to occur. Once the people returned and drank the now "modified" contents, the rest, as they say, "is history" (may be the first *"kegger"* was really a *"stomacher"*. So it is highly possible that the first alcoholic drink was mead (beer made from the sugar in honey).

Similar serendipitous discoveries that other natural materials could be fermented to produce the same taste as in mead probably followed in a relatively short period of time. Considering the fact that the wine trade is ancient, alcoholic products were one of the early trade goods that humans bartered. Today the world produces >43 x 109 gal of alcoholic drinks per year. From a purely biological perspective ethanol is considered a powerful narcotic. That is, as well as being a "mind-altering" substance, it is highly addictive and the data indicate that addiction is strongly influenced by genetic components. Ethanol is a potent biological toxin that damages all organs in the body. Brain development in the human fetus is very susceptible to irreparable damage by ethanol. If humans had not had such a long history of ethanol usage and its use was introduced into our society today, it is likely that it would be categorised as an illegal drug. It has, however, become such an intimate component of most cultures, both socially and commercially, that it is unlikely that its use can be seriously curtailed.

Ethanol were introduced to society today it would be considered an illegal drug. Because of the harmful and sometimes fatal effects that ethanol has, it would be considered as dangerous as cocaine or other illegal drugs. Society over a period of time has come to accept the nature of ethanol. Society has fooled themselves into believing that they don't have a problem with alcohol addiction, but what would happen if alcohol were banned tomorrow and became illegal?

Absolutely anything done in excess is bad for a person but when done in moderation it is okay. Alcohol fits perfectly into this situation because too much leads to social problems and alcoholism while small doses are enjoyable in social settings. For this reason laws on drinking and driving are a good idea. These rules are there to protect individuals and remind them moderation is acceptable but overindulging is not.

MICROBES IN CHEMICAL INDUSTRY

Microbes have been used for about 100 years to produce industrial chemicals for human use. The microbes are cultivated under rigorously controlled environmental conditions conducive to optimum production of the given product in rather humongous fermenters. Fermenters are tanks that may hold 1,000 of gallons, or more, of a culture. They must be made of materials, usually stainless steel, that can be heat sterilised and which will not react with the microbes or with the desired products. They must be able to be tightly sealed to prevent contamination and yet must contain numerous openings for monitoring the progress of the fermentation and for controlling the internal environment. All industrial microbial processes deal with similar problems:

- Finding the *least expensive medium* in which to grow the microbe so as to maximise yield and profits.
- Often this is a waste product from another industrial process, such as corn steep liquor, sugar processing wastes or whey.
- Maintaining strain purity and developing better strains for improving the yield.
- A single mutation may decrease the yield by a significant percentage or result in undesirable substances being produced. The industrial research laboratories constantly seek better strains for the production of their product.
- Preventing contamination by other microbes and by viruses (phage) that live on the microbe involved.
- The media must be sterilised prior to being inoculated with the desired organism and purity must be maintained throughout the production process. A small quantity of a contaminant may produce an enzyme that can destroy the product in 1,000s of gallons of medium. For many microbes, viruses present a constant danger as a single virus can infect and destroy the desired microbe in an entire tank. The sterilisation of large containers and huge quantities of media represent both an engineering and microbial challenge.

- Developing rapid and efficient methods for purification of the desired produce in a stable form that is safe to use.
- The products of many fermentations are often unstable in the impure form or subject to unwanted modifications if they are not purified quickly. The final growth mixture may contain dangerous substances from which the desired product must be separated. As every step in the purification results in a lose of the product, the search for more efficient purification procedures is never ending.
- Always striving to improve yield by modifying the strain, nutrients or environmental conditions.
- As product yields are exquisitely sensitive to subtle modifications in the nutrient and the environmental conditions, these are constantly monitored For example, the pH, oxygen content, nitrogen/phosphorous ratio etc. may be adjusted during the production process.
- Safe and inexpensive disposal of the massive quantities of waste products remaining after the product is formed.
- The waste products of these large fermentations present major waste disposal problems as they are rich in organic matter that are highly polluting if released untreated into the environment. However, the cost of treatment cuts into the profit margin and increases the cost of the product.

Simple organic chemicals like ethanol, acetic acid (vinegar) acetone, butyric acid and lactic acid are readily made either by organic chemical synthesis or by microbial fermentation. The method of choice depends upon the price of the raw materials and on the availability of industrial facilities to carry out either process. That is, in some cases it is cheaper to manufacture ethanol by fermentation and in other cases by chemical conversion from petroleum or natural gas.

Immediately procceding the first world war the process of acetone-butanol fermentation by bacteria was discovered. When the war began England found itself cut off from a supply of acetone (at this site go to "search", type in acetone & follow the steps until you reach the pathway), a crucial ingredient in the making of gunpowder. Chaim Weismann, a Jewish biochemist was put in charge of developing the microbial process for the commercial production of acetone. His success made such an important contribution to the war effort that the British government offered him any reward he chose. Being an ardent Zionist, he asked that the British support the formation of a Jewish State in Palestine.

Today acetone and butanol are more cheaply made from petroleum, but as these natural resources run low in the next century we may have to return to the microbiological technology. The following is a partial list of organic chemicals made commercially by microbes:

— 2,3, butainediol; buttery taste
— Enzymes
— Organic acids such as citric, lactic, ascorbic (vitamin C), acetic. These are utilised both as foods, and in industrial chemical processes.
— Polysaccharides
— Poly-beta hydroxybutyric acid
— Methane
— Hydrogen
— Biological pesticides

Starting with several mutant strains of *E. coli*, researchers have manipulated them by using different growing conditions to show that they can be persuaded to produced pure D- or L-lactate. The work shows that the central fermentation metabolism of *E coli* can be changed to the production of an indigenous fermentation product, D-lactate, or to the production of a nonindigenous one, L-lactate. Propylene glycol (PG) is widely used in the drug industry to help manufacture drugs and preserve the moisture of foods. PG is manufactured from propylene, which is a nonrenewable resource originating from petroleum. Researchers enlisted the bacterium *Escherichia coli* to produce PG from glucose. They did so by engineering the bacteria to overproduce two enzymes that help the bacterium grow in the absence of oxygen.

MICROBES IN PHARMACOLOGICAL INDUSTRY

Another contribution of microbes to winning a war came through the serendipitous discovery of penicillin by the English microbiologist A. Fleming in 1929. Fleming who was known as a bit of a character for painting pictures on petri dishes using different colored microbes, observed that a mold contaminating a plate of *S. aureus* was excreting something that inhibited the growth of that pathogen. He surmised that it might be used to fight bacterial infections and began to investigate it. Although he made little progress on it, and finally gave up, others began to investigate its possibilities and eventually a tiny amount of penicillin was isolated and given to a policeman suffering from a fulminating infection of *S. aureus*. He began to recover when the supply ran out and he died.

The amounts of penicillin were so small in those early days that it was reisolated from the urine of patients and used again. However, clinical tests looked so promising that when the second WW came along the U.S. took over the investigation and the development of penicillin became, after the development of the atomic bomb, the *second highest research priority in the war effort*. From this followed the antibiotic era and the huge pharmacological industry that operates world wide to day.

The revolution in molecular biology offers the possibility of yielding a whole new range of pharmacologically active microbiological produces through the application of genetic engineering technology. Under the proper conditions the cloned genes can be made to direct the synthesis of their protein product. In this way a substance that has a specific effect on another gene or gene product, but which is normally made in tiny amounts in a target organism can be made in commercially large quantities which can then be used for therapeutic purposes. For example, although clots are constantly forming in our bodies, they are dissolved before they do serious damage by special "clot-dissolving enzymes". In the case of strokes or embolisms where life threatening clots form in the brain or lungs, a cloned version of one of these clot-dissolving enzymes has been shown to be effective in saving lives and minimizing damage from strokes. However, their low concentration and difficulty of isolation have, until recently, made these clot-busting enzymes too rare and expensive to use widely. Now these enzymes are now being made through genetic engineering technology in large enough quantities so as to become a standard treatment for stroke victims.

The following is a partial list of microbial produced commercial pharmacological and related biotech products:

- Vitamins
- Amino acids
- Nucleic acids
- Antibiotics
- Alkaloids
- Steroids
- Proinsulin
- Insulin
- Human growth hormone
- Somatostatin
- Interferons
- Platelet-derived growth factor
- Fibroblast growth factor
- Tumor Necrosis Factor
- Other cytokines are coming on line all the time

Biotech industries, which produce the *biological materials* listed above from genetically engineered microbes, plants or animals have developed rapidly in the last few years. Most of these industries, some of which are worth millions of dollars, did not exist 15 years ago and new biotech industries are appearing all the time as scientist find how to produce new biological products using genetically engineered microbes.

Biotech industries are expected to be one of the fastest growing industries in the next century, particularly as the human population ages and as the information from the human genome project comes on line. However, a word of caution is advisable here. Many biotech industries fail because they could not make a commercially viable product due to some unexpected hitch developing along the way or they find that the market they thought was there really isn't or they were beaten out by a better or less expensive product. Many of the products that work under controlled laboratory conditions, fail to perform up to expectations in the real commercial world. It is a very fluid situation and one should invest one's money in such enterprises carefully.

Enzymatic Actvities

Metabolism in hyperthermophiles is accomplished by enzymes that demonstrate an extreme thermal stability. They maintain their native form at conditions, which will cause their mesophilic counterparts to denature and inactivate. Functional folding of proteins and enzymes is initially determined by their amino acid sequence (primary structure), and the generated, mutually attracting and repulsing forces result in the energetically most favorable, native structure (secondary, tertiary and quaternary). While it was initially presumed that protein thermostability was a result of only additional hydrogen bonds and salt bridges, detailed comparisons between homologs from mesophilic and thermophilic microorganisms throughout the years showed it to be due to different combinations of several stabilising features. The core of homologs of mesophilic and thermophilic microorganisms is often observed to be similar at the amino acid level, suggesting that gain in stabilisation is achieved in regions that are less conserved. Direct comparison of the amino acid composition of thermostable proteins to homologs from mesophiles reveals a higher contents of Ala, Arg and Tyr. In addition, the residues Asn, Gln and Cys occur less frequently in the sequences of enzymes from thermophiles, because of their sensitivity to chemical deamination or oxidation at high temperatures. Structural factors, such as surface loop deletion, increased occurrence of hydrophobic residues with branched side chains, and an increased proportion of charged residues also contribute to a higher intrinsic thermostability of proteins. Other stabilising mechanisms include an increased number of ion-pair networks and salt bridges in thermostable proteins, compared to their mesophilic counterparts.

observed enzymes of hyperthermophilic origin, compared to homologs from mesophiles. An abundance of disulfide bonds in intracellular proteins of thermophiles, in particular the archaeal Pyrobaculum aerophilum and Aeropyrum pernix, underlines its role in protein thermostability. For enzymes from hyperthermophiles an increased number of subunits has been observed, compared to counterparts from mesophiles. Multimerization is believed to be a thermostabilising factor. Besides enhancing enzyme activity, metal ions, like Co, Mg, Caand Mn, are known to play a role in protein stabilisation.

Thermostabilisation can also be achieved by external factors. Thermophiles, in contrast to mesophilic organisms, are known to accumulate compatible solutes, such as mannosylglycerate and di-myo-inositol-phosphate, which are believed to play a stabilising role intracellularly. Likewise, chaperones are proteins that assist in the folding of other proteins, preventing them from aggregation. These so-called thermosomes have been observed in increased quantities in heatshocked thermophiles.

Starch-degrading Enzymes

Starch (or glycogen in animals) is composed of a-glucose units, linked by α-1,4- and α-1,6- glycosidic bonds, thereby forming the two high-molecular-weight components amylose (20%) and amylopectin (80%). The starch-conversion is an important industrial process in which thermostable enzymes are applied. One of the most important applications is the enzymatic production of High Fructose Corn Syrup (HFCS), which is added to several softdrinks as a sweetener. Both the liquefaction and the saccharification steps in starch-conversion are preferably carried out at elevated temperatures.

The saccharolytic bacterium Rhodothermus marinus was found to have high levels of a-amylase activity when grown on starch. An α-amylase from *T. maritima* with high specific activity was produced in *E. coli*, although addition of calcium ions for its activity was required. Pullulanases (type I and II) are capable of hydrolyzing the α-1,6-bonds in starch. Whereas type I only debranches, type II is also able to degrade α-1,4-linkages. Both types have been identified in *T. maritima* and characterised.

The first glycosidase that has been isolated from the genus Fervidobacterium is a type I pullulanase from *F. pennavorans* Ven5. It was produced in *E. coli* and did not require calcium ions, like most of the reported pullulanases. The remaining oligosaccharides, like maltose and maltotriose, are attacked by the exo-active a-glucosidases (α- 1,4) and glucoamylases (α-1,4 and α-1,6), although the latter is rare in thermophiles. Thermostable bacterial α-glucosidases have been identified in *T. ethanolicus* and *T. maritima*.

Cellulose-degrading Enzymes

Cellulose consists of glucose units linked via ß-1,4-glycosidic bonds and is nature's most abundant plant polymer. Together with hemicellulose and lignin it forms a network in the cell wall of plants. Due to this incorporation and since a great part of the cellulose is composed of hydrolysis-resistant crystalline regions, alkaline pretreatment at high temperature is requisite.

Consequently, thermostable cellulases have gained considerable interest. Complete hydrolysis entails the synergistic action of endo- and exoglucanases, and ß-glucosidases. Most cellulosedegrading thermozymes have been isolated from bacterial thermophiles, although several glucanases of archaeal origin have been characterized. Two thermostable endoglucanases from *T. maritima,* CelA and CelB, have been characterised and their optimal activity lies at 95°C. In sequence comparison and enzymatic features they appeared to be analogous to two cellulases from the related genus *T. neapolitana.*

An endoglucanase, isolated from A. aeolicus, was found to be active on carboxymethylcellulose. Similar to *T. maritima* and *T. neapolitana,* growth of the organism on β-1,4-linked polymers has not been reported yet. In recent years, the possibilities have been explored to produce thermostable cellulases in plants for an efficient degradation of endosperm cell walls of barley. Examples of these transgenic plants are a thermostable (1,3-1,4)-β-glucanase, a hybrid from parental enzymes from Bacillus macerans and B. amyloliquefaciens and produced in barley, and a 1,4-β-glucanase from *Acidothermus cellulolyticus* produced in chloroplasts of tobacco plants.

Xylan-degrading Enzymes

Like cellulose, hemicellulose forms a fraction of the cell wall, where it is mainly composed of the heterogeneous polymer xylan. The main chain of xylan consists of β-1,4-linked xylose residues, which can be degraded by endo-active xylanases and exo-active β-xylosidases. The application of thermostable xylan-degrading enzymes in the paper industry, involving the bleaching process, is promising. The initial step in paper production involves a heat-treatment to open up the cell wall, removing 90% of the lignin from the wood pulp.

The specific removal of the 10% residual lignin by thermostable xylanases instead of the traditional chemical treatment could reduce the amount of chlorine derivatives presently used. Up till now, not many thermostable versions of these enzymes have been characterised, since the number of thermophiles or hyperthermophiles which are known to be able to grow on this substrate is rather limited. Most of the xylanases described so far have been isolated from Thermotoga spp. that have the ability to grow on xylan, such as *T. maritima, T. neapolitana,* and

T. thermarum, where they are tightly bound to the toga. With a high thermostability and a pH optimum of around 6.0 they meet the prerequisites to be used in the paper pulp bleaching process, as was already tested for one of the T. maritima xylanases. Nearly all xylanase-containing Thermotoga spp. also contain the corresponding β-xylosidases, to realise complete xylan degradation.

Pectin-degrading Enzymes

Pectin is a branched heteropolysaccharide present in cell walls, where it forms a matrix, intertwining the other polymers, celluloses and hemicelluloses. The largest part of the pectin molecule consists of a-1,4-linked D-polygalacturonate, which can be partially methylated. Degradation of this polymer can occur via hydrolysis (hydrolases, belonging to family 28 of the glycoside hydrolases) or β-elimination. Pectinases are widely applied in the fruit juice extraction and clarification processes (acidic pectinases), as well as in the textile industry, where they are used for retting and degumming of fiber crops (alkaline pectinases). The information on growth on pectin at high temperatures is rather limited.

The only thermophiles for which growth on pectin has been reported are Caldicellulosiruptor and Thermoanaerobacterium strains. Recently, it was found that nearly all Thermotoga species degrade pectin. Only one archaeon, Desulfurococcus amyloliticus, has been reported to grow on pectin. As a consequence, relatively few thermostable pectinolytic enzymes have been characterised. Hence, their industrial potential may have been overlooked, since most attention has been given to other thermozymes, such as the cellulases.

MICROBES IN AGRICULTURE INDUSTRY

Other interactions with beneficial microbes can be of direct benefit to agricultural plants and animals. A classic example of mutualism in action is the partnership between legumes and bacteria called Rhizobia. The bacteria take up residence in plant roots, receiving nutrients. In exchange, they fix nitrogen from the air into a form that the plants can use, replacing a need for nitrogencontaining fertilizer. There are other cases of microbes helping a host organism scavenge essential nutrients, or fend off pathogens. In the intestinal tract of ruminants, a complex mixture of bacteria enables the animal to extract sufficient nutrients from a diet of grasses.

Even the best-recognised and most-studied forms of mutualism are not understood well enough to be effectively controlled or expanded to cover hosts previously unknown to benefit from a particular interaction. Scientists have been unsuccessful in getting Rhizobia to form a mutualistic relationship with wheat roots, for instance. For the few classic examples of mutualism in agricultural systems, there are likely to be many more interactions taking place in obscurity.

Study of interactions between organisms that boost agricultural success is a field rich with opportunities. More knowledge of microbial ecology and mutualistic interactions will pave the way for advances that enhance agricultural organisms' nutrient use, pathogen resistance, and hardiness. Microbial ecology will likely be found to have impacts on agricultural systems beyond those currently recognised.

Complex interactions between plants and the consortia of microbes found in soil probably extend beyond resisting pathogens and scavenging nutrients. Properly tuned interactions could help improve drought resistance and salt tolerance of plants and have other growth-promoting activities. Understanding and managing microbial ecology will have major benefits for stressed agricultural systems.

The massive scale of human agricultural and food production enterprises brings with it an array of problems that microbiology can help address. Any technological advances that increase resistance to pathogens or nutrient scavenging will also contribute to reduced use of pesticides and fertilisers. This represents a corresponding reduction in pollutants. Other pollutants are a direct consequence of agricultural production itself, rather than production practices.

Waste produced by animals, particularly when produced in high densities, frequently represents a serious environmental and health hazard. Animal manure is typically accumulated in bulk and some of the material is used as fertilizer on agricultural fields. Technology to harness microbes for digestion of animal waste could alleviate some of the environmental and health hazards generated by large-scale animal rearing operations. Microbes may also be harnessed for the remediation of agricultural chemicals or for mitigating greenhouse gases.

Microbial digestion, another form of fermentation, can be harnessed to produce alternative fuels. Fermentation of animal wastes can create flammable gases, such as methane. Devising bioreactors that efficiently convert animal waste on a large scale would help eliminate an environmental and health hazard, while also satisfying growing energy needs. Fermentative processes also produce fuels, such as ethanol from plant material.

The inefficiency of this type of fermentation for fuel production has kept it from being widely adopted. Improvements in fuel generating technology would allow the gradual replacement of highly polluting fossil fuels with more environmentally friendly fuel sources. However, continued removal of plant waste from fields may have unintended effects, such as altering the composition and characteristics of the soil, affecting microbial populations and subsequent plant growth. Thus, this practice needs to be examined and followed over multiple years to monitor its effects.

The role of beneficial organisms in promoting the health of agricultural plants and animals extends beyond combating pathogens. Research into how beneficial microorganisms can promote growth, improve stress tolerance, and aid in the uptake of nutrients are research areas ripe for discovery and innovation. Research into these complex and often delicate interactions between different organisms should ultimately pay off by revealing ways to assure that agriculture can become heartier and less environmentally taxing.

The same communities of microbes that benefit agricultural health and efficiency are likely to be disturbed by some of the practices of industrialised agriculture. One way to fortify agriculture against disease and stress is to supplement systems with probiotic and biocontrol organisms, but a complementary and sometimes alternative approach is to protect beneficial organisms that may already be present in the environment.

Research in microbial ecology will help to determine how to preserve a balance in microbial communities that favours agriculture. Heavy pesticide and fertilizer use, in particular, are two practices that should be studied using a holistic or integrated approach to determine their impact on microbial ecology within the context of tradeoffs between risks and benefits. More knowledge in this area will help determine optimal tradeoffs such that the benefits of use outweigh the disruption caused by chemical inputs into our agricultural systems.

Microbial communities are both vulnerable to, and contribute to, removal of pollutants. Research into how multiple organisms work in partnership to degrade complex molecules is essential to increase options for dealing with the byproducts of industrialised agriculture. Understanding this aspect of microbial ecology may also lead to improvements in waste disposal and energy generation through fermentation, as well as bioremediation.

Organisms that cannot currently be cultivated in a laboratory setting are likely to be pivotal in advances in probiotics, biocontrol, and microbial ecology. Research efforts targeted on identification of these previously uncharacterised organisms, whether through new cultivation methods or by indirect detection approaches made possible by genomic knowledge, will strengthen our ability to use beneficial organisms effectively.

Research offers opportunities for understanding more completely the impacts of technologies that are already used in agriculture and food production. Questions remain concerning if genetically modified organisms interact differently in the environment as compared to their non-engineered counterparts. Whether or not there are negative impacts associated with the use of GMOs needs to be considered and balanced within the broader context of sustained use and potential benefits.

Increased attention to food hygiene, which has occurred in the developed world over the last century, has dramatically reduced incidence of foodborne infection. However the void of microbially-based immune stimulation has been proposed as a culprit in weakening immune systems in animals and people. Further research into this area will determine whether this is a real phenomenon and, if so, how to balance tradeoffs between a microbiologically safer food supply and maintaining a healthy immune system throughout life.

Food and agriculture microbiology research provides potential solutions to problems that cut across many fields. The scientific principles and the implications of food and agricultural research are increasingly linked to public health and the environment, topics that historically have received more public attention. For example, interventions against Avian Influenza, which has been devastating poultry production, may also prevent dissemination of another human influenza epidemic. Providing biologically-based alternatives for protection of crops and animals against pathogens will help reduce pesticide and antibiotic use, both of which have public health and environmental implications.

Despite the need for continued advances in agriculture and food microbiology, and the proven track record of agricultural research, research support for these fields over the last few decades has been lean and is, in fact, decreasing. Reversing the decline in funding and recognition of the value of agricultural research requires fundamental changes, in addition to an infusion of financial support. The major barriers to advancing agriculture and food research are institutional and perception based.

The profile and priority of agricultural research needs to be raised. Designated research centers of excellence, similar to those in the biomedical and defense arenas, would make strides in this respect. There are numerous institutions that provide a backbone of exceptional scientific work for U.S. agriculture, but their programs run in aging facilities and with limited financial resources.

Within U.S. research institutions, agriculture is too often a subordinate priority. One way to reverse this would be to raise the institutional overhead rate that is currently allowed on USDA grants from its uniquely low level to a level on par with that provided by other funding agencies, such as the National Science Foundation (NSF) or the National Institutes of Health (NIH). With limited overhead capital, administrators and investigators are restricted in their efforts to build strong programs and recruit personnel to pursue state-of-the-art agricultural research. Increases in overall research expenditures for U.S. Department of Agriculture (USDA) programs would then be needed to maintain even the current level of direct funds for research.

A healthy agricultural research community depends on an influx of young scientific talent. Trouble recruiting and maintaining graduate students is impacting programs and will ultimately affect the field. Several measures can be taken to alleviate this problem. A program of prestigious and remunerative fellowships for graduate students and postdoctoral fellows would provide some needed recruiting leverage. Internships involving industry, non-governmental organisations (NGOs), and government agencies would have mutually beneficial value. Such internships would infuse awareness and technical knowledge of agricultural science to institutions and provide the visiting scientists with networking and training opportunities.

INDUSTRIAL USES OF MICROBIAL BIOSENSORS

Biomolecules such as enzymes, antibodies, receptors, organelles and microorganisms as well as animal and plant cells or tissues have been used as biological sensing elements. Among these, microorganisms offer advantages of ability to detect a wide range of chemical substances, amenability to genetic modification, and broad operating pH and temperature range, making them ideal as biological sensing materials. Microorganisms have been integrated with a variety of transducers such as amperometric, potentiometric, calorimetric, conductimetric, colorimetric, luminescence and fluorescence to construct biosensor devices. Several reviews papers and book chapters addressing microbial biosensor development have been published. The intent of this review is to highlight the advances in the rapidly developing area of microbial biosensors with particular emphasis to the developments since 2000.

Enzymes are the most widely used biological sensing element in the fabrication of biosensors. Although purified enzymes have very high specificity for their substrates or inhibitors, their application in biosensors construction may be limited by the tedious, time-consuming and costly enzyme purification, requirement of multiple enzymes to generate the measurable product or need of cofactor/coenzyme. Microorganisms provide an ideal alternative to these bottle-necks.

The many enzymes and co-factors that co-exist in the cells give the cells the ability to consume and hence detect large number of chemicals; however, this can compromise the selectivity. They can be easily manipulated and adapted to consume and degrade new substrate under certain cultivating condition. Additionally, the progress in molecular biology/ recombinant DNA technologies has opened endless possibilities of tailoring the microorganisms to improve the activity of an existing enzyme or express foreign enzyme/protein in host cell. All of these make microbes excellent biosensing elements.

Microbe Immobilisation Methods

The basis of a microbial biosensor is the close contact between microorganisms and the transducer. Thus, fabrication of a microbial biosensor requires immobilisation on transducers with a close proximity. Since microbial biosensor response, operational stability and long-term use are, to some extent, a function of the immobilisation strategy used, immobilisation technology plays a very important role and the choice of immobilisation technique is critical. Microorganisms can be immobilised on transducer or support matrices by chemical or physical methods.

Chemical Methods

Chemical methods of microbe immobilisation include covalent binding and cross-linking. Covalent binding methods rely on the formation of a stable covalent bond between functional groups of the microorganisms' cell wall components such as amine, carboxylic or sulphydryl and the transducer such as amine, carboxylic, epoxy or tosyl. To achieve this goal, whole cells are exposed to harmful chemicals and harsh reaction condition, which may damage the cell membrane and decrease the biological activity.

This method has therefore not been successful for immobilisation of viable microbial cells. Cross-linking involves bridging between functional groups on the outer membrane of the cells by multifunctional reagents such as glutaraldehyde and cyanuric chloride, to form a network. Because of the speed and simplicity, the method has found wide acceptance for immobilisation of microorganisms.

The cells may be cross-linked directly onto the transducer surface or on a removable support membrane, which can then be placed on the transducer. The ability to replace the membrane with the immobilised cells is an advantage of the latter approach. While cross-linking has advantages over covalent binding, the cell viability and/or the cell membrane biomolecules can be affected by the cross-linking agents. Thus cross-linking is suitable in constructing microbial biosensors where cell viability is not important and only the intracellular enzymes are involved in the detection.

Physical Methods

Adsorption and entrapment are the two widely used physical methods for microbial immobilisation. Because these methods do not involve covalent bond formation with microbes and provide relatively small perturbation of microorganism native structure and function, these methods are preferred when viable cells are required.

Physical adsorption is the simplest method for microbe immobilisation. Typically, a microbial suspension is incubated with the electrode or an immobilisation matrix, such as alumina and glass bead, followed by rinsing with buffer to remove unadsorbed cells. The microbes are immobilised due to adsorptive interactions such as ionic, polar or hydrogen bonding and hydrophobic interaction. However, immobilisation using adsorption alone generally leads to poor long-term stability because of desorption of microbes.

The immobilisation of microorganisms by entrapment can be achieved by the either retention of the cells in close proximity of the transducer surface using dialysis or filter membrane or in chemical/biological polymers/gels such as alginate, carrageenan, agarose, chitosan, collagen, polyacrylamide, polyvinylachohol, poly(ethylene glycol), polyurethane, etc.. A major disadvantage of entrapment immobilisation is the additional diffusion resistance offered by the entrapment material, which will result in lower sensitivity and detection limit. Microbial biosensor can be classified based on the transducers into electrochemical, optical and others.

Types of Electrochemical microbial Biosensor

There are three types of electrochemical microbial biosensors:

- amperometric,
- potentiometric, and
- conductometric.

Amperometric Biosensor

Amperometric microbial biosensor operates at fixed potential with respect to a reference electrode and involves the detection of the current generated by the oxidation or reduction of species at the surface of the electrode.

Amperometric microbial biosensors have been widely developed for the determination of biochemical oxygen demand (BOD) for the measurement of biodegradable organic pollutants in aqueous samples. The conventional standard method for the determination of BOD measures the microorganisms' oxygen consumption/respiration over a period of 5 days and is reported as BOD_5. While BOD_5 is a good indicator of the concentration of organic pollutants in water, it is extremely slow and hence not suitable for process control. To address this limitation, several BOD biosensors based on amperometric oxygen electrode transducer modified with microorganisms degrading/metabolising organic pollutants have been reported.

The microbial strains used as biological sensing element include Torulopsis candida, Trichosporon cutaneum, Pseudomonas putida, Klebsiella oxytoca AS1, Bacillus subtilis, Arxula adeninivorans LS3, Serratia marcescens LSY4, Pseudomonas sp., P. fluorescens, P. putida SG10, Thermophilic bacteria, Hansennula anomala and yeast. Because any given strain provides a narrow substrate spectrum, singlestrain-BOD-biosensor has limitations in analysing complex samples. This bottleneck can be alleviated by employing a mixture of two or more microorganisms to broaden the substrate and hence analyte spectrum with a stable performance.

As the most extensively investigated microbial biosensor, the first commercial BOD biosensor was produced by Nisshin Denki (Electric) in 1983. Since then, several more BOD biosensors have been commercialised by DKK Corporation, Japan; Autoteam FmbH, Germany; Prufgeratewerk Medingen GmbH, Germany; Dr. Lange GmbH, Germany; STIP Isco GmbH, Germany; Kelma, Belgium;LARAnalytik and Umweltmesstechnik GmbH, Germany; Bioscience, Inc., USA; USFilter, USA.

While most of the research and development in BOD biosensors has focused in identifying different microorganisms that can determine BOD of a specific waste, research efforts have also been directed at improving the amperometric transducer itself. For example, a miniaturised oxygen electrode based on thick-film screen-printing was recently developed to replace the bulky Clark dissolved oxygen electrode transducer. The widely used thick-film screen-printing technique was used to print the platinum-working electrode, Ag/AgCl reference electrode and platinum auxiliary electrode of the amperometric oxygen electrode on an inert substrate.

The oxygen electrode was then modified with A. adeninivorans LS3 by entrapment in poly(carbamoyl) sulfonate (PCS) gel and successfully applied for rapid (~100 s) and stable (up to 2 months) BOD determination. Similarly, to extend the dynamic range of the BOD sensor, which in the case of dissolved oxygen electrode is limited by the solubility of oxygen in the sample, a ferricyanidemediated microbial biosensor using a novel yeast strain for BOD measurement was developed. Recently, an amperometric transducer array featuring four individually addressable platinum electrodes was constructed and modified with two microbial strains with different substrate spectra for the measurement of BOD and poly cyclic aromatic hydrocarbons (PAH) simultaneously.

Besides BOD biosensor, amperometric microbial biosensors have also been applied for measurement of several other chemicals. Because of its importance in fermentation industry and clinical toxicology, microbial biosensors for ethanol has garnered the second most research attention after BOD. Different microorganisms metabolising ethanol such as *Trichosporon brassicae, Acetobacter aceti, Candida vini, Gluconobacter suboxydans, C. tropicalis, Aspergillus niger, Saccharomyces ellipsoideus, G. oxydans* and *Pichia methanolica* have been immobilised on oxygen electrode to fabricate ethanol biosensors.

While these biosensors posses good sensitivity and stability, they usually have poor selectivity. Thus, there is a great interest to develop selective microbial ethanol biosensor. An improved selectivity for ethanol determination in presence of glucose was achieved by replacing oxygen with ferricyanide as the electron acceptor mediator for *G. oxydans* immobilised on a glassy carbon electrode by cellulose acetate membrane which also restricted the availability of glucose to the cells by size exclusion.

Sugars are important ingredients of different media and sensors for determination of sugars are therefore highly desired. Microbial biosensors for sugars have ranged from the simple modification of Clark and microfabricated oxygen electrode with *S. cerevisiae* and *E. coli* K12 mutants, respectively, to modification of graphite electrode with *G. oxydans* in conjunction with hexacyanoferrate (III) as a mediator.

Phenol and substituted phenols have received considerable attention in waste analysis program due to their high toxicity to mammals, humans and plants. A variety of amperometric microbial biosensors have been reported for these EPA Priority chemicals. p-Nitrophenol (PNP) degrading bacterial Arthrobacter JS 443 and Moraxella sp. isolated from PNP contaminated sites in the U.S. have been immobilised on oxygen and carbon paste electrodes using polycarbonate membrane and Nafion, respectively, and by directly mixing in the carbon paste to fabricate biosensor for PNP. Other microbial biosensors for phenols include Rhodococcus erthropolis modified Clark oxygen electrode for 2,4-dinitrophenol (2,4-DNP) and *P. putida* DSM 50026, a well-known phenol degrading microorganisms, modified thick-film and screenprinted graphite electrodes for phenols.

Neurotoxic organophosphate (OP) compounds have found wide applications as pesticides and insecticides in agriculture and as chemical warfare agents in military practice. Amperometric biosensors based on genetically engineered Moraxella sp. and *P. putida* with surface-expressed organophosphorus hydrolase (OPH) have been developed for sensitive, selective and cost-effective detection of OPs. These biosensors relied on the amperometric detection of PNP generated from hydrolysis of OP compounds by surface-displayed OPH or oxygen consumed and electrochemically active intermediates formed during the further mineralisation of the PNP by the cells.

The inhibition of bacterial respiration and hence the decrease of oxygen consumption rate, has been utilised to fabricate cyanide biosensor. Whole-cell biosensors consisting of dissolved oxygen electrode modified with Nitrosomonas europaea, Thiobacillus ferrooxidans, Saccharomyces cerevisiae and Pseudomonas fluorescens were reported for batch and continuous cyanide monitoring.

Other amperometric microbial biosensors based on monitoring of cell respiration include biosensor for surfactants, representing a widespread group of organic pollutants, using surfactant-degrading bacteria, hydrogen peroxide by coupling immobilised living Acetobacter peroxydans and for acetic acid using Fusarium solani.

Over the last two decades, the microbiologically influenced corrosion (MIC) of metallic materials has received great attention. Astable, reproducible and specific microbial biosensorwas developed for monitoring MIC of metallic materials in industrial systems based on Pseudomonas sp. isolated from corroded metal surface and immobilised on acetylcellulose membrane at oxygen electrode.

A linear relationship between the biosensor response and the concentration of sulphuric acid was established. The biosensor response time was 5 min and was dependent on many parameters such as pH, temperature, corrosive environment and immobilised cell loading. The same group also used Acetobacter sp. to develop amperometric microbial biosensor for monitoring microbiologically influenced corrosion caused by fungal species.

Another application of amperometric microbial biosensors is the detection of heavy metal ions for environmental control. A microbial biosensor to detect Cu^{2+} by an amperometric method has been developed using recombinant *S. cerevisiae* containing plasmids with Cu^{2+}-inducible promoter fused to the lacZ gene. In the presence of Cu^{2+}, the recombinant strains are able to utilise lactose as a carbon source and lead to the oxygen consumption change, which can be detected by using oxygen electrode.

A novel promoter-based electrochemical biosensor for on-line and in situ monitoring of gene expression in response to cadmium has also been described. A cadmium-responsive promoter from *E. coli* was fused to a promoterless lacZ gene, and then the β-galactosidase activity was monitored using screen-printed electrode in the presence of cadmium. This whole-cell biosensor could detect nanomolar concentrations of cadmium on-line or in-site within minutes.

Potentiometric Microbial Biosensor

Conventional potentiometric microbial biosensors consist of an ion-selective electrode (pH, ammonium, chloride and so on) or a gas-sensing electrode (pCO_2 and pNH_3) coated with an immobilised microbe layer. Microbe consuming analyte generates a change in potential resulting from ion accumulation or depletion. Potentiometric transducers measure the difference between a working electrode and a reference electrode, and the signal is correlated to the concentration of analyte. Due to a logarithmic relationship between the potential generated and analyte concentration, a wide detection range is possible. However, this method requires a very stable reference electrode, which may be a limitation of these transducers.

The simplest potentiometric microbial biosensor is based on the modification ion selective electrode. Several microbial biosensors based on modification of glass pH electrode with genetically engineered *E. coli* expressing organophosphorus hydrolase intracellularly and on the outer surface of cells and wild-type OP degrading bacteria Flavobacteium sp. have been reported. The principle of detection is based on the detection of the protons released by OPH catalysed hydrolysis of OP and correlating to the concentration of OPs.

Similarly, recombinant *E. coli* harboring the plasmids encoding for β-lactamase and penicillinase synthesis immobilised on pH electrode using gluten and acetylcellulose membranes entrapment, respectively, were developed for monitoring penicillin. A new type of solid state silicon-based light addressable potentiometric sensor for monitoring hydrogen ion was integrated to the auxotrophic bacteria *E. coli* WP2 (requiring tryptophan for its growth) to fabricate a potentiometric microbial assay for tryptophan.

While pH electrodes are the most widely applied ion selective electrode for microbial biosensors, other ion selective electrodes have also been utilised. For example, an ammonium ion selective electrode was coupled with urease-yielding Bacillus sp. isolated from soil to develop a disposable microbial biosensor for monitoring the presence of urea in milk. Similarly, a chloride ion selective electrode was modified with TCE degrading bacterium Pseudomonas aeruginosa JI104 for TCE monitoring in batch and continuous modes in wastewaters. Apotentiometric oxygen electrode with immobilised *S. ellipsoideus* was also successfully used to produce a microbial biosensor for the determination of ethanol with an extended response range. Based on the same format, sucrose biosensor based on an immobilised *S. cerevisiae* was also described.

Conductimetric Biosensor

Many microbe-catalysed reactions involve a change in ionic species. Associated with this change is a net change in the conductivity of the reaction solution. Even though the detection of solution conductance is non-specific, conductance measurements are extremely sensitive. Recently, a single-use conductivity and microbial sensor were developed to investigate the effect of both species and concentration/ osmolarity of anions on the metabolic activity of *E. coli.* This hybrid sensing system combines physico-chemical and biological sensing and greatly increases the ease with which comparative data could be assimilated.

Microbial Fuel Cell

Microbial fuel cells (MFCs) have been studied as a BOD sensor for a long time. Since Karube et al. reported a BOD sensor based on MFC using the hydrogen produced by Clostridium butyricum immobilised on the electrode in 1977, a variety of MFC BOD sensors with use of electron-mediator have been developed. Even though the addition of mediators in these biosensors can enhance the electron transfer, these biosensors have poor stability because of the toxicity of mediators. Recently, mediator-less microbial fuel cells have been exploited to fabricate novel BOD sensors for continuous and real-time monitoring. Furthermore, Kim et al. reported that the performance of a microbial fuel cell as BOD sensor was improved using respiratory inhibitors.

Optical Biosensor

The modulation in optical properties such as UV-vis absorption, bio- and chemi-luminescence, reflectance and fluorescence brought by the interaction of the biocatalyst with the target analyte is the basis for optical microbial biosensors. Opticalbased biosensors offer advantages of compactness, flexibility, resistance to electrical noise, and a small probe size.

Bioluminescence Biosensor

Bioluminescence is associated with the emission of light by living microorganisms and it plays a very important role in realtime process monitoring. The bacterial luminescence lux gene has been widely applied as a reporter either in an inducible or constitutive manner. In the inducible manner, the reporter lux gene is fused to a promoter regulated by the concentration of a compound of interest. As a result, the concentration of the compound can be quantitatively analysed by detecting the bioluminescence intensity.

In the constitutive manner, the reporter gene is fused to promoters that are continuously expressed as long as the organism is alive and metabolically active. This kind of reporter is good for evaluating the total toxicity of contaminant. Both types of reporters have been shown to be useful for biosensor development.

Heavy metal-mediated toxicity in the environment is dependant on bioavailable metal concentrations. Bioluminescent microbial biosensors have been extensively investigated to monitor bioavailable metal. Ralstonia eutropha AE2515 was constructed by transcriptionally fusing cnrYXH regulatory genes to the bioluminescent luxCDABE report system to fabricate a whole cell biosensor for the detection of bioavailable concentration of Ni^{2+} and Co^{2+} in soil.

Several optical biosensors consisting of bacteria that contain gene fusion between the regulatory region of the mer operon (merR) and luxCDABE have been developed to quantitatively response to Hg^{2+}. The mer promoter is activated when Hg^{2+} binds to MerR, then result the transcription of the lux reporter gene and subsequent light emission. Bioavailable copper in soil is also monitored by using engineered *P. fluorescens* through mutagenesis of *P. fluorescens* containing copper-induced gene and Tn5::luxAB promoter probe transposon.

In order to monitor nutrients in an aquatic ecosystem, a biosensor for monitoring phosphorus bioavailablity to Cyanobacteria (Synechococcus PCC 7942) was developed. The reporter strain Synechococcus harbors the gene coding the reporter protein luciferase under the control of an inducible alkaline phosphatase promoter, which can be induced under phosphorous limitation and shows improvement to conventional phosphorus detection methods. Bioluminescent microbial biosensors using the inducible reporter gene have also been developed for the measurement of bioavailable naphthalene, tributyltin and halogenated organic acids.

The environmental problems caused by industrial and agricultural pollution have increased the demand for the development of pollutant and toxicity detection methods. The fusion of reporter genes to promoters that are induced when cell are stressed by toxic chemicals are one promising approach that has been used to fabricate biosensor for such application. Recombinant *E. coli* bearing fabA'::lux fusion and plasmid pUCD607 containing the full luxCDABE cassette have been constructed as biosensors for water pollutant detection. The online pollutant and toxicity test, using bioluminescence-based biosensors, was proved to be sensitive and reliable.

Lux-marked rhizobacterium P. fluorescens has been developed to evaluate the pollution-induced stress, which influences rhizobacterium carbon flow based on the fact that bioluminescence output of biosensor is directly correlated with metabolic activity and reports on carbon flow in root exudates. Furthermore, lux-marked whole cell biosensors for evaluation of interactive toxicity of chlorophenol and toxicity assessment of a wastewater treatment plant treating phenolics-containing waste have been reported, respectively. These biosensors responded to tested pollutants fast and enable a rapid toxicity test possible.

Genotoxicants is a class of hazards, which can cause DNA damage. An optical-fiber bioluminescent microbial sensor to detect the DNA damage hazard-mitomycin C by the induction of a selected promoter and the subsequent production of bioluminescent light through a recombinant lux reporter was reported. Bioluminescence production was shown to be dosedependent. *E. coli* containing plasmid-borne fusion of the recA promoter-operator region to the Vibrio.scheri lux genes has also been reported for genotoxicant detection. When the recombinant *E. coli* strains are challenged with DNA damage hazards, they increase their luminescence. Furthermore, this study was expanded by investigating and demonstrating the luminescence response of these strains to ultraviolet radiation, which can cause DNA damage. Another lux-based Psedumonoas aeruginosa biosensor was fabricated to quantify bacteria exposure to UV radiation in biofilm.

Fluorescence Spectroscopy and Biosensor

Fluorescence spectroscopy has been widely applied in analytical chemistry. It is a sensitive technique that can detect very low concentrations of analyte because of the instrumental principles involved. At low analyte concentrations, fluorescence emission intensity is directly proportional to the concentration. Fluorescent materials and green fluorescent protein have been extensively used in the construction of fluorescent biosensor.

GFP-based biosensor

Like bioluminescent reporter lux gene, gfp gene coding for the green fluorescent protein (GFP) has also been widely applied as reporters and fused to the host gene that allows reporter activity to be examined in individual cells. Because GFP is very stable and not known to be produced by microorganism indigenous to terrestrial habits, it provides great advantage and flexibility when evaluating reporter activity. The primary disadvantage of GFP as a reporter protein is the delay between protein production and protein fluorescence.

The GFP-based microbial biosensor has been shown to be useful in assessing heterogeneity of iron bioavailability on plant. In this sensor, ferric iron availability to cells was assessed by quantifying the fluorescence intensity of cells containing a plasmid-borne transcriptional fusion between an ion-regulated promoter and gfp.

plasmid-borne transcriptional fusion between an ion-regulated promoter and gfp. Recently, Wells et al. developed an ultrasensitive biosensor for arsenite by using laser-induced fluorescence confocal spectroscopy to measure arsenite-stimulated enhanced green fluorescent protein synthesis of genetically engineered E. coli bioreporter cell, which has an inherent single-molecule detection capability.

A recombinant soil bacterium Sinorhizobium meliloti has been constructed by fusing the gfp gene to the melA promoter, which is induced on exposure to galactose and galactosides. Using this fusion strain, a biosensor was developed to determine the concentration of galactosides. Similarly, gfp reporter gene has also been used to develop biosensors for various applications, such as detecting bioavailable toluene and related compounds and N-acyl homoserine lactones in soil, measuring water availability in a microbial habitat, monitoring cell populations and so on.

Besides green fluorescent protein, other fluorescent materials have also been used in the construction of microbial biosensor. Recently, fiber-optical microbial sensors for determination of BOD were reported. The biosensors consisted of either a layer of oxygen-sensitive fluorescent materials that are made up of seawater microorganisms immobilised in poly(vinyl alcohol) sol-gel matrix and an oxygen fluorescence quenching indicator with linear range of 4-200 mg/l, or an immobilised P. putida membrane attached to an optical fiber sensor for dissolved oxygen from ASR Co. Ltd. with detection limit of 0.5 mg/l.

Colorimetric Biosensor

A sensitive biosensor based on colour changes in the toxinsensitive coloured living cells of fish was reported. In the presence of toxins produced by microbial pathogens, the cells undergo visible colour change and the colour changes in a dose-dependant manner. The results suggest this cell-based biosensor's potential application in the detection and identification of virulence activity associated with certain air-, food-, and water-borne bacterial pathogens.

A simple fiber-optic based microbial sensor to detect organophosphates based on the absorbance of PNP formed from the hydrolysis of organophosphates by the genetically engineered E. coli expressing organophosphorus hydrolase on the cell surface. This biosensor can be easily extended to other organophosphates such as coumaphos through the monitoring of its hydrolysis product coumarin.

A colorimetric whole cell bioassay for the detection of common environmental pollutants benzene, toluene, ethyl benzene and xylene (BTEX), found at underground fuel storage tanks, using recombinant E. coli expressing toluene dioxygenase and toluene dihydrodiol dehydrogenase was reported. The bioassay was based on the enzyme catalysed conversion of the BTEX components to their respective catechols followed by the reaction with hydrogen peroxide in presence of horseradish peroxidase to colrimetric products that can be monitored at 420 nm.

Sensors Based on Baroxymeter

As a new application, baroxymeter has been developed as a portable wastewater direct toxicity assessment device based on manometric bacterial respirometry. Respirometry was measured as the pressure drop in the headspace of a close vessel due to oxygen uptake by the microorganism in contact with sample. This microbial pressure sensor showed good reproducibility and comparable responses with other reported methods.

Sensors Based on Infrared Analyser

A new method was reported for monitoring inhibitory effects in wastewater treatment plants based on continuous measurement of the microbial respiration product CO_2. Activated sludge microbes are used as the biological elements and their respiratory activity is inhibited by the presence of toxic compounds, resulting in a decrease in CO_2 concentration which was analysed by using a CO_2 infrared analyser. Based on the measurement of CO_2 concentration in the off gas produced during degradation of carbon compound by microbial respiration activities, a microbial biosensor was developed to monitor the extent of organic pollution in wastewater both off-line in a laboratory and online in a wastewater treatment plant.

FUTURE OF INDUSTRIAL MICROBIOTECHNOLOGY

Industrial application of biotechnology is emerging as a spin-off from developments in other fields such as the pharmaceutical sector. This emergence is largely because industrial biotechnology has not received the same level of public policy attention as has biotechnology in other sectors. There are other structural factors that influence the diffusion of industrial biotechnology. These include the dominance of physical and chemical technology as a source of concepts for the design of industrial plants limits the scope for introducing biological processes.One of the main advantages of industrial biotechnology is the prospect for the controlled production of biological catalysts. These biocatalysts are more specific and selective than their non-biological counterparts. As a result, they offer greater potential for cleaner industrial production.

In other words, biocatalysts generate fewer by-products and can start with relatively less purified feedstocks. And because they are self-propagating, they can be used in applications such as waste treatment.But despite these advantages, biocatalysts are generally fragile (requiring large amounts of water) and have low volumetric productivity. Over the years, however, incremental technological innovations and new bioreactor designs have helped to improve the industrial performance of biocatalysts.

With incremental improvements in biocatalysts and the emergence of new design concepts, biotechnology's capacity to diffuse in the industrial sector will be enhanced. This prospect is enhanced by the growth in the biological sciences, as well as complementary fields such as chemistry and informatics. The use of biomass for energy and industrial uses has been on the agenda of many governments for nearly two decades. Much of the interest was triggered by the oil crises of the 1970s. Although interest in the field waned with the decline in energy prices, advances in the biological sciences have continued to enhance the prospects for technological improvement and wider application.

In addition to energy, living plants can be used to produce chemicals such as citric acid, lysine and lactic acid. Genetic modification offers new possibilities for using plants as a source of raw materials for chemicals or even finished products. As enzyme technology improves, attention is shifting to other areas of bioprocessing by tapping the potential in the world's splendor of microbial life. Much of this world remains underutilized largely because microorganisms have so far been poorly studied and documented. With the advent of DNA sequencing, microorganisms will become an important addition to industrial activities through scientists' discovery of new biocatalysts.

The field of genomics is therefore likely to extend its influence from medicine and agriculture to industrial production. Methods such as forced evolution and rational design will increasingly be used to discover new enzymes for industrial use. In addition, methods such as gene shuffling are helping firms to optimize their bioprocessing activities. It is expected that the genomes of major industrial microorganisms will be sequenced in the coming years, and this will add significantly to the genetic library of industrial biotechnology resource.

The evolution of market opportunities for industrial biotechnology is difficult to predict, partly because of the nascent nature of the industry, poor understanding of its structure, and a lack of concerted efforts to improve the policy environment for the diffusion of biotechnology products. What is likely to emerge, however, is a scenario dominated by niche markets in a wide range of sub-sectors. Furthermore, the blurring of boundaries between agriculture, health and industry makes it difficult to predict potential areas of market expansion. Even though the life science industries model is currently being questioned, the generic nature of the technology suggests that firms that have established a lead in pharmaceutical or agricultural biotechnology are likely to become equally important players in industrial biotechnology.

REFERENCES

Allenby, Braden R. "Integrating Environment And Technology: Design For Environment" in Allenby, Braden R., Richards, Deanna, (ed.) *Greening Industrial Ecosystems*. Washington DC. National Academy Press Office. 1994 .

Biotechnology Industry Organization. *Biotechnology in Perspective*. Washington, D.C.: Biotechnology Industry Organization. 1990.

Olson, Steve. *Biotechnology: An Industry Comes of Age*. Washington. D.C.: National Academy Press. 1986.

Bibliography

Alexander, M. *Biodegradation and Bioremediation*. Academic Press. New York, 1994.

Allenby, Braden R. "Integrating Environment And Technology: Design For Environment" in Allenby, Braden R., Richards, Deanna, (ed.) *Greening Industrial Ecosystems*. Washington DC. National Academy Press Office. 1994 .

Andria M. Costello, Ann J. Auman, Jennifer L. Macalady, Kate M. Scow, and Mary E. Lidstrom, "Estimation of Methanotroph Abundance in a Freshwater Lake Sediment", *Environmental Microbiology*, 2002.

Astrid A. Van de Graaf et al, "Anaerobic Oxidation of Ammonium Is a Biologically Mediated Process", *Applied and Environmental Microbiology*, 1995.

Banks, J.G. and R.G. Board, "Some factors influencing the recovery of yeasts and moulds from chilled foods", *Intl. J. Food Microbiol.*, 1987.

Biotechnology Industry Organization. *Biotechnology in Perspective*. Washington, D.C.: Biotechnology Industry Organization. 1990.

Bryant, T.N., Lee, J.V., West, P.A. and Colwell, R.R., "A probability matrix for the identification of species of Vibrio and related genera", *Journal of Applied Bacteriology*, 1986.

Carson, J., Wagner,T., Wilson,T. and Donachie, L., "Miniaturized tests for computer-assisted identification of motile Aeromonas species with an improved probability matrix", *Journal of Applied Microbiology*, 2001.

Cookson, J. T. *Bioremediation Engineering: Design and Application*. McGraw-Hill. New York. 1994.

Cox, R.P. and Thomsen, J.K., "Computer-aided identification of lactic acid bacteria using the API 50 CHL system", *Letters in Applied Microbiology*, 1990.

D'Aoust, J.-Y., "Effective enrichment-plating conditions for detection of Salmonella in foods", *J. Food Prot.*, 1984.

Dawson, C.A. and Sneath, P.H.A., "A probability matrix for the identification of vibrios", *Journal of Applied Bacteriology*, 1985.

Dekker, J., and G. Comstock. "Ethical and Environmental Considerations in the Release of Herbicide Resistant Crops." *Agriculture and Human Values*. Summer 1992.

Edgington, Stephen M. "Environmental Biotechnology." *Bio/Technology*. December 1994. p. 1338-42.

Environmental Defense Fund. *Genetically Engineered Foods: Who's Minding the Store?* New York. NY: Environmental Defense Fund. 1995.

Feltham, R.K.A., Wood, P.A. and Sneath, P.H.A., "A general-purpose system for characterizing medically important bacteria to genus level", *Journal of Applied Bacteriology*, 1984.

Fiksel, Joseph, and Vincent T. Covello. (eds.) *Biotechnology Risk Assessment: Issues and Methods for Environmental Introductions*. N.Y.: Pergamon Press. 1986.

Fowle, John R., III. (ed.). *Application of Biotechnology: Environmental and Policy Issues.* Boulder. Colorado: Westview Press for the American Association of Advancement of Science. 1987.

Gendel, Steven, et al. (eds.) *Agricultural Bioethics: Implications of Agricultural Biotechnology.* Ames: Iowa State University Press. 1990.

Ginzburg, Lev. R. *Assessing Ecological Risks of Biotechnology.* Boston: Butterworth-Heinemann. 1991.

Gould, W.A., "Tomato Production", *Processing and Quality Evaluation,* AVI Publishing Co., Inc., Westport, Connecticut, 1974.

Hanson, J.R., J. L. Macalady, D. Harris and K.M. Scow, "Linking toluene degradation with specific microbial populations in soil", *Applied and Environmental Microbiology,* 1999.

Hauschild, A.H.W. and R. Hilsheimer, "Enumeration of food-borne Clostridium perfringens in egg yolk-free tryptose-sulfite-cycloserine agar", *Appl. Microbiol.,* 1974.

Hettinger, N. "Owning Varieties of Life: Biotechnology, Intellectual Property, and Environmental Ethics." Center for Biotechnology Policy and Ethics. Texas A&M University. College Station. Texas. 1994.

Holmes, B. and Costas, M., "Identification of Enterobacteriaceae by computerized methods", In: Board, R.G., Jones, D. and Skinner, F.A. (Eds.) *Identification methods in applied and environmental microbiology,* Oxford: Blackwell Scientific Publications, 1992.

Jarvis, B. and A.P. Williams, *Methods for detecting fungi in foods and beverages,* In: Beuchat, L.R. (ed.), Food and beverage mycology. Second edition, Avi Publishing Co., Westport, Connecticut, 1987.

Jetten et al, "Microbiology and application of the anaerobic ammonium oxidation (anammox) process", *Environmental Biotechnology,* 2001.

Margaret G. Mellon. *Biotechnology and the Environment : A Primer on the Environmental Implications of Genetic Engineering.*

Mislivec, P.B., L.R. Beuchat and M.A. Cousin, "Yeasts and Moulds", In: *Compendium of Methods for the Microbiological Examination of Foods,* APHA, Washington, D.C., 1992.

Newell, S.Y., "Fungi in marine/estuarine waters", Pages 1394-1400 in: Bitton, G., ed. *The Encyclopedia of Environmental Microbiology,* Wiley, New York, 2002.

Olson, Steve. *Biotechnology: An Industry Comes of Age.* Washington. D.C.: National Academy Press. 1986.

On, S.L.W., Holmes, B. and Sackin, M.J., "A probability matrix for the identification of Campylobacters Helicobacters and allied taxa", *Journal of Applied Bacteriology,* 1996.

Organisation for Economic Cooperation and Development (OECD). *Biotechnology, Agriculture and Food.* Paris. OECD. 1992.

Peeler, J. T., G. A. Houghtby, and A. P. Rainosek, "The most probable number technique", *Compendium of Methods for the Microbiological Examination of Foods,* 3rd Ed., 105-120, 1992.

Peters, Pamela. *Biotechnology: A Guide to Genetic Engineering.* Dubuque. IA: Wm. C. Brown Publishers, 1993.

Rittmann, B.E., and P.L. McCarty. *Environmental Biotechnology: Principles and Applications.* Boston, MA: McGraw-Hill. pp. 720-721. 2001.

Turner A. P. F. (ed.). *Biosensors: Fundamentals and Applications.* Oxford University Press. Oxford. 1987.

Index